Almanac of Soviet Manned Space Flight

Dennis Newkirk
foreword by James Oberg

Almanac of Soviet Manned Space Flight

Library of Congress Cataloging-in-Publication Data
Newkirk, Dennis.
  Almanac of Soviet manned space flight: a revealing
launch-by-launch history of the red star in orbit/Dennis
Newkirk.
      p. cm.
  Includes bibliographical references.
  ISBN 0-87201-848-2
  1. Astronautics—Soviet Union—History.   I. Title.
TL789.8.S65N48 1989
629.45'00947—dc20   89-23608
                                                          CIP

Contents

☆

6
Mars Precursors (1986–1989)

Appendix
Specifications of Soviet Space Launch Vehicles, Ballistic Missiles, and Rocket Engines

Index

Foreword

Sleuthing the secrets of the Soviet space program has required a strange and rare collection of skills. Whereas NASA overwhelms researchers with a torrent of data, interviews, photographs, and multifaceted minutiae, the Soviets traditionally have attempted to strictly control the space information available to the outside world, so as to establish images that often have little relation to reality. An investigator must conduct a wide-ranging survey, must be able to recognize the significant details that slip past the secrecy, must balance them against sound engineering judgment, and must then assess their significance in the "big picture."

Metaphorically, one is working with scattered pieces to several jigsaw puzzles. Many of the pieces are damaged, and some are counterfeit. Some might belong in different puzzles. Some are faded, folded, stretched by time into new shapes.

Under these circumstances, what hope is there for a reasonably accurate image? Dennis Newkirk is working in the grand tradition of earlier sleuths who have succeeded in laying out the broad outlines of the known, of pinpointing detailed chronologies and technical specifications, and of sketching the boundary of the unknown. Much Soviet data are eventually published, even if in obscure places, sometimes by mistake. Independent Western observers, particularly radio listeners and naked-eye skywatchers, regularly add crucial information. And most important, the technical feasibility of spaceflight engineering (as practiced in other spacefaring nations) gives a touchstone against which competing hypotheses can be evaluated.

The result here is a reliable reconstruction of what seems to have been happening in the Soviet space program, as if the Soviets had never attempted to cover up their embarrassments and politically awkward realities. These days, under the glasnost tide, new official Soviet revelations repeatedly confirm the estimates made by the tenacious sleuths whom Newkirk has joined. This validates old techniques and supports new probes into hitherto impenetrable darknesses.

Why should such an effort even be required? The "game" often appears tedious, trivial, and dead-ended. Years of efforts by dozens of dedicated Western sleuths often condense to a few extra details of one Soviet mission's backup crew, or one unsuccessful precursor test with an ad hoc cover story, or a blueprint of one canceled space prototype. Who benefits from such labors?

Ask any sleuth, ask Dennis Newkirk or myself or our colleagues around the world, and the first answer is personal satisfaction. We climb the mountain of Soviet space secrecy "because it is there." They are trying to hide things; our human reaction is to want to expose these things. And our successes have been considerable, as this book clearly details.

Moreover, eventual Soviet admissions have been prodded by Western revelations, and merely waiting patiently would be in vain without the pressure of publication. By making many things secret, the Soviets have made them even more "juicy" for the Western news media, and this has guaranteed widespread publication of the very things the Soviets would like to withhold. Recent developments show considerable maturity in Moscow, and current (if only temporary?) trends are to be more candid about current activities, about a portion of planned missions, and about selected chapters of the past. But the underlying cause of much of today's glasnost is the bitter lesson that such secrets will sooner or later be dug out by Western space sleuths anyway. Today's Soviet honesty is a tribute to the failure of their past policies of dishonesty, and that failure was brought about by conscious actions of the Western sleuths.

Another reason is that history requires a contemporary analysis of events. Future generations will have neither the knowledge of the environment, nor the access to all printed and verbal material, nor the accumulated wisdom of today's sleuths. Centuries from now, when the names of sports heroes, actors, preachers, even presidents, and yes, of countries, too, are all forgotten, the human activity for which this century will be known will be the breakout into space. For those scholars in unknown, unborn languages, our finest bequest— besides the very fact of spaceflight—will be our documentation of how it was done.

There are benefits to our own generation as well. An appreciation of how the Soviets conduct their manned space program has value to the American manned space program, both to learn techniques of advantage and to identify blind avenues to be avoided. The Soviets are putting the best of their aerospace industry into spaceflight, so any assessment of their space activities is automatically a measure of their highest capabilities in aerospace technology. Such assessments of capabilities allow speculative inquiries into actual intent, and give, via views of space hardware with well-defined applications, unprecedented views into the minds of Kremlin policymakers.

Both nations will be pursuing parallel manned space programs into the next century, into the unforeseeable future. These activities will be both in competition and in cooperation. Rational and successful planning for American space activities absolutely requires sound technical appraisals of the Soviet past, present, and future in space, and readers will find what they need in Newkirk's impressive catalog. Beyond the hardware, they will detect the thrill of the successful sleuth, and they can be confident—as the Soviets remain anxious—that such dogged, skillful investigators continue to assault the boundaries of what the Soviets would like to allow us to know about their space efforts, to the complete picture of what we must know in order to chart our course into the next millennium. This book is one such map.

James Oberg
Dickinson, Texas
(Author of *Red Star in Orbit*
and other works)

Acknowledgments

This book is a compilation of information from more than 50 books, magazines, and other sources. Much of the information about the flights from the 1960s to the mid 1970s comes from Nicholas Johnson's *Handbook of Soviet Manned Spaceflight*, and Reginald Turnill's *Spaceflight Directory*. Information on flights from the 1970s to the present have been collected from many magazines, articles, the Foreign Broadcast Information Service translations of Soviet publications, and Soviet news broadcasts. Another important source was the British Interplanetary Society's *Spaceflight* magazine and their *Journal*, which have had the most consistent coverage of Soviet manned spaceflight during the last 15 years. Interesting stories revealed in James Oberg's many books and articles on spaceflight have also been placed in the mission summaries. The Soviets supply much of the basic information on the day-to-day activities and events of their spaceflights. This information is given almost daily in normal news broadcasts during missions. Soviet magazine and newspaper articles are also translated by the U.S. government and are available at most large libraries.

I would also like to thank James Oberg, Nicholas Johnson, and Loyd Swenson for contributing their comments and information during the preparation of this book. Lastly, without the support of my family and Jennifer Stone's encouragement this book would not have been possible.

Dennis Newkirk
Fairfax, Virginia

Introduction

The purpose of this book is to present a history of Soviet manned* spaceflight activities, and highlight the fact that the Soviets may soon become the major power off the Earth. Just as in the late 1950s, the United States is facing the threat of being perceived as second rate in the pioneering field of spaceflight because of Soviet domination in space activities.

The Soviets now have the same capabilities as the U.S. possessed in the early 1970s. If they choose, they can soon send people to the moon and the planets. They increasingly talk of solar power satellites, satellite power relay, satellite assembly and repair stations in orbit, and multiple manned space complexes. These systems have been advocated for the U.S. manned space program since the 1970s, but they had no political support. U.S. space projects were allowed to dwindle over the 1970s into a few very specialized projects, constantly underfunded and behind schedule. Meanwhile, the Soviet program expanded and accumulated years of spaceflight experience. In the United States, debate continues over the future direction of the U.S. space program while the Soviets are making steady progress in space.

Soviet scientists believe that the most important part of future space exploration will be manned orbital spacecraft. From Earth orbit space stations, space industries can be developed, and unmanned and manned flights through the solar system can be prepared. The Soviets plan to establish a large permanent space station in Earth orbit around 1995, serviced by Soviet space shuttles. They can also be expected to launch a manned Mars fly-by mission in the late 1990s, resume manned lunar explorations by 2000, and land people on Mars soon after that. During these activities, they can also begin the massive space manufacturing projects previously mentioned. Whether projects like solar power satellites can be economical is not yet known, but the Soviets are investigating the possibilities. These projects would also undoubtedly bring the Soviets world-wide recognition.

---

*The term "manned" is used throughout this book as the traditional reference to any piloted spaceflight of either men or women.

Sensitivity to this issue by the U.S. government is shown by recent plans for a Mars Sample Return mission. Such a mission had been advocated by many in the space science community for years, but significant funds for NASA never materialized until the Soviets publicized their own plans for such a mission in 1987. This decision was the first taken by the U.S. in a new round of the space race for world prestige.

This book covers all known Soviet manned and man-related flights. Man-related flights are flight tests of man-rated spacecraft, flights directly supporting manned missions, and research and development flights directly related to manned spacecraft development. The book is divided into chapters covering each major space effort from the first manned flights to the first shuttle flight. Despite being broken up into chapters, the flights are still listed in chronological order within chapters, and from chapter to chapter. At the beginning of each chapter is an orientation to the time period covered and the programs in play at the time.

In addition to the chronological listing of spaceflights, an illustration of the spacecraft is provided. The illustration includes all other spacecraft directly involved in the mission.

Chapter 1 covers the period from 1960 through 1966, when the Soviets were in the space spectacular business. Each flight of this period broke previous records or introduced new activities to manned spaceflight. The Soviets continued the series of space firsts that started the space race with the launch of Vostok 1 and the first man in space. The Soviets managed to stay one step ahead of the U.S. in spaceflight by flying the first multi-man crew and performing the first spacewalk. During this period, the U.S. moon race challenge was accepted and preparations for a manned lunar flight and eventually lunar landing began.

Chapter 2 covers the period from 1966 through 1968, when the Soviets began development of the techniques needed for the planned manned lunar missions. They performed development flights similar to the U.S. Gemini program and made tests of their lunar spacecraft. The Soviets and U.S. both experienced their first crew fatalities, delaying both equally in the moon race. By the end of this period, the Soviets lost almost all hope of winning the moon race after the successful circumlunar flight of Apollo 8 in December 1968.

Chapter 3 covers the period from 1969 through 1973, when the spacecraft and boosters developed for the moon race were eventually tested and the program was terminated. The Soviets turned to Earth orbital space stations to continue their manned space program and investigate the military uses of manned spaceflight. But this was a time of many failures for the Soviets, delaying their space station program for three years and stopping the super booster program for 15 years.

Chapter 4 covers the years 1973 to 1976, when the Soviet military began experimenting with manned spaceflights and the first successful space station missions were flown. During this period, the Soviets began preparing for their second-generation space stations, and began developmental test flights of their first space shuttle program in hopes of beating the U.S. shuttle to flight and reducing costs of supplying the second-generation space stations.

Chapter 5 covers the period from 1976 through 1985, when the Soviets introduced a second-generation space station, enabling longer missions by continual resupply using robot cargo spacecraft and large space station expansion modules. Using these spacecraft, the Soviets continued to extend the time limits that people could work in space. They also ended work on a small space shuttle after unsuccessful tests and began development of a new heavy booster and large space shuttle to support a permanent space station in the 1990s.

Chapter 6 covers the period from 1986 to 1989, when the Soviets began operating their third-generation manned space station. The new station allowed work to start on preparing a manned Mars flight and began testing large-scale production and research in space. The Soviets also made the first test flights of their large booster and space shuttle.

Despite the great detail presented here, it must always be remembered that much of manned space program activities remain unknown. Included in the unknown activities are tests of launch escape systems, landing systems, and possible tests of unsuccessful spacecraft designs. Also, any book on U.S. spaceflight would have to include ballistic spaceflights of rocket planes like the X-15. Only as this book was being completed did the Soviets begin releasing the first vague reports of similar Soviet activities in the 1960s. This indicates that the Soviets have never released as much information about their space program as in the past few years, but there is much more that has not been made public.

Flight Notes

Altitudes listed in mission titles are for the initial orbit. They are usually provided by Royal Aircraft Establishment or Soviet news reports.

The term "aborted" is used if the flight failed to accomplish its major goal but accomplished lesser goals. The term "failed" is used if the spacecraft is seriously damaged or the crew killed.

Flights that fail or are aborted that are later followed by a similar successful flight are designated with the letter A. For example, the aborted Soyuz T-10A mission was followed by Soyuz T-10B.

Because the Soviets did not start publicly naming their space launch vehicles until very recently, this book uses the well known "Sheldon" naming system, originated in the 1960s by Library of Congress spaceflight researcher Dr. Charles Sheldon. See the Appendix for equivalent Soviet names and U.S. DOD space launcher (SL) designations.

Conversions

Time

All dates are given in Moscow time unless otherwise stated. All mission times are accurate to at least ± 5 minutes. Launch and landing times listed are for the spacecraft and not necessarily the crew launched by that spacecraft.

Eastern Daylight Time = GMT − 4 hr
Eastern Standard Time = GMT − 5 hr
Greenwich Mean Time = Moscow Standard Time − 3 hr
Greenwich Mean Time = Moscow Summer Time − 4 hr
Moscow Summer Time = GMT + 4 (from the last Sunday in March, until
⠀⠀⠀⠀⠀⠀⠀⠀⠀⠀⠀⠀⠀⠀⠀⠀⠀⠀⠀the last Sunday in September)
Baykonur time = Moscow + 2 hr

Measurement

This book uses metric units of measurement, so some simple English conversions are listed below.

1 kilometer (km) = 0.6 mile
1 kilogram (kg) = 2.2 pounds
1 meter (m) = 1.09 yard
1 liter (l) = 1.057 quart

Chapter 1

Man into Space

The space age had its beginnings in World War II. The wartime development of short-range ballistic missiles and atomic weapons led to the development of large rockets, which enabled the long-held dream of spaceflight to be realized. The German V-2 ballistic missile laid the groundwork for future rockets that would be used in spaceflight. The V-2 reached an altitude of around 80 km on its wartime missions, taking it beyond most of the Earth's atmosphere (about the same height as flights of the manned U.S. X-15 rocket plane in the 1960s). Orbital spaceflight using V-2 technology was possible soon after the war. The V-2 and similar missiles (today classed as sounding rockets) made flights out of the atmosphere on military development and scientific research missions in the U.S. starting on April 16, 1946, and in the U.S.S.R. starting on October 27, 1947 (see Table 1-1).[1] Both countries used captured German equipment and scientists as a resource to aid the development of domestic rocket development establishments. The U.S. eventually formed the Germans into a new U.S. rocket development organization, while the Soviets returned most captured Germans to their homes after learning their methods and technology.

While paralleling rocket research in the U.S., the Soviet Union also matched the United States by developing its own atomic bomb. Work in both countries continued past the atomic bomb in the early 1950s to development of the thermonuclear or hydrogen bomb. The Soviets tested their first hydrogen bomb only nine months after the first U.S. test in 1952. The Soviets had matched the U.S. in bombs, but could not match its ability to deliver those weapons over long distances. By 1951, the U.S. had long-range B-36 bombers and had begun developing the Atlas intercontinental ballistic missile (ICBM), by which to deliver nuclear weapons to Soviet territory in case of war.[2] At the time, the Soviets could not match the capabilities of the U.S. bombers with their own, and doubted their ability to stop the planes with air defenses. With the Soviets unable to counter the U.S. aircraft delivery system offensively or defensively, they leapfrogged past development of long-range bombers to a spaceflight delivery system using ICBMs.[3]

Table 1-1
Soviet High-Altitude Biological Research Flights

Date	Rocket	Payload Weight	Altitude	Payload
1951 – 1952	V-2/SS-1 Scunner?	?	96 km	6 flights carried 9 dogs total
1952 – 1956	V-2/SS-1 Scunner?	?	96 km	9 flights carried 9 dogs total
1956 – 1957	V-2/SS-1 Scunner?	?	96 km	4 flights carried 3 dogs total
May 16, 1957	SS-2 Sibiling	2196 kg	211 km	5 dogs
Aug. 27, 1958	SS-3 Shyster	1690 kg	452 km	2 dogs
Sept. 19, 1958	SS-3 Shyster	1515 kg	473 km	
Oct. 31, 1958	SS-3 Shyster	1515 kg	473 km	
July 2, 1959	SS-2 Sibiling	2000 kg	241 km	2 dogs, 1 rabbit
July 10, 1959	SS-2 Sibiling	2200 kg	211 km	2 dogs
June 15, 1960	SS-2 Sibiling	2100 kg	221 km	2 dogs, 1 rabbit
June 24, 1960	SS-2 Sibiling	2100 kg	212 km	
Sept. 16, 1960	SS-2 Sibiling	2100 kg	210 km	
Sept. 22, 1960	SS-2 Sibiling	2100 kg	210 km	

Sources: Congressional Research Service, The Library of Congress, Soviet Space Programs 1976–80, Manned Space Programs and Life Sciences, Part 2. Government Printing Office: Washington, 1984, p. 486. Congressional Research Service, The Library of Congress, Soviet Space Programs 1980–87, Part 1. Government Printing Office: Washington, May 1988, p. 31.

The ICBM

Josef Stalin had initiated serious study of ICBMs soon after the war, but it was Nikita Khrushchev who in 1954 gave final approval to a missile designed by Sergei Korolev the previous year. The missile was a direct descendant of the German V-2, and much of its technology was conceived by German scientists and engineers during development of the V-2 at Peenemunde. Korolev, however, took the German technology to a much greater scale than anyone else, including Wernher von Braun, then working for the U.S. Army building a V-2 offspring called the Redstone.

Korolev's missile, which would later be known in the West as the SS-6, was much larger than the V-2 and was as powerful as 20 V-2's put together. The SS-6 was designed as an elegant multistage rocket, which made up for other relatively crude aspects of the design (see Figure 1-1). What the Soviets lacked in advanced Western technology, they made up for in brute force. While in the U.S., development of the monocoque, pressurized propellant tanks on the Atlas enabled greater performance with smaller missiles, Korolev's SS-6 was made of thick-walled, heavy steel tanks, making it much larger and less efficient than its American counterparts.

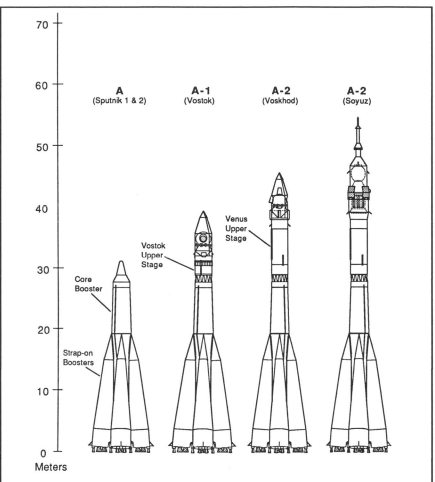

Figure 1-1. The A-type booster derived from the SS-6 ICBM has evolved over 30 years to launch Soviet manned spacecraft of the Vostok, Voshkod, and Soyuz types.

A-Type Rocket

The SS-6 or A-type rocket consisted of 5 parts, a core stage 2.95 meters in diameter and 28 meters long, which was surrounded by 4 strap-on boosters, each 19 meters long and 3 meters in diameter. The booster was 10.3 meters in diameter at the base, from tail fin to tail fin. *(continued on next page)*

(continued)

The A-type used an RD-107 rocket engine in each strap-on booster. The RD-107 had four main nozzles with two steering vernier engines, which gimballed on one axis; the main engines did not gimbal. Each RD-107 engine consisted of 4 combustion chambers, fed by a single turbo-pump mounted above the chambers. The core stage used an RD-108 engine, which was the same as the RD-107, but with four steering verniers. Development of the RD-107 and RD-108 engines started in 1954 at the Gas Dynamics Laboratory in Leningrad. The booster is generally considered a two-stage booster with the strap-on boosters comprising the first stage and the rest of the core stage the second stage. The A-type designation was given to the SS-6 when it was used for satellite launch missions.

The A-1 type booster was used to increase the lifting capacity for launches of Vostok spacecraft. The A-1 consisted of an A-type with an upper or third stage, 2.65 meters in diameter and 2.5 meters long. The upper stage weighed about 1,400 kg empty and had a single RD-7 engine with a thrust of 90,000 kg.

The A-2 type was used to launch Voskhod, Soyuz, and other spacecraft heavier than the A-1's could launch. It was the same as the A-1 but replaced the old third stage with a new Venus upper stage measuring 8 meters long and 2.6 meters in diameter. The Venus stage used a single RD-461 type engine and four steering verniers which was very similar to the RD-108. By 1988, more than 1200 A-type boosters had been launched.

The hydrogen bomb, for which the SS-6 was designed, began as a relatively heavy device. The Soviets and the Americans later developed lighter H-bombs that required much smaller rocket boosters, and U.S. development of the Atlas had been keyed to the lightweight bomb being developed. But, the Soviets decided to design SS-6 for the heavier warhead. This later turned out to be fortunate for the space effort because the SS-6 could launch larger satellites than any U.S.-designed ICBM, including the Atlas.

Though it could launch heavier payloads, the SS-6 made a poor ICBM. Its range was barely enough to reach the northern United States, and it was completely unprotected from attack while sitting on the launch pad.[4] Only four of the missiles were ever put into operational status.[5] Ironically, more than 1,000 of the rockets would be used for more than 30 years as a reliable space booster.

Sputnik

Korolev, the designer of the SS-6, had always been more interested in spaceflight than ICBMs, and he tailored the rocket for spaceflight in addition to its military mission. On hearing of the U.S. Vanguard program, which was to launch a small satellite in 1957, Korolev tried to get permission in 1956, to launch a satellite first. After the initial proposal failed, Korolev was officially put in charge of the project in the summer of 1957.

Khrushchev saw the missile as a great propaganda tool to use against the West, both as a weapon and for spaceflight. By matching the United States militarily, the Soviet Union assured its place as a world superpower. By surpassing all other nations in spaceflight, the U.S.S.R. hoped to gain politically and economically around the world by influencing Third World and other nonaligned nations. Khrushchev also hoped to make political gains inside the Soviet Union by using the missile force to increase his influence while lessening any opponents' power.[6] To do this, Khrushchev created in 1960, a totally new branch of the armed forces, called the Strategic Rocket Force, equal in level with the Soviet Army, Navy and Air Force.[7,8] The Strategic Rocket Force was responsible for development and operation of all rockets, both military and scientific, with scientific contributions coming from the Soviet Academy of Sciences.

The Soviets announced in September 1956, that they would also launch a satellite as the U.S.S.R.'s unexpected contribution to the International Geophysical Year (IGY). The Soviets even provided the frequencies the satellite would transmit on in May 1957, so that amateur radio operators around the world could listen as it passed overhead. The world media and the general public in the West considered the possibility of a Soviet satellite as highly unrealistic. At the time, the Soviets did little to change this belief, releasing no information on the rocket or launch facility.

The SS-6 ICBM's first successful test flight was on August 3, 1957. After the successful test, Korolev was then given final approval for launch of Sputnik 1 after the next test flight.[9] Interestingly, the U.S. Atlas ICBM was launched only four months after the first SS-6. While it did lack some power in comparison, the U.S. was clearly not far behind in military and spaceflight capability. The Soviets, however, were the first into space when they launched Sputnik 1 on October 4, 1957. They had been planning to launch a satellite to research aspects of the space environment and the Earth that would match and surpass all planned U.S. experiments for IGY satellites. To be in time to beat the U.S. Vanguard project, the Soviets had to replace the more sophisticated satellite with the simple spherical Sputnik satellite when development fell behind schedule. The satellite they intended to be first was later launched as Sputnik 3 on May 15, 1958.

Sputnik 2, launched on November 3, 1957, in some ways was more significant than its predecessor. While Sputnik 1 had weighed 83.6 kg, or about ten times more than the planned U.S. Vanguard satellites, Sputnik 2 carried a dog, Laika, around the Earth in a pressurized container on top the the SS-6 core booster stage. The combined weight of the container and final booster stage was over 6,500 kg. It was this demonstration of great weight disparity between the Soviet and U.S. capabilities that generated much of the early concern that in turn fired up the space race.[10]

Sputnik 2 showed the world that the Soviets could launch comparatively very heavy satellites or nuclear bombs. It also showed the desire to put living organisms into orbit, perhaps including people. Soviet experimental flights of ballistic sounding rockets had been carrying animal payloads since 1951. By September 1960, more than 29 flights carrying dozens of animals had been conducted at altitudes from 96 to 473 km, well out of the Earth's atmosphere (see Table 1-1). These flights tested life-support systems and biological effects of spaceflight that could be directly applied to the man-in-space program. In the United States, similar test flights were also conducted.[11]

As expected, following the orbital flight of Laika, the Soviets began serious plans to put a man into orbit. Khrushchev hoped such a flight would have great propaganda value, considering the effect of the first few Sputniks. The propaganda value was working in the U.S. Public concern resulted in congressional hearings from 1957 into 1958, slowly making space an important issue for politicians to deal with and use to their advantage.

The man mainly responsible for the Sputniks, Premier Khrushchev, wasted no time in the exalting the U.S.S.R.'s ICBM and space booster. Khrushchev claimed the missiles were being produced by the dozens. This added insult to injured U.S. national pride, and so Khrushchev had pushed the United States into a public competition that far surpassed anything the Soviets could match for years to come.

U.S. Reaction

The Sputnik 1 and 2 flights had an intensifying public impact in the western nations. While the Soviets had captured the imagination of the western press, no massive changes were made to government programs. Political pressure for an increased space program was slow to build. To appease public demands, the National Aeronautics and Space Administration was formed from the existing National Advisory Committee for Aeronautics on October 8, 1958, more than a year after Sputnik 1. Initially, NASA had very vague goals, but by the end of the year the goal of placing a man in space was finally agreed to by the President and the Congress. Increased space research was seen as necessary to match future Soviet developments.[12] The Soviets were already months ahead of the United States with their man-in-space project, and the United States had entered the next round of the space race with little hope of beating a Soviet cosmonaut into space.

Part of the U.S. government's problem in finally reaching the decision to create NASA was the division of military and civilian programs. The military space program was already well established and, overall, things were not really as bad as they might have seemed. There also has been some speculation that the government was actually waiting for either the Soviets to launch the first Earth satellite or the scientific Vanguard project to clear the way for U.S. military satellites. The argument is that some in the United States feared Soviet reaction by flying a militarily developed satellite over their territory before the right of over-flight was established. The validity of this argument was demonstrated by the downing of a U-2 spy plane over the U.S.S.R. on May 1, 1960.[13]

U.S. intelligence agencies in the late 1950s were pressing for development of reconnaissance satellites to avoid the need for U-2 spy plane missions over unfriendly territory. The U.S. Air Force had been developing the Discoverer series of photo-reconnaissance satellites, and their Thor-Agena booster since at least 1956. The Discoverer satellites, first launched in 1959, were among the first and were the heaviest U.S. satellites of the period. However, they were not well covered by the popular press even though they were not classified at the time.[14] Instead, the press publicized the Explorer, Pioneer, and unique Score satellites as a meager response to the Soviet space program. The Score satellite was an almost exact copy of the Sputnik 2 flight. Score placed a 2,000-kg Atlas core stage, and an attached 150-kg radio transmitter payload into orbit. The effort nearly matched the Soviet mission, but clearly showed that the U.S. was still a little behind in weight-lifting ability. Still, the main focus had shifted from the unmanned space probes to the man in space goal, and the efforts of both countries to be first.

Vostok Program

The design of the Soviet Vostok manned spacecraft began in early 1958. As with the Sputnik satellites before it, Korolev was the chief designer of the Vostok. After stormy engineering discussions, the design was complete by April 1958 (see Figure 1-2). The possibility of designing a spacecraft for sub-orbital flight was considered, but was rejected in favor of a spacecraft capable of orbital flight lasting several days. The Vostok was designed as a fully automatic spacecraft. Its main purpose was to give more space spectaculars to Khrushchev for propaganda. In addition, the flights gave cosmonauts and workers valuable flight experience. The A-1 booster used to orbit the Vostok was an SS-6 ICBM with an upper stage added to lift the heavy spacecraft into a low orbit. Initial construction of the spacecraft began in early 1959 at the Star City space center near Moscow (see page 9).

After the first few orbital test flights (named Korabl Sputnik), the spacecraft's systems were reviewed. During tests of the capsule's ejection seat, a test pilot

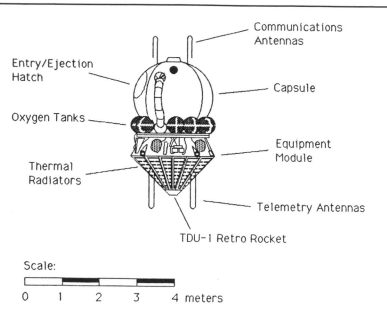

Figure 1-2. The Vostok spacecraft was the world's first manned spacecraft. Early tests carried dogs and simulated humans, as did early NASA Mercury test flights.

Vostok Spacecraft

This spacecraft consisted of a spherical capsule and an equipment module. The spacecraft's total weight was 4,730 kg and the total length was 4.9 meters. The equipment module was 2.6 meters long and the 2,500-kg capsule was about 2.3 meters in diameter. The capsule contained the cosmonaut's ejection seat, life support equipment, flight controls, and communication system. The cosmonaut's ejection seat contained emergency supplies and a small oxygen supply for use after ejection. The capsule's systems were self-contained including batteries to provide power for its systems for up to ten days. The equipment module weighed 2,050 kg fueled (1,900 kg empty) and contained systems for telemetry, heat regulation, and engine systems. The engine system consisted of the attitude control system that could change the spacecraft's attitude but not the orbital parameters, and the liquid fuel TDU-1 retrorocket. The TDU-1 could produce 1,583 kg thrust for 45 seconds to lower the orbit into the atmosphere for reentry.

had been killed after being hit by the hatch during ejection. The redesigned spacecraft included a new ejection system with a two-second delay between hatch jettison and seat ejection. A new analog computer control system and changes in the horizon sensor systems were also incorporated. Korolev approved the design changes in August 1960.[15] Work on the redesign continued from September to December 1960.

A major disaster for the Soviet space program occurred on October 24, 1960. Final preparations for the launch of the new SS-7 ICBM were underway on October 23 when a fuel leak developed. Normally, the rocket would be drained of propellants before repairs could begin, but the Soviets apparently felt comfortable enough with the problem to fix the booster without unloading propellants, because a similar problem had been solved without unloading during a Vostok launch a few months before on July 23. Workers welded the leak shut and launch was rescheduled for the next day. On October 24,

---

Soviet Space Centers

The main centers concerned with Soviet manned space flight are the Baykonur Cosmodrome and the Gagarin Training Center at Star City. Star City (Zvezdny Gorodok) is near Moscow and includes facilities for cosmonaut training and mission control. It is the Soviet equivalent to NASA's Johnson Spaceflight Center. The Baykonur Cosmodrome, located east of the Aral Sea near Lenninsk, was built to launch the first ICBM and later became the Soviets' manned launch facility. It is many times the size of NASA's Kennedy Space Center and Cape Canaveral. Another launch center at Plesetsk is the world's busiest space launch site, launching only unmanned satellites. There are also many small tracking and communications facilities based on ships, satellites, and within the U.S.S.R.:

Ground Tracking Stations:	Dzhusaly
	Yavpatoriya
	Ussurisk
	Ulan-ude
	Petropavlovsk-Kamchatskiy
	Tbilisi
	Zvenigorod
Original Space Tracking Ships (1961):	*Krasnodar*
	Il'ichevsk
	Aksay
	Dolinsk (continued on next page)

(continued)

Military Missile Tracking Ships:	*Chazhma*
	Chumikan
	Chukotka
	Sakhalin
	Spassk
	Marshal Nedelin
	Marshal Krylov
	Kamchatka
	SSV 33
Academy of Sciences Ships:	*Academic Sergey Korolev*
	Cosmonaut Yuri Gagarin
	Cosmonaut Vladimir Komarov
	Cosmonaut Pavel Belyayev
	Cosmonaut Vladislav Volkov
	Cosmonaut Viktor Patsayev
	Cosmonaut Georgy Dodrovolsky
	Kegostrov
	Morzhovets
	Bezhitsa
	Ristna
	Nevel
	Chazhma
	Borovichi
	Chumkin

30 minutes before the planned launch, Field Marshal Mitrofan Nedelin, commander of the Soviet Strategic Rocket Force, observed the technicians as they finished reconnecting the still-fueled-boosters' electrical system, when the second stage accidentally ignited. The flames from the second stage rocket engine burned into the first stage causing an explosion, killing many technicians and the field marshal. While not directly connected to the Vostok program, the loss of skilled and experienced space center workers must have been costly, and almost certainly delayed Vostok.

Another accident in the Vostok development occurred on March 23, 1961. During a ground test in an isolation chamber, cosmonaut Valentin Bondarenko was killed in a flash fire in a pure oxygen atmosphere.[17] The Soviets did not reveal the accident until 1986. This was very similar to the disastrous Apollo 1 fire that killed Ed White, Gus Grissom, and Roger Chafee while testing their capsule in 1967.

Figure 1-3. This photo of the A-type launch pad at the Baykonur Cosmodrome was taken by a U-2 aircraft in 1959. The launch pad was the only one in service at Baykonur until the late 1960s, when two more A-type launch pads were built. The flame trenches under the launch pads are 45 meters deep, and displace 1 million cubic meters volume. The booster sits on a concrete platform structure that extends over the deepest part of the flame pit. (Source: U.S. Central Intelligence Agency.)

After making two completely successful test flights in a row with the redesigned spacecraft and an aborted launch that was successfully recovered, the Soviets launched Yuri Gagarin into orbit, beating NASA's first sub-orbital Mercury flight. The Vostok 1 flight proved to be a great propaganda success for the Soviets. So much so, that a month later after the 15-minute sub-orbital Mercury flight of Alan Shepard, President Kennedy was forced to respond by committing the U.S. to the next round of the space race, which would end on the moon.

Four months after Vostok 1, Vostok 2 was launched with Gherman Titov on board. During the flight, Titov experienced some space sickness and as a result, the next Vostok mission was postponed for a year to study the phenomenon. Medical concerns over the effects of weightlessness on the human body were many, in both the U.S. and U.S.S.R. The only previous weightlessness experiments available were 20 to 30 seconds in aircraft. Doctors had voiced concern over just how long a person could survive. But as flights grew longer and longer, the doctors continually pushed back the safety limits. The question of a limit remains open today, and keeps getting pushed back year after year as the Soviets continually set records in flight duration. The problem at the time was immediate space sickness due to confusion in the body's vestibular system, which senses orientation to gravity.

The question of space sickness did not affect the U.S. space program until the late 1960s when the spacecraft were large enough to move around in extensively. The early Soviet capsules, while small, did allow room to float around enough to affect the vestibular system and cause sickness in some people. After resumption of flights, the Soviets performed two double spacecraft flights, and put the first woman in orbit. These six manned Vostok missions used up the spacecraft's potential for spectacular flights by orbiting men and a woman for durations that reached the safe limit. By orbiting two spacecraft at a time, the Soviets got a little more propaganda value from the last four missions, but little more could be gained from further Vostok flights.

Voskhod Program

Khrushchev wanted to continue with a highly visible space program, but the Vostok could not be used directly for the manned lunar effort that President Kennedy had proposed. The Soviet's moon ships being developed for flight to the moon could not be ready to fly for at least three years. Until then, Khrushchev ordered Chief Designer Korolev to beat the U.S. in launching the first multiman crew. In the United States, the Gemini, a two-seater spacecraft, was being planned to develop flight techniques needed for lunar flight. To surpass Gemini for Khrushchev, Korolev revised plans to modify the Vostok to carry three people and changed its name to Voskhod.[18]

While delaying initial flights of the manned Soyuz moon ship to support the interim Voskhod, the Soviets probably went ahead with development of Soyuz-compatible upper stages to be used in a Soyuz circumlunar mission. During the lull in manned spaceflight activity in between Vostok and Voskhod, flight tests of these Earth escape stages may have taken place under the Polyot name. Only two flights called Polyot were ever flown, and no later flights resembled the Polyot missions.

The Polyots were the first maneuverable satellites, having large rocket engines capable of restarting in orbit. This feature would be necessary to send a probe to the moon after being parked in Earth orbit. The Soviets' plan for the lunar flight was to launch a booster stage and fuel it in orbit. After the refueling, a manned Soyuz moon ship would dock and be boosted to the moon by the booster stage (this is described in more detail in Chapter 2). Whether the Polyots were involved in the manned lunar program or not is still not clear, but the first Soyuz missions were certainly designed to support use of such a booster stage.

Meanwhile, test flights of the interim Voskhod spacecraft were being prepared. The Voskhod was a modified Vostok in which the ejection seat was removed, and the cabin configured to carry three people. An optional inflatable airlock, used for extra vehicular activities (EVA), was also designed to be attached to the side of the capsule (see Figure 1-4). These modifications made the Voskhod too heavy to be orbited by the A-1 type booster. The extra weight required a more powerful upper stage for the A-type booster. Fortunately, a larger upper stage, called the Venus, was already being developed for Soyuz and other unmanned uses.

The Voskhod program also introduced the possibilities of flying doctors, scientists, women, engineers, and journalists.[19] With each was the opportunity to claim new firsts in space for the U.S.S.R. Before the program was canceled, two journalists were selected for spaceflights that never materialized. They were Yaroslav Golovanov and Yuri Letunov.[20] Valentina Tereshkova was also to make a Voskhod spacewalk in a repeat of Leonov's Voskhod 2 spacewalk.[21] Despite the intention to fly relatively untrained people on Voskhods, the Voskhod was probably the most dangerous manned program ever undertaken, since the crew could not escape from the spacecraft in case of an early launch failure. Ejection seats could no longer be used because of the limited space in the capsule, and new abort tractor rockets being developed for the next generation Soyuz spacecraft were not ready. But the Soviets probably did not consider it any more dangerous than the Vostok since they thought that ejecting early in a launch was not likely to succeed even though it was possible. It was not until the U.S. shuttle mission STS-5 that another manned spacecraft would be flown with no provisions for escape early in the launch. Fortunately, there were no Voskhod launch failures.

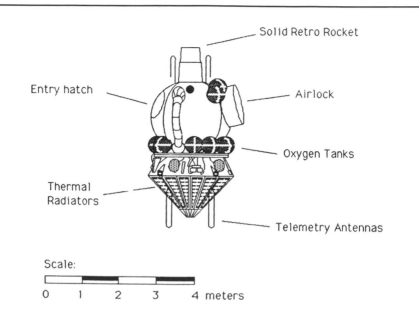

Figure 1-4. The Voskhod was a modified version of the Vostok spacecraft, and could carry a crew of two or three people in spacesuits. It could be equipped with an inflatable airlock for EVA.

Voskhod Spacecraft

The Voskhod weighed about 5,300 kg, depending on the particular mission and configuration. The added weight required the use of the more powerful A-2 booster for launch, which also allowed higher orbital altitudes. Planned high-orbit missions required a larger solid propellant retrorocket, which was added to the top of the capsule. Voskhod was the first manned spacecraft to have no escape system to save the crew from a malfunctioning booster early in a launch. A tractor rocket escape tower was not developed until 1966 for the first flights of the Soyuz spacecraft. Because ejection seats were not used on the Voskhod, landing rockets were tied to the capsule's parachute lines. The rockets fired just before the capsule hit the surface, slowing the impact to tolerable levels.

As the Soviets made relatively minor changes to produce the Voskhod, the United States was making major changes in its second manned spacecraft, Gemini. Its design was much different from Mercury, although it retained the same general appearance. To allow easier servicing, spacecraft components were highly modular and designed for easy replacement. On-board computers, reusability, docking and extensive maneuvering in orbit were also important new features. These would be critical developments for landing on the moon. During the middle 1960s, only two Voskhod flights took place, while 12 Gemini missions developed the technology and skills to fly beyond Earth orbit. The Soviets began falling behind the United States in spaceflight technology even as they made space spectaculars with the Voskhod in Earth orbit.

The first Voskhod mission carried the first three-man crew, beating the first flight of the two-man Gemini. The second flight was dedicated to the first spacewalk, or extra vehicular activity (EVA), upstaging by less than three months the Gemini 4 EVA of Ed White. The Voskhod program then came to a sudden end.

Despite the risk and limited propaganda value of further Voskhod flights, it has been reported that a Voskhod-type spacecraft launch failed in July 1965. Whether this was a manned Voskhod or a similar research or military reconnaissance satellite is not publicly known. There is speculation that a Voskhod had been planned for launch in early 1966, and that the next Voskhod mission would have been a long duration flight of two men for about a week. There has also been speculation about a possible Voskhod mission with an all-female crew, and even a joint flight of two Voskhod's similar to the joint Vostok flights, but the Voskhod program was reportedly canceled before 1966.[22]

The end of the Voskhod program apparently was not the failure of a mission, but that the political pressures to fly the risky Voskhod's had been relieved, after Khrushchev's downfall. It was also clear that the political and propaganda value of a week-long flight would have been minor at best since the U.S. Gemini program had already flown a record-setting two-week flight, and any joint flights would have been only repeats of the joint Vostok missions.

The flight of Kosmos 110 is included in this listing, because it initially was probably planned as a precursor to a third Voskhod flight before the Voskhod program was terminated. Kosmos 110 became the first in a long series of Soviet biological research flights lasting more than 20 years. Kosmos 110 was also notable as the first flight to use the 51.6° inclination orbit, a characteristic of nearly all future man-related flights. The new inclination enabled heavier payloads to be launched than to a 65° orbit, but it also brought the rockets' flight paths closer to the Chinese border than before. Figure 1-5 shows a time line of all Soviet man-related flights.

In related spacecraft development in the early 1960s, the Soviets used the Vostok design as a reconnaissance satellite to counter similar work by the U.S.

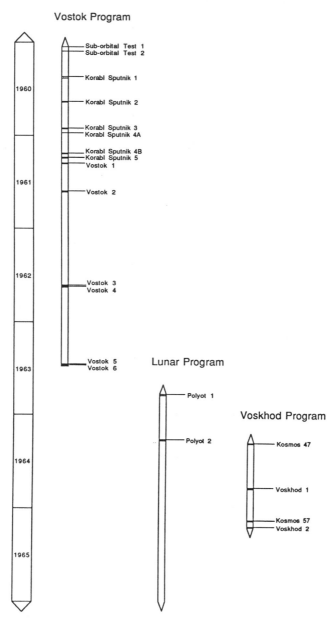

Figure 1-5. The Vostok, Voskhod, and Lunar programs were operating from 1960 to 1965. All known flights related to these programs are shown here and are described in this chapter.

Reconnaissance Versions of Vostok and Voskhod Spacecraft*

Soviet reconnaissance satellites are classed into different generations and different sub-types depending on the capabilities of the spacecraft. Vostok and Voskhod derivative spacecraft were divided into first-, second-, and third-generation reconnaissance satellites. Subsequent versions of reconnaissance spacecraft were based on the Soyuz manned spacecraft and are listed in Chapter 4.

First Generation: These reconnaissance satellites were based on the Vostok spacecraft, and were flown on flights between Kosmos 4 to 157. Mission duration was up to 18 days although 8 days was the most common. As with the Vostok, they could not maneuver in orbit and their utility was limited by this factor. They are no longer in use.

Second Generation: These satellites were based on the Voskhod spacecraft and flew in similar orbits to a manned Voskhod. The basic Voskhod had been improved and modified for two types of reconnaissance missions usually lasting 8 days. They are no longer in use.

Low resolution: Flew missions from Kosmos 120 to 344
High resolution: Flew missions from Kosmos 22 to 355

Third Generation: The third-generation satellites were based on the Voskhod design with an enlarged equipment module including in-orbit maneuvering capability, larger telescopes and film supplies. Flew missions from Kosmos 208 to 1004. Some flights also conducted military experiments, and some were used for publicized Earth resources work. An experiment module was carried on top of the capsule housing scientific, military and weather sensing instruments. On high resolution flights, this module was used to house a maneuvering engine.

Morse code type: Flown from Kosmos 251 to 632. No longer in use.
Low resolution: Flown from Kosmos 470 to 1597. No longer in use.
Medium resolution: These flights usually lasted 12 to 24 days. When launched from Baykonur they went into 70° or 70.4° orbits. When launched from Plesetsk they went into 72.9° or 82.6° orbits. First flight was Kosmos 867. They are still in use.
High resolution: These flights usually lasted 8 to 14 days and were launched from Plesetsk into 72.9, 82.3 or 62.8° orbits. First flight was Kosmos 364. They are still in use.

---

* Peebles, Curtis, *Guardians: Strategic Reconnaissance Satellites*, Novato, CA: Presidio Press, 1987, pp. 154, 157.

The capsule in this case carried a large camera and film supply. The capsule returned to Earth with the film and camera after a short mission, usually about a week in duration. During the 1960s, the Soviets extended the lifetime of the satellites and eventually changed to the Voskhod design for longer mission duration and more capabilities. Modern versions of the same spacecraft were still flying in the late 1980s as bio-satellite and commercial material processing flights.

Sub-orbital Test 1
Launched: January 20, 1960
Landed: January 20, 1960
Altitude: ?

This was the first spaceflight directly involved in the Vostok program. The flight was to test the Vostok capsule heatshield during reentry heating. The landing point of the capsule is not known, but was probably the Kamchatka peninsula where Soviet ICBM tests usually terminate. For those flights, the A-type boosters were targeted to impact 1,700 km southeast of Hawaii.[23] No further details of the test are known.

Sub-orbital Test 2
Launched: January 31, 1960
Landed: January 31, 1960
Altitude: ?

This flight was probably similar to the January 20 flight to test reentry heating on the capsule heatshield. No further details of this test are known.

Korabl Sputnik 1
(Sputnik 4)—Failed
Launched: May 15, 1960, 3:00 a.m.
Reentry: October 15, 1965, 12:21 a.m.
Altitude: 312 × 369 km @ 65°

Korabl Sputnik means spaceship satellite in Russian. This was the first orbital test flight of a Vostok and carried a simulated man in a pressure suit in the ejection seat. The flight was to test the automatic systems that would be used for a manned flight.

The spacecraft's weight at launch was 4,540 kg. The launch sequence is shown in Figures 1-6 and 1-7. To avoid confusion during the flight, the Soviets used a tape of a Russian choir to test the communication system, instead of a single voice. Stories about lost and dead cosmonauts were all too common in the early 1960s to add to the outrageous rumors by using a recording of a single voice that could be mistaken for a real pilot. There is absolutely no evidence, including recently released Presidential papers, that any of the reports of cosmonauts dying in flight in the early 1960s were in any way factual. All Korabl Sputnik flights transmitted live television from the capsule showing any reactions of the animal specimens.[24]

After the spacecraft performed the planned tests, the landing sequence was initiated on the 64th orbit. The capsules pointing system used an infrared sensor to sense the Earth's horizon. The horizon was used to determine the orientation angle needed to fire the retrorocket to reduce orbital velocity and place the capsule on a path entering the atmosphere. The sensor failed and the spacecraft pointed in the wrong direction during the retrofire sending the ship into a 290 × 675-km orbit.[25] It was later determined that the sensor had actually failed three days before the landing attempt, but the failure went undetected.

After the misguided retrofire, the capsule separated from the equipment module as programmed in a normal reentry sequence, leaving the capsule and equipment module drifting in a useless orbit. The capsule's self-contained systems continued to operate for eight days after the failed reentry attempt. The Vostok capsule was designed to be self-supporting for eight days in case the retrorocket failed to fire. In a normal manned mission, the orbit was intentionally low, which would force the spacecraft to reenter due to atmospheric drag after a maximum of eight days. The capsule finally reentered in 1965 and was destroyed on landing since power for landing systems was long since depleted.[26]

Korabl Sputnik 2

(Sputnik 5)

Launched: August 19, 1960, 11:38 a.m.
Landed: August 20, 1960, 2:02 p.m.
Altitude: 297 × 324 km @ 65°

On July 23, the first attempt to launch this flight was aborted on the launch pad. When the command to ignite the A-1 boosters' engines was given, there

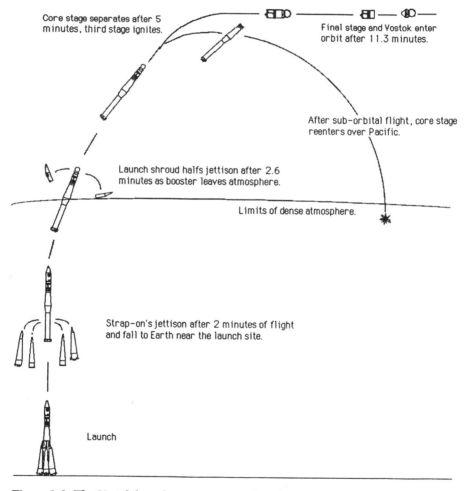

Core stage separates after 5 minutes, third stage ignites.

Final stage and Vostok enter orbit after 11.3 minutes.

After sub-orbital flight, core stage reenters over Pacific.

Launch shroud halfs jettison after 2.6 minutes as booster leaves atmosphere.

Limits of dense atmosphere.

Strap-on's jettison after 2 minutes of flight and fall to Earth near the launch site.

Launch

Figure 1-6. The Vostok launch sequence started with ignition of the booster core and strap-on boosters at launch. About 2 minutes later, the strap-on boosters and launch shroud were jettisoned followed by core stage cutoff after 5 minutes. The upper stage then boosted the Vostok into orbit and the spacecraft separated for orbital flight.

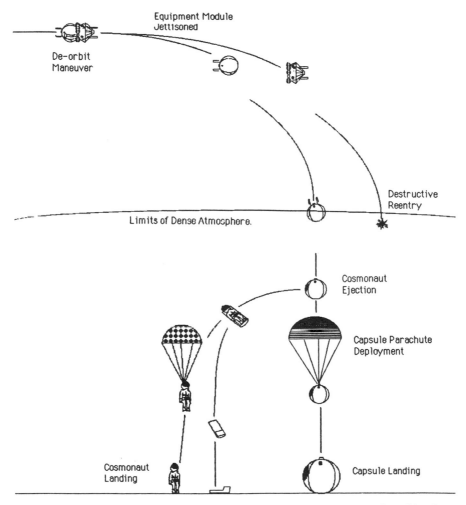

Figure 1-7. Vostok landing sequence. All Soviet manned spacecraft capable of landing are designed to land on the ground or water. The Vostok capsule landed by parachute, but landing on the ground was too harsh for humans. All cosmonauts ejected from the capsule and landed by parachute separately.

was no response. None of the engines ignited and there was no damage to the launch pad or the booster.[27]

When finally launched, this second orbital test flight of the Vostok spacecraft carried the dogs Belka and Strelka, and a variety of other specimens including 40 mice, two rats, and plants. Live television of the dogs was transmitted to mission control during the flight to monitor their reactions to the flight.[28] The animals were carried in a container strapped into the ejection seat so they could be recovered without the risk of being killed by the capsule's hard landing. The capsule was equipped with a parachute so it could be recovered with its information and experiment results intact, but the landing was very hard in comparison to the ejection seat. The capsule landed less than 10 km from the target point.[29,30]

Korabl Sputnik 3

(Sputnik 6)—Failed
Launched: December 1, 1960, 10:26 a.m.
Reentry: December 2, 1960, 12:12 p.m.
Altitude: 166 × 232 km @ 65°

This flight tested the redesigned Vostok spacecraft and its new systems like the new computer control system.[31] The flight was limited to a low altitude orbit, as normal manned missions would be, to ensure orbit decay within the eight-day limit of capsule resources in event of a retrorocket malfunction.

Again, as on the Korabl Sputnik 1 flight, the retrofire burn was not at the correct attitude and the capsule was destroyed during a very steep reentry into the Earth's atmosphere.[32,33] The cosmonauts were not unduly concerned, because they could always override the automatic system in event of a failure and orient the spacecraft themselves. The failure did change the flight plans for Vostok 1 from 6–18 orbits to 1 orbit.

Korabl Sputnik 4A

Aborted
Launched: December 21, 1960
Landed: December 21, 1960
Altitude: ?

This flight was to test the redesigned spacecraft systems again after the Korabl Sputnik 3 failure. The flight was proceeding normally until well into the first stage burn when one of the A-1's strap-on boosters prematurely shut down. The booster continued flying a normal launch profile seemingly unaffected by the anomaly. The strap-on boosters separated at the programmed time and the booster continued to fly toward orbit. When the third stage ignited, the booster was well below the velocity needed to achieve orbit because of the early strap-on shutdown and the command was given to abort the flight. The spacecraft would have been well out of the atmosphere at the time of the abort. To perform the abort, the Vostok separated from the firing third stage, turned around, and fired its retrorocket in the direction of flight. The capsule then separated from the equipment module in preparation for reentry into the atmosphere, on a steep ballistic path toward the western U.S.S.R. The capsule reentered and landed normally and its cargo of animals was recovered in good condition.[34] Despite the failure to reach orbit, the spacecraft systems performed perfectly during the abort. There are reports that another test flight of the Vostok may have been attempted in February 1961, but no details are known.[35]

Korabl Sputnik 4B

(Sputnik 9)
Launched: March 9, 1961, 9:29 a.m.
Landed: March 9, 1961, 11:16 a.m.
Altitude: 173 × 239 km @ 64.9°

This flight was a rehearsal for Vostok 1 and another attempt at testing the redesigned systems after two consecutive failed missions. It carried a simulated man in a pressure suit in the ejection seat, the dog Chernushka, mice, and a guinea pig in boxes attached to the cabin's walls. The flight was successful and even the animals apparently survived the hard capsule landing.[36]

Korabl Sputnik 5

(Sputnik 10)
Launched: March 25, 1961, 9:00 a.m.
Landed: March 25, 1961, 10:47 a.m.
Altitude: 164 × 230 km @ 64.9°

This flight was the final rehearsal before Vostok 1. It carried a simulated man in a pressure suit in the ejection seat and the dog Zvezdochka in a container attached to the cabin's wall. Both were recovered successfully clearing the way to the first manned spaceflight less than a month later.[37, 38]

Vostok 1

First Man in Space
Launched: April 12, 1961, 9:07 a.m.
Landed: April 12, 1961, 10:55 a.m.
Altitude: 169 × 315 km @ 65.07°
Crew: Yuri Gagarin
Backup: Gherman Titov
Call sign: Cedar

The booster and spacecraft were moved from the assemble building to the launch pad on April 11, a day before the launch.[39] There are reports that U.S. intelligence detected the booster for Vostok 1 being moved to the launch pad and knew that a manned flight was imminent. President Kennedy was informed and reportedly drafted a message of congratulations to the Soviets before the flight began.[40]

The night before the launch Gagarin and his backup, Titov, slept in a building near the launch pad, and unknowingly had their sleep monitored by sensors under their beds. Both reportedly moved little in the night. Gagarin and Titov prepared for the launch by putting on pressure suits and rode out to the launch pad in a bus. One of the engineers working for Chief Designer Korolev, Konstantin Feoktistov, had checked and double-checked the capsule's switch settings and systems before Gagarin's arrival. Feoktistov would soon become a cosmonaut and later would design the civilian Salyut space stations.

Gagarin boarded the capsule 90 minutes before launch. The countdown was delayed at least once due to a faulty valve. During the countdown, music was played over the radio for Gagarin's enjoyment.[41] This would be a standard practice for Soviet launches for decades to come. Just before the launch, he was promoted from lieutenant to major.[42]

As NASA was still preparing for its first sub-orbital manned flight, the Soviets surprised most of the world by launching the first man into orbit (see Figure 1-8). At launch, the spacecraft weighed 4,725 kg. Gagarin's pulse rate reached a maximum of 158 during the launch.[43] A maximum acceleration of up to six gravities was produced during the upper stage burn.[44] The Soviets announced the launch during the flight and named it only as Vostok. After subsequent flights, the mission was commonly referred to as Vostok 1.

The flight was limited to one orbit to provide information on weightlessness during a substantial period, but not too long as to pose extra risk. The spacecraft's automated systems worked well and Gagarin's main task was to observe them. During the flight, Gagarin surprised many Soviet scientists by his ability to see small objects on the ground (much like Gordon Cooper did during the Mercury Atlas-9 flight in 1962). Gagarin also tested some food and drinks during the flight. He did not have a camera to take pictures of the Earth, but he used a

Figure 1-8. Vostok 1 Launch. The Vostok capsule is visible through the round opening in the launch shroud. The opening allowed for ejection in case of an emergency early in the launch, although the chances of a successful ejection were thought to be low. Note the ice formed over the liquid oxygen tanks which makes the booster appear white in areas. (Source: Tass from Sovfoto.)

tape recorder to note his observations, which were transmitted at high speed when within range of a tracking station or ship. During normal orbital flight, the capsule was put into a slow tumble to even out thermal heating on the spacecraft's exterior.[45]

The Vostok capsule was a completely automatic spacecraft, as demonstrated by the Korabl Sputnik test flights. There were provisions for the capsule to be manually controlled during the only major maneuver during the flight—orientation for retrofire. A manual system for pointing the spacecraft named Vzor was available, but the system was normally locked with a combination lock. Soviet doctors believed that the cosmonaut could possibly become mentally disturbed by the effects of spaceflight and interfere with the orientation system, hence the lock. An envelope with the combination was taped to the cabin wall for use in an emergency by, it was hoped, a rational cosmonaut.[46] Fortunately, there were no ill effects of this type.

As the spacecraft came around the world over South America and across Africa, the capsule performed the retrofire maneuver correctly. The capsule then separated from the equipment module to begin its fall into the atmosphere (see Figure 1-7). At 10:35 a.m., the capsule reentered the atmosphere and began to tumble, but the motion eventually damped out. The Vostok capsule fell in a ballistic path experiencing up to eight gravities of deceleration.

At 7,000 meters altitude, the capsule hatch was blown off and Gagarin ejected, landing by parachute. The capsule also landed by parachute nearby, but without braking rockets the landing was too harsh for people. The capsule and Gagarin landed southwest of Engels in the Saratov region. The official landing site was near Smelovka, Saratskaya. A 40-meter-tall titanium monument now marks the official spot.[47]

After the mission, the Soviets were very vague about the actual landing and whether Gagarin landed in the capsule, because the Federation Aeronautique Internationale (FAI) rules of aerospace world records require that a pilot must be in control of the vehicle (or at least in it) from the moment of take-off until landing. Ejecting from the capsule would violate these rules and not give the U.S.S.R. credit for the first manned space flight. After questioning the Soviets about the flight and the landing for two days, and still not getting answers from the Soviets, the FAI gave in, even though at the time the Soviets had released very little proof of an actual manned flight.[48]

There is no doubt that Gagarin used the ejection seat during the landing like all latter Vostok cosmonauts. The descriptions of the landing site of Vostok 1 are different depending on the source. The Soviets said that the capsule (with Gagarin inside) landed in a plowed field, while it was also written that he landed in a pasture.[49] The latter was probably true and the deception was necessary from the Soviet viewpoint due to the technicality in the FAI rules. The Soviets later released many statements admitting that Gagarin ejected, as was normal procedure for landing in a Vostok capsule.[50]

Analysis of the landing time for Gagarin and the capsule also support that the ejection seat was used. In 1986, a doctor for the cosmonauts also said that Gagarin landed like all the other Vostok cosmonauts. The doctor was in one of four teams of parachutists that were prepared to aid the cosmonaut after landing. The team was not needed for the Vostok 1 landing and were waved off after ground teams reached Gagarin. If the rules had been strictly adhered to, the U.S. would today hold the records for the first man in space and the first man in orbit, instead of the U.S.S.R.

Vostok 2

First Day-Long Flight

Launched: August 6, 1961, 9:00 a.m.
Landed: August 7, 1961, 10:18 a.m.
Altitude: 166 × 232 km @ 64.9°
Crew: Gherman Titov
Backup: Andrian Nikolayev
Call sign: Eagle

After the successful flight of Gagarin, Soviet spaceflight doctor V.I. Yazdivsky and cosmonaut commander General Kamanin wanted the next manned flight to last three orbits to cautiously gather information on weightlessness.[51] But, political pressures, the pilot, and the engineers wanted a one-day flight, and got it. The spacecraft's weight at launch was 4,713 kg. Minor improvements were made to the spacecraft including an improved capsule air conditioning system.

During the flight, Titov tested the spacecraft's attitude control system twice, tested specially prepared food on the third and sixth orbits, and tested exercise methods. He experienced some symptoms of space sickness on the fifth orbit. After sleeping for five orbits, starting on orbit seven, the symptoms diminished. Television from inside the capsule was relayed to mission control during the flight.[52] He also photographed the Earth using a hand-held Zritel camera. During the flight, Premier Khrushchev promoted Titov from captain to major and made him a full Communist Party member.[53] After this mission, the next Vostok flight was postponed a year to study the space sickness experienced by Titov.

Vostok 3
First Double Flight

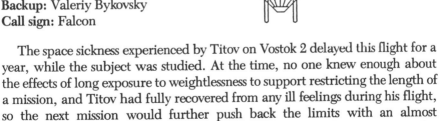

Launched: August 11, 1962, 11:30 a.m.
Landed: August 15, 1962, 9:54 a.m.
Altitude: 166 × 218 km @ 65°
Crew: Andriyan Nikolayev
Backup: Valeriy Bykovsky
Call sign: Falcon

The space sickness experienced by Titov on Vostok 2 delayed this flight for a year, while the subject was studied. At the time, no one knew enough about the effects of long exposure to weightlessness to support restricting the length of a mission, and Titov had fully recovered from any ill feelings during his flight, so the next mission would further push back the limits with an almost four-day-long flight. Boris Volynov was also trained as backup cosmonaut for the flight.[54] The Vostok 3 spacecraft weighed 4,722 kg at launch.

Nikolayev was the first cosmonaut to unstrap himself and float around in the small cabin. He reported none of the space sickness symptoms encountered by Titov, which affect only about half of space travelers. The cabin's two television cameras were used to observe his reactions to weightlessness.[55] During the flight, his electro-encephalogram, electro-oculogram, and galvanic skin reactions were monitored. The Soviets broadcast the first live pictures of a cosmonaut from space during the flight. The flight was proceeding normally when Vostok 4 was launched unexpectedly.[56]

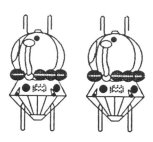

Vostok 4

Launched: August 12, 1962, 11:02 a.m.
Landed: August 15, 1962, 10:09 a.m.
Altitude: 169 × 222 km @ 65°
Crew: Pavel Popovich
Backup: Vladimir Komarov
Call sign: Golden Eagle

Popovich arrived at the Baykonur Cosmodrome on August 8, to prepare for the flight and watch the assembly of the 4,728-kg Vostok 4 spacecraft with its booster (see Figure 1-9). He also must have watched as Nikolayev was launched the day before his own launch. Boris Volynov also trained as backup cosmonaut for the flight as for the Vostok 3 flight.[57]

The Vostok 4 flight was launched without any advanced word during the Vostok 3 flight. The Soviets said that Vostok 4 used the same launch pad as Vostok 3 a day earlier. At the time there was only one A-type launch pad at the Baykonur Cosmodrome. The Soviets reported that the main goal of the joint

Figure 1-9. The Vostok is shown here in the A-type booster assembly building at Baykonur. The spacecraft is mated to its upper stage and is being lowered to a horizontal position before being slid into its launch shroud and mated to the booster rocket. (Source: Tass from Sovfoto.)

flight was to gain experience of tracking and planning rendezvous between two orbiting spacecraft.

This would have been an important task in the original Soviet manned lunar mission program. The Vostok did not have the ability to rendezvous, but the spacecraft did come within 6.5 km of each other in very similar orbits.[58] Popovich made radio contact with Nikolayev an hour after launch and was sighted by Nikolayev shortly afterwards.[59] While the spacecraft were within sight of each other the cosmonauts could use their radios to talk to each other.

Both cosmonauts said they could see the other, describing each other's ship as a bright point of light. The cosmonauts communicated together by radio often as both followed the same schedule of tests. Both made the first live television broadcasts from space shown on both Soviet and European television. The space sickness issue raised by Titov's flight was dispelled when only Popovich experienced some mild disorientation during the flight.

During Vostok 4's final orbit, it was separated by 2,850 km from Vostok 3 along their orbits. With the joint mission completed, both spacecraft landed at approximately the same time about 193 km apart, south of Karaganda, four days after the first was launched.[60,61]

Vostok 5

Double Flight
Launched: June 14, 1963, 3:00 p.m.
Landed: June 19, 1963, 2:00 p.m.
Altitude: 162 × 209 km @ 65°
Crew: Valeriy Bykovsky
Backup: Boris Volynov
Call sign: Hawk

The Soviets said that the Vostok 5–6 flight was to study the biological effects of prolonged spaceflight on humans and to perfect the spacecraft's systems.[62] The flight was a virtual repeat of the Vostok 3–4 flight, except that Vostok 6 would launch the first woman into space.

The launch had to be delayed for one day due to bad weather at the Baykonur Cosmodrome launch site.[63] After the booster's upper stage separated, Bykovsky turned the spacecraft around and observed it drift away, as U.S. astronauts did on Mercury flights.

During the first day of the mission, Bykovsky made normal reports of his and the spacecraft's condition. He also ate normal food during the flight, including beef and chicken.[64] Vostok 6 was launched on the second day of the flight as Vostok 5 flew over the Baykonur Cosmodrome launch site.

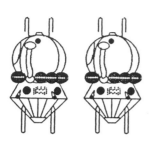

Vostok 6

First Woman in Space

Launched: June 16, 1963, 12:30 p.m.
Landed: June 19, 1963, 11:20 a.m.
Altitude: 168 × 218 km @ 65°
Crew: Valentina Tereshkova
Backup: Irina Solovyova
Call sign: Seagull

Valentina Tereshkova, a textile factory worker and amateur parachutist, became the first woman in space on Vostok 6.[65] There were four other women in training for the flight, and were never identified publicly until 1988.[66] The women reportedly continued ground support jobs at Star City after the flight of Tereshkova. Some reports state that Tereshkova was originally the backup, and was substituted for some reason at the last moment. Her backups were officially listed to be Irina Solovyova and V. Ponomareva.[67] To ensure that women could withstand the long flight, two simulated flights were conducted on the ground lasting twelve and six days.

The Vostok 6 spacecraft weighed 4,713 kg at launch. There are some reports that Tereshkova experienced severe space sickness during the flight, but there are also reports that the flight was extended, and was initially planned for only one day.

The closest approach between the Vostok 5 and Vostok 6 spacecraft was 5 km.[68] While the spacecraft were within line of sight of each other, the cosmonauts were able to talk to each other over the radio. During the flight, both cosmonauts were shown on television in both the Soviet Union and western countries.[69] Both cosmonauts tested the spacecraft's control systems and took pictures of the Earth, moon, and stars through the three port holes in the cabin. The joint flight ended uneventfully as Vostok 6 landed first, 625 km northeast of Karaganda. Vostok 5 landed at 2:00 p.m., 540 km northwest of Karaganda.[70]

During the flight, Tereshkova spent more time in space than all the Mercury astronauts combined.[71] After the flight, Tereshkova was taken to the International Congress of Women, which was meeting in Moscow.[72] The Soviets presented her as an example of the equality of women in the U.S.S.R. Since before the flight, she was Cosmonaut Nikolayev's fiancee and they were married five months later. She then officially retired from the cosmonaut corps to raise their child, but in reality she later resumed training for a later canceled Voskhod flight that was to include an EVA. The women cosmonauts continued to train until 1969, when they were finally disbanded.[73] Tereshkova and Nikolayev were divorced in 1982.[74] It would be 19 years until the next woman, also a Soviet, would fly in space.

Polyot 1
First Maneuverable Satellite
Launched: November 1, 1963, 11:53 a.m.
Reentry: October 16, 1982
Altitude: 339 × 592 km @ 59.92°

It is theorized that the two Polyot flights could have been tests of the original Soviet lunar flight plan to launch a manned Soyuz around the moon. This plan and the Soyuz manned lunar spacecraft are described in Chapter 2. The plan would have required launching a manned Soyuz, a Soyuz B booster stage, and three Soyuz V tanker spacecraft. The size and complexity of this mission plan would lead to its demise shortly after the intensity of the U.S. space program was recognized. The Soviet change in lunar mission plans is described in Chapter 2.

The Soyuz B and V data here are from analyses by Phillip S. Clark and Ralph F. Gibbons of Soviet design studies. The Soyuz B was to be launched into orbit with empty fuel tanks and be fueled by three Soyuz V propellant tankers in orbit. The Soyuz B was 4.7 meters long, 2.5 meters in diameter, and weighed about 1,950 kg. It would be launched with a maneuvering unit, 3.1 meters long, attached to the aft end, and weighing 3,500 kg fueled. The Soyuz B would have to use the maneuvering unit to dock with the Soyuz V tankers. The Soyuz V tanker would have been 4.2 meters long, 2.5 meters in diameter, and weighed 6,100 kg fueled. The tanker would carry 4,155 kg of propellant for transfer to the Soyuz B. It would require three flights of the Soyuz V, one carrying only UDMH (unsymmetrical dimethyl hydrazine) and two carrying nitrogen tetroxide propellants to fully fuel the stage. A manned Soyuz would then be launched to dock with the Soyuz B upper stage, and be boosted out of Earth orbit to fly around the moon.[75]

The Polyot flights are believed to have been tests of the Soyuz B rocket body weighing about 1,590 kg, without a maneuvering unit, and with 1,000 kg of fuel in its tanks that would normally be launched without fuel. Premier Khrushchev announced Polyot 1 as the first maneuverable Earth satellite and the beginning of a new era of spaceflight. Polyot 1 maneuvered within 24 hours of launch, after several engine firings, to a final orbit at 343 × 1,437 km @ 58.55°.[76]

Polyot 2
Launched: April 12, 1964, 12:22 p.m.
Reentry: June 8, 1966
Altitude: 242 × 485 km @ 59.92°

Polyot 2 was acclaimed similarly to the first mission. During the second day in orbit the spacecraft maneuvered from its initial orbit to 310 × 500 km @ 58.06°.[77]

The Soyuz circumlunar plan using the Soyuz B and V was canceled about the time of the Polyot flights in favor of using a Proton launch vehicle for launching men in a Zond spacecraft (a light-weight version of the Soyuz) on the same type of circumlunar mission. In the final analysis, the Polyot missions were then probably flown as technology demonstrations for future restartable upper stages to be used in the lunar program.

Kosmos 47

First Test of Voskhod

Launched: October 6, 1964, 10:12 a.m.
Landed: October 7, 1964, 10:30 a.m.
Altitude: 174 × 383 km @ 64.62°

This was the first flight of the modified Vostok or Voskhod spacecraft, and it was a rehearsal for the Voskhod 1 flight.[78] This was also the first man-related flight to be flown under the Kosmos label. Many such flights would use the Kosmos name during the next two decades. The Kosmos name was used by the Soviets to help disguise military flights by mixing them with scientific and civilian missions also given Kosmos names.

Voskhod 1

First Three-man Crew

Launched: October 12, 1964, 10:30 a.m.
Landed: October 13, 1964, 10:47 a.m.
Altitude: 177 × 377 km @ 64.9°
Crew: Vladimir Komarov, Boris Yegorov, and
     Konstantin Feoktistov
Backups: Boris Volynov, Georgiy Katys, and Vasili Lazarev
Call sign: Ruby

This was the first manned flight of the Voskhod and the first to use the A-type booster and Venus upper stage (see Figure 1-10). The Venus upper stage was painted a black and white checked pattern for some Voskhod flights. The spacecraft's weight at launch was 5,320 kg.

Voskhod 1 was manned by a most extraordinary crew. Komarov was one of the most experienced and skilled of the cosmonauts. Yegorov was the first physician in space. Originally, Yegorov was the backup for Ilyin who was

Figure 1-10. A Voskhod spacecraft on its A-2 type booster is shown here seconds before launch. The launch pad support arms are about to fall away from the booster as it lifts off. The spacecraft has no emergency abort systems to save the crew during the first few minutes of the flight. The capsule could separate from a malfunctioning booster only after launch shroud separation. (Source: Tass from Sovfoto.)

removed from the flight late in the training process.[79] The other crewman, Feoktistov, was one of the top scientist-engineers in the Soviet space program and was the personal choice of Korolev from the engineers in his design bureau to make the flight. Feoktistov was later the designer of the civilian Salyut space stations and a prime force in the Soviet space program to this date. Yegorov and Feoktistov were specialists and not career cosmonauts (similar to current NASA space shuttle payload and mission specialists). They were given only four months to prepare for the flight.[80] During the next year, two journalists, Yaroslav Golovanov and Yuri Letunov, were also selected for spaceflights, which never materialized.[81]

For many years there were many theories as to how the Soviets managed to squeeze three cosmonauts into such a small spherical capsule. Pictures published in 1987 clearly show the seating arrangement.[82] The three men were fitted into the Voskhod by positioning three seats next to each other with the middle seat raised a few inches above the other two so the seat frames at the cosmonauts shoulders could overlap slightly. Yegorov was in the middle seat with mission commander Komarov in the right seat at the attitude controls. This arrangement was cramped and did not allow enough room for the cosmonauts to wear pressure suits although they were reportedly carried on board. This seems unlikely since there would be very little room to put on a pressure suit in the cramped capsule. Instead, the cosmonauts wore gray flight suits and white helmets.[83]

Both Yegorov and Feoktistov were said to have experienced some space sickness throughout the flight. This might be due to the nature of the medical experiments performed by the crew investigating space adaptation and space sickness. They included excessive head movements that, as NASA Skylab astronauts later found, promoted space sickness.[84]

The objective of the flight was to test the Voskhod spacecraft and carry out scientific and medical experiments. Komarov tested the attitude control system on sixth and seventh orbits and experimented with electrostatic ion attitude control thrusters. Feoktistov tested a horizon sensor navigational aid, and Doctor Yegorov observed the crew's reactions to weightlessness. There were televised periods from the spacecraft shown in the U.S.S.R. of the crew performing experiments.[85] During the flight, Komarov said the crew was eager to extend the mission by one day, but Korolev would not change the flight plan.

At 9:55 a.m., October 13, Komarov oriented the Voskhod for retrofire. After retrofire, at 10:19 a.m., the capsule separated from the equipment module and reentry began shortly afterwards. The parachutes opened at 5,000 meters altitude, at a velocity of 220 meters per second. The capsule landed at 10:47 a.m., 420 km northwest of Karaganda (52°02′N, 68°08′E). The capsule's landing rockets, which were tied to the parachute lines worked so well the crew reportedly did not feel touchdown. The crew landed in a somewhat changed country though—the man that was mostly responsible for their spaceflight, Premier Khrushchev, had been removed from office during the mission, and Brezhnev and Kosygin greeted the cosmonauts when they returned to Moscow.[86,87]

Kosmos 57

Launched: February 22, 1965, 10:41 a.m.
Reentry: March 31, 1965, 1:42 p.m.
Altitude: 165 × 427 km @ 64.74°

This was a test of a Voskhod with an airlock attached to the side of the capsule. There are reports that the spacecraft may have started tumbling when it separated from the upper stage. The spacecraft exploded during its second orbit, breaking into 168 detectable pieces, at about the time the airlock should have deployed.[88] All the debris reentered by April 6, 1965. Despite this malfunction, the Soviets went ahead with the manned launch of Voskhod 2 a month later.

Voskhod 2

First Spacewalk

Launched: March 18, 1965, 10:00 a.m.
Landed: March 19, 1965, 12:02 a.m.
Altitude: 167 × 475 km @ 64.79°
Crew: Pavel Belyayev and Aleksei Leonov
Backups: Dmitri Zaikin and Yevgeniy Khrunov
Call sign: Diamond

The main objective of the Voskhod 2 flight was to perform the first EVA (extravehicular activity or spacewalk). The Soviets decided to include an EVA in a Voskhod flight when they learned that NASA planned for a "stand up" EVA during the Gemini program. The Soviets wanted to be first to perform an EVA to add to the list of space firsts, so Voskhod 2 was planned and flown before the Gemini EVAs. The Soviet EVA was little more than a stunt compared to the U.S. plans, which were to test manned maneuvering units and tools in space. In preparing for the flight, Leonov trained for long periods in isolation, under observation. He also practiced EVAs in vacuum chambers and on a Tu-104 aircraft flying brief periods of weightlessness.[89] Some Soviet psychiatrists believed that the experience of EVA and having no stationary references would be the most dangerous part of the mission.[90] As they later found out, there were much more important physical dangers. A third cosmonaut, Viktor Gorbatko, also trained as a backup crewman for the flight.[91]

Both cosmonauts wore EVA type spacesuits for the entire mission. The space suits may have been one of the first pieces of moon hardware to be tested by the Soviets. The spacesuits were totally self-contained units (like the Apollo PLSS suits), and in that respect more advanced than anything the U.S. had yet developed. The suit was very similar to the Vostok spacesuit, consisting of a pressure suit covered by a white outer garment for thermal protection.[92]

The spacecraft's weight at launch was 5,682 kg. When deployed, the airlock was one meter in diameter and two meters long. The airlock was necessary because the Voskhod carried only enough oxygen to replenish what the cosmonauts consumed in normal breathing. There would not be enough to refill the cabin after depressurizing for an EVA. Also, the equipment in the cabin was not designed for exposure to vacuum and low temperatures.[93] If the EVA was canceled, the flight would probably have lasted for two or three days.

During the second orbit, Leonov began preparing for the EVA by breathing oxygen for an hour to prevent getting the bends. The space suit pressure was only 304 mm Hg, less than half of the Voskhod cabin pressure.[94] The pressure difference would cause any nitrogen in the blood stream to boil out causing the bends. Breathing oxygen before an EVA was a way of clearing the nitrogen out of the body and was a practice used on both U.S. and Soviet spacecraft for at least the next 25 years. Leonov then got into the airlock, pressurized his spacesuit, checked for suit leaks, adjusted his helmet, and tested the oxygen life support system.

Belyayev then closed the airlock hatch and Leonov gradually depressurized the airlock. Leonov would be connected to the spacecraft during the EVA by a 5.35-meter tether, which also served as a communication link to the spacecraft.

Leonov's only duties during the EVA were to attach a camera to the end of the airlock to film his movements during the EVA and to photograph the spacecraft. After leaving the airlock at 11:35 a.m., Leonov placed the camera on the end of the airlock and tried to photograph the Voskhod, but he found that he couldn't bend to reach the still camera that was attached to the leg of his spacesuit (see Figure 1-11).[95] When he tried to bend his arms or legs, he lowered the volume inside the spacesuit, which increased the air pressure in the suit. The increasing pressure made any further bending increasingly difficult.

Figure 1-11. Aleksei Leonov is shown here outside the Voskhod 2 airlock during the world's first spacewalk. He wore a modified Vostok pressure suit, which caused great problems when he tried to reenter the airlock. (Source: Tass from Sovfoto.)

During the EVA, Belyayev held the ship steady using attitude control rockets, and said he could feel the ship move when Leonov bounced against it (he bounced against the capsule five times). Belyayev wore an identical spacesuit and could have gone outside to assist Leonov in an emergency (provided the outer hatch was closed first). There were airlock controls both inside the airlock and in the capsule.

When it was time to end the EVA, Leonov found that it was just as difficult to bend to get into the airlock as it had been to reach his camera. After several futile attempts at bending his legs into the airlock, he had to dangerously reduce air pressure in the suit to 190 mm Hg to reduce the ballooning effect. He found he still could not bend enough to go in feet first, and instead, went head first into the airlock. He somehow managed to turn around in the flexible airlock tube to reach the outer hatch and close it.[96] The EVA lasted 23 minutes with 11 minutes in the airlock. Leonov later said that he was on the verge of heat stroke and covered with sweat after the EVA. He lost 6 kg (12 pounds) of body weight that day. After the EVA, and with the objective of the flight behind them, the cosmonauts performed some minor experiments like testing their color perception (reduced 25.5%, average), and Leonov wrote notes about the EVA.

The crew then prepared for landing by first jettisoning the airlock from the side of the capsule. On the 16th orbit while preparing for a normal automatic retrofire, they discovered that the automatic guidance system had failed and they would have to use the manual solar orientation system.[97] This meant that Belyayev would manually orient the ship for retrofire on the 17th orbit. Belyayev put the spacecraft in the correct attitude for the retrofire by 11:19 a.m., and as Korolev counted down the seconds for him over the radio, he initiated retrofire at 11:36 a.m. Because of the inaccuracies of the manual procedure, the capsule overshot the landing area by 3,200 km and landed in a snow covered forest, 180 km northeast of Perm.

The capsule's radio beacon antenna broke off while landing among the trees, making it more difficult for the rescue forces to locate them. Eventually, they were located and supplies were air dropped to them 2.5 hours after landing. Since there were no clearings nearby for helicopters to land, it would be the next day until rescue forces reached the capsule. The cosmonauts were forced to retreat to the capsule after meeting timber wolves that night and had to hold the hatch closed to stop the wolves, while trying to rest in the very uncomfortable space suits.[98] After skiing to recovery helicopters the next day, they were airlifted back to Star City.[99]

After the flight, the Soviets portrayed EVA as simple and without difficulty, even though Leonov experienced great difficulty. This lead NASA to be somewhat unprepared for its EVAs, during the Gemini program, endangering astronauts' lives.

Kosmos 110

First Bio-satellite
Launched: February 22, 1966
Landed: March 16, 1966
Altitude: 175 × 512 km @ 51.9°

This flight was the first Bio-satellite mission and carried the dogs Verterok and Ugolyok. The flight may also have been the test flight for a later canceled Voskhod 3 mission.

After launch, the ship was maneuvered to a 190 × 882-km orbit, into the Van Allen radiation belts to test the effects of the additional radiation on the dogs. The dogs were carried in separate compartments in the capsule, and they were fed through tubes to their stomachs. Anti-radiation drugs were given to one dog through tubes to an artery. Both dogs' condition worsened during the flight, because they could not move or exercise in any way for the 22 days of the mission. After firing of the solid retrorocket attached to the front end of the capsule, the equipment module separated and the capsule went on to land successfully.

The Soviets recovered the capsule and found that the dogs experienced substantial bone calcium loss during the flight and could not move normally until ten days after landing. Two dogs used as controls were put through the same tests and confinement on Earth during the flight for comparison purposes. The Soviets reportedly were alarmed by the poor physical condition of the dogs after landing.[100]

This flight was also the first use of a 51° inclination orbit by a Soviet man-related spacecraft. From Baykonur, the 51° orbit takes the most advantage of the Earth's rotation while launching as near the Chinese border as considered prudent. The Soviets did not want to risk an international incident with the Chinese by having a manned spacecraft land there in case of a launch abort.

The Bio-satellite series of biological science flights continued over the years using almost the same version of the Voskhod spacecraft. Later Bio-satellites, however, were more like the military reconnaissance version of the Voskhod (described earlier). The other Bio-satellite flights to date were Kosmos 605, 690, 782, 936, 1129, 1514, 1667, and 1887. Since these flights were not directly related to manned spacecraft development, they are not included here.

References

1. Ordway, Fredrick I. and Sharpe, Mitchell R., *The Rocket Team*. Cambridge: MIT Press, 1979, pp. 333, 353.

2. McDugall, Walter, ...*the Heavens and the Earth, A Political History of the Space Age*. New York: Basic Books Inc., 1985, p. 105.
3. Baker, David, *The Rocket*. London: New Cavendish Books, 1978, pp. 115–116.
4. Oberg, James E., "Korolev and Khrushchev and Sputnik," *Spaceflight*, Vol. 20, No. 4, April 1978, p. 148.
5. Peebles, Curtis, *Guardians: Strategic Reconnaissance Satellites*. Novato, CA: Presidio Press, 1987, p. 68.
6. Oberg, James E., *Red Star in Orbit*. New York: Random House, 1981, pp. 28–29.
7. Oberg, "Korolev and Khrushchev and Sputnik," op. cit. p. 147.
8. Burrows, William E., *Deep Black: Space Espionage and National Security*. Random House: New York, 1986, p. 99.
9. McDugall, op. cit. p. 61.
10. Congressional Research Service, The Library of Congress. *Soviet Space Programs 1976–80, Unmanned Space Activities, Part 3*. Government Printing Office: Washington, 1985, p. 801.
11. Congressional Research Service, The Library of Congress. *Soviet Space Programs 1976–80, Manned Space Programs and Life Sciences, Part 2*. Government Printing Office: Washington, 1984, p. 486.
12. McDugall, op. cit. pp. 172–200.
13. Congressional Research Service, The Library of Congress. *Soviet Space Programs 1976–80, Unmanned Space Activities, Part 3*. Government Printing Office: Washington, 1985, p. 815.
14. McDugall, op. cit. p. 143.
15. Foreign Broadcast Information Service, U.S.S.R., Space, JPRS-USP-84-003, June 1984, Joint Publications Research Service, pp. 46, 48.
16. Oberg, James E., *Uncovering Soviet Disasters*. Random House: New York, 1988, pp. 178–181.
17. Bond, Peter, *Heroes in Space: From Gagarin to Challenger*. Basil Blackwell: New York, 1987, p. 17.
18. Clark, Phillip S. and Gibbons, Ralph F., "The Evolution of the Soyuz Program," *Journal of the British Interplanetary Society*, Vol. 36, No. 10, Oct. 1983, p. 434.
19. Foreign Broadcast Information Service, U.S.S.R., Space, JPRS-USP-86-005, Sept. 1986, Joint Publications Research Service, p. 62.
20. *Spaceflight*, Vol. 30, Feb. 1988, p. 72.
21. Kidger, Neville, "Above the Planet," *Spaceflight*, Vol. 31, Feb. 1989, p. 45.
22. Clark, Phillip S., "Soviet Launch Failures," *Journal of the British Interplanetary Society*, Vol. 40, No. 10, Nov. 1987, p. 527.

23. Johnson, Nicholas L., *Handbook of Soviet Manned Spaceflight.* American Astronautical Society: San Diego, 1980, pp. 27, 28.
24. Congressional Research Service, The Library of Congress, *Soviet Space Programs 1976–80, Manned Space Programs and Life Sciences, Part 2.* Government Printing Office: Washington, 1984, pp. 488–89.
25. Peebles, Curtis, "Setting Out for Space," *Journal of the British Interplanetary Society*, Vol. 40, No. 2, Feb. 1987, p. 89.
26. Johnson, op. cit. p. 28.
27. Clark, op. cit. p. 528.
28. Grahn, S., "Cosmosvision," *Spaceflight*, Vol. 17, No. 12, Dec. 1975, p. 440.
29. Johnson, op. cit. p. 29.
30. Congressional Research Service, The Library of Congress, *Soviet Space Programs 1976–80, Manned Space Programs and Life Sciences, Part 2.* Government Printing Office: Washington, 1984, p. 488.
31. Johnson, Nicholas L., op. cit. p. 30.
32. Peebles, op. cit. p. 89.
33. Congressional Research Service, The Library of Congress, *Soviet Space Programs 1976–80, Manned Space Programs and Life Sciences, Part 2.* Government Printing Office: Washington, 1984, p. 489.
34. Peebles, op. cit. p. 90.
35. Clark, op. cit. p. 528.
36. Congressional Research Service, The Library of Congress, *Soviet Space Programs 1976–80, Manned Space Programs and Life Sciences, Part 2.* Government Printing Office: Washington, 1984, p. 489.
37. Johnson, op. cit. p. 32.
38. Congressional Research Service, The Library of Congress, *Soviet Space Programs 1976–80, Manned Space Programs and Life Sciences, Part 2.* Government Printing Office: Washington, 1984, p. 489.
39. Bond, op. cit. p. 13.
40. Daniloff, N., *The Kremlin and the Cosmos.* New York: Alfred A. Knopf, 1972, p. 138.
41. Bond, op. cit. p. 14.
42. Time-Life Books Inc., *Life in Space.* Little, Brown and Company: Boston, 1983, p. 39.
43. Johnson, op. cit. pp. 34, 38.
44. Oberg, *Red Star in Orbit*, op. cit. p. 52.
45. Congressional Research Service, The Library of Congress, *Soviet Space Programs 1976–80, Manned Space Programs and Life Sciences, Part 2.* Government Printing Office: Washington, 1984, p. 490.
46. Oberg, *Red Star in Orbit*, op. cit. p. 53.
47. Bond, op. cit. p. 12.

48. Mallan, L., "The Russian Spacemen Who Weren't There!" *Science & Mechanics*, June 1966, pp. 32–33.

49. Oberg, *Red Star in Orbit*, op. cit. p. 55.

50. Foreign Broadcast Information Service, U.S.S.R., Space, JPRS-USP-84-003, June 1984, Joint Publications Research Service, p. 47.

51. Daniloff, op. cit. p. 108.

52. Grahn, op. cit. p. 440.

53. Time-Life Books Inc., op. cit. p. 51.

54. *Spaceflight*, Vol. 31, No. 2, Feb. 1989, p. 57.

55. Bond, op. cit. p. 20.

56. Johnson, op. cit. p. 46.

57. *Spaceflight*, Vol. 31, No. 2, Feb. 1989, p. 57.

58. Johnson, op. cit. pp. 46, 47.

59. Ertel, Ivan D. and Morse, Mary L., *The Apollo Spacecraft, A Chronology, Vol. I*. NASA SP-4009, Government Printing Office: Washington D.C., 1969, p. 181.

60. Bond, op. cit. pp. 20–22.

61. Furniss, T., *Manned Spaceflight Log*, 2nd Ed., London: Janes Publishing Co. Ltd., 1986, p. 18.

62. Morse Mary L. and Bays, Jean K., *The Apollo Spacecraft, A Chronology, Vol. II*. NASA SP-4009, Washington D.C.: Government Printing Office, 1973, p. 64.

63. Bond, op. cit. p. 24.

64. Johnson, op. cit. p. 50.

65. Congressional Research Service, The Library of Congress, *Soviet Space Programs 1976–80, Manned Space Programs and Life Sciences, Part 2*. Government Printing Office: Washington, 1984, p. 495.

66. *Spaceflight*, British Interplanetary Society, Vol. 30, Feb. 1988, p. 72.

67. *Spaceflight*, Vol. 31, No. 2, Feb. 1989, p. 57.

68. Johnson, op. cit. p. 52.

69. Morse, Mary L. and Bays, Jean K., *The Apollo Spacecraft, A Chronology, Vol. II*. NASA SP-4009, Washington D.C.: Government Printing Office, 1973, p. 64.

70. Furniss, op. cit. p. 23.

71. Congressional Research Service, The Library of Congress, *Soviet Space Programs 1976–80, Manned Space Programs and Life Sciences, Part 2*. Government Printing Office: Washington, 1984, p. 495.

72. Daniloff, op. cit. p. 109.

73. Kidger, Neville, *Spaceflight*, Vol. 30, July 1988, p. 301.

74. Bond, op. cit. p. 25.

75. Clark and Gibbons, op. cit. pp. 436–437.

76. Clark, Phillip S., "The Polyot Missions," *Spaceflight*, Vol. 22, No. 9–10, Sept.–Oct. 1980, p. 312.
77. Ibid.
78. Johnson, op. cit. p. 69.
79. *Spaceflight*, British Interplanetary Society, Vol. 30, Feb. 1988, p. 27.
80. Bond, op. cit. p. 66.
81. *Spaceflight*, Vol. 30, Feb. 1988, p. 72.
82. Hart, Douglas, *The Encyclopedia of Soviet Spacecraft*. New York: Bison Books, 1987, pp. 145, 146–7 (photos).
83. Turnill, R., *Spaceflight Directory*. London: Frederick Warne Ltd., 1977, p. 327.
84. Bond, op. cit. pp. 68, 71.
85. Brooks, Courtney G. and Ertel, Ivan D., *The Apollo Spacecraft, A Chronology, Vol. III*. NASA SP-4009, Government Printing Office, Washington D.C., 1976, p. 7.
86. Johnson, op. cit. p. 75.
87. Bond, op. cit. p. 70.
88. Johnson, op. cit. p. 69.
89. Bond, op. cit. p. 73.
90. Johnson, op. cit. p. 76.
91. *Spaceflight*, Vol. 31, No. 2, Feb. 1989, p. 57.
92. Bond, op. cit. p. 75.
93. Johnson, op. cit. p. 76.
94. Bond, op. cit. p. 75.
95. Johnson, op. cit. p. 78.
96. Congressional Research Service, The Library of Congress, *Soviet Space Programs 1976–80, Manned Space Programs and Life Sciences, Part 2.* Government Printing Office: Washington, 1984, p. 498.
97. Bond, op. cit. p. 79.
98. Crook, Wilson W., "Of Space Walks and Wolves: The Incredible Flight of Voskhod 2," *Space World*, Jan. 1987, pp. 12–13.
99. Johnson, op. cit. pp. 82–83.
100. Johnson, op. cit. p. 71.

Chapter 2

The Moon Race

The series of Soviet space spectaculars that started with Sputnik had continually increased the attention of the U.S. public. After Gagarin's Vostok flight on April 12, 1961, President Kennedy was forced by political and public pressure to propose a major new space effort, with the goal of landing a man on the moon. The secretive Soviets were also planning to win that goal first. The Soviets were totally convinced that their years of spaceflight experience and superiority in heavy lift launch vehicles would let them achieve any space goal before another country. But, the Soviets were overconfident and underestimated the U.S. effort. This led to the Soviets wasting valuable time in the early 1960s, while the U.S. built a massive space industry infrastructure of research centers, factories, and launch complexes.

Soon after Kennedy's announcement, Soviet Premier Nikita Khrushchev was briefed by Valentin P. Glushko, a top Soviet rocket engine designer, on the current Wernher von Braun plan to send men to the moon. The von Braun plan was a method called Earth Orbit Rendezvous (see Figure 2-1). EOR meant assembling a spacecraft in Earth orbit that would then fly on to the moon. This could require as many as 15 launches of individual components, and had not yet been accepted as the final NASA plan. Glushko mentioned another possible method—a direct ascent to the moon (see Figure 2-1) using a giant rocket 15 times bigger than the newest Soviet A-2 type. At that time, the United States had yet to even launch a booster as powerful as the A-2. After hearing these plans, Khrushchev concluded that President Kennedy's moon landing goal was only propaganda and delayed any action on a moon program.[1]

In July 1962, NASA chose a relatively new plan, Lunar Orbit Rendezvous (LOR, see Figure 2-1), as the method for the Apollo moon landing missions. This would require fewer and smaller boosters, save a planned $1.5 billion, and could be accomplished six to eight months sooner than EOR or direct ascent.[2] The drawback to LOR was the development of not one, but two specialized spacecraft, and perfecting untried methods of rendezvous and docking around the moon before 1970.

43

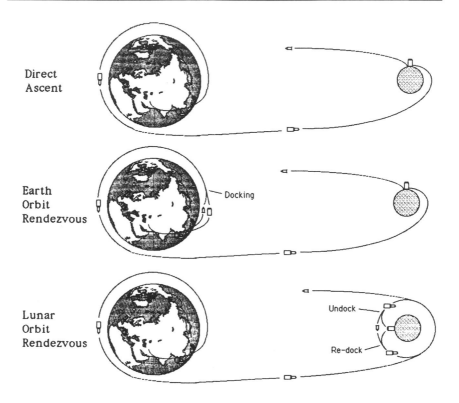

Figure 2-1. Lunar mission profiles. Direct ascent requires the largest booster and requires landing a massive spacecraft on the moon, but offers the simplest flight plan by not requiring rendezvous. Earth orbit rendezvous uses smaller boosters but more than one launch, and requires rendezvous and assembly in Earth orbit before leaving for the moon. Lunar orbit rendezvous requires a booster slightly smaller than direct ascent, but also requires rendezvous of two spacecraft around the moon after a landing. LOR also requires development of two independent spacecraft, one to land on the moon and one to return to Earth.

By this time, it must have become clear to the Soviets that the massive work begun by NASA was a serious challenge. The Soviets were concerned about the U.S. effort, but still thought that the goal of landing on the moon before 1970 was very unrealistic. Not wanting to waste resources unnecessarily, the Soviets began their own program that would only send people around the moon. This was their first manned lunar effort. Over the next few years their lunar effort went through major changes as the threat of the U.S. challenge grew more ominous.

The Lunar Program

Over the years, rumors and details about Soviet space flights and developments have been slowly made public from various sources around the world. Even though the details of the manned lunar program are not perfectly clear, there is no doubt that it was real, despite Soviet denials of the early 1970s. There is much evidence for a lunar program, including published statements from the Soviets (see Figure 2-2). The Soviets have said that Gagarin was training extensively for a mission in 1968, before his death. However, after he had served as back-up to Komarov on the ill-fated Soyuz 1, in 1967, he was apparently not assigned to another Soyuz Earth orbit mission. Cosmonaut Valentina Tereshkova, during a trip to Cuba, said that Gagarin would command the first lunar landing mission (see Table 2-1).[3] In the 1960s, the Soviets also built a fleet of ships to provide communications for Earth orbit and lunar expeditions that was very similar to the NASA Deep Space Network. In May 1967, cosmonaut Belyayev was certain he would make a circumlunar flight very soon.[4] These are only a few examples of the statements that indicate that the manned lunar effort was a major program.

The Soviets' manned lunar program is difficult to explain since it ultimately comprised three different plans to reach the moon. After joining the moon race in 1962, when the NASA effort was recognized, the Soviets apparently adopted an EOR plan to send men around the moon. They evidently felt that this would be enough of an effort for the immediate future, and believed that the NASA lunar landing effort was far from being realized by the 1970 goal date. The Soviets' plan was to launch a large empty rocket booster into orbit, and then fuel it using an automatic tanker spacecraft. Then a three man spacecraft would dock to the fueled booster and be launched out of Earth orbit on a lunar fly-by mission. Gagarin said that a plan like this was being worked on, in September 1963.[5]

Table 2-1
Possible Lunar Program Cosmonauts

Mission	Commanders	Flight Engineers (SLM Pilots)
Earth orbit test	Filipchenko	Kubasov
Lunar orbit tests	Leonov?	Sevastyanov
First lunar landing	Gagarin	Makarov
Second landing	Leonov?	Grechko
Third landing		Rukavishnikov

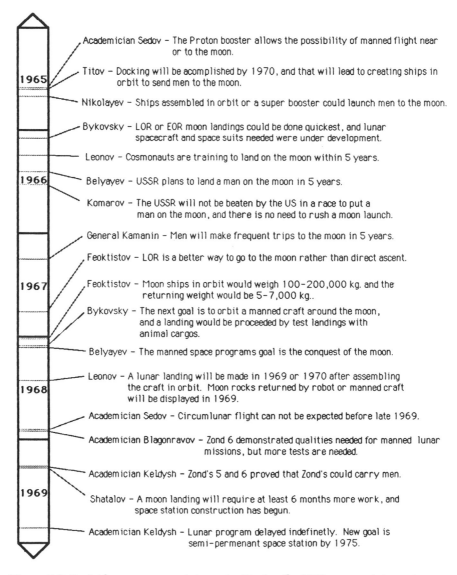

Figure 2-2. Soviet lunar program statements. During the 1960s, many statements were made showing the evolution of Soviet lunar landing plans from the then traditional EOR and direct ascent method, to LOR or EOR, and then abandonment of the program after the first U.S. landing on the moon.

The First Lunar Program

For the manned lunar spacecraft, the Soviets apparently adopted a 1961 General Electric proposal for the Apollo spacecraft that could carry three men on a journey around the moon. The spacecraft proposal included a mission module, a bell-shaped reentry module, and a propulsion module.[6] This was very similar to what the Soviets eventually built and named the Soyuz. Details of the design of the Soyuz are in many areas very similar to the GE Apollo proposal. The Soyuz was the only spacecraft from the fly-by plan that was ever flown operationally (see Figure 2-3).

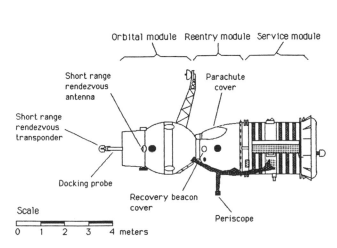

Figure 2-3. The Soyuz spacecraft was originally designed to dock to a booster stage in orbit and be propelled around the moon. The Soyuz was later modified to fly crews to Earth-orbiting space stations.

Soyuz Spacecraft

The Soyuz weighed about 6,600 kg and was used to carry a three-person crew into a low Earth orbit. It consisted of a near spherical orbital module, a bell-shaped reentry module, and a service module. The spacecraft was about 9 meters long from rear antenna to docking probe (exact length depended on the type docking system) and 10 meters wide with solar arrays deployed, with 9 cubic meters habitable volume. The Soyuz used the same A-2 booster as the Voskhod for launch. *(continued on next page)*

(continued from page 47)

The Soyuz that flew between 1966 and 1970 was originally designed for the Soyuz B-V circumlunar mission described in Chapter 2. The spacecraft was never used for that mission because the plan was abandoned in favor of a much larger lunar landing effort to match the U.S. Apollo program. Different variants of the original Soyuz were developed for use into the 1990s and are described in later chapters.

The forward third of the Soyuz comprised the orbital module. The orbital module was roughly spherical measuring 2.25 meters in diameter and 3.1 meters long including the docking system (later changed to 2.65 meters for Soyuz 10 and 11). The module served as a sleeping quarters, laboratory, air lock and cargo hold.

The Soyuz reentry module comprised the middle element of the spacecraft. It was 2.2 meters long and 2.3 meters in diameter. It was bell-shaped to provide aerodynamic lift to enable 3- to 4-gravity reentries. The capsule could also perform ballistic reentry from lunar distances which resulted in 10- to 16-gravity decelerations. The Soyuz carried up to three people, but did not allow enough room for the crew to wear pressure suits for launch or landing. The Soyuz reentry capsule could land on the ground or water, although a ground recovery in the Soviet Union was preferred for easy crew recovery. Small landing rockets fired at 2 meters altitude to ease landing shock on either land or water.

The rear of the Soyuz was the service module, consisting of a pressurized section for instrumentation and an unpressurized engine compartment. It was 2.3 meters long, and 2.3 meters in diameter with a flared base 2.72 meters in diameter, and weighed 2,650 kg fueled. The attitude control system consisted of eighteen 10-kg and twelve 1-kg hydrogen peroxide thrusters mounted on the forward and rear ends of the module. The engine compartment contained 4 tanks of propellant and a torus shaped tank, which was always flown empty to allow launch on the A-2 booster. In 1964 design studies, the torus tank was jettisonable, but this was never conclusively demonstrated in flight. The main propellant was nitric acid and UDMH (unsymmetrical dimethylhydrazine). The KTDU-35 main engine had a thrust of 409 kg. Situated next to the main engine was the dual nozzle backup engine, which had 403 kg of thrust. Either engine could be fired many times for a total of 500 seconds, but the backup engine was restartable only if it fired more than 90 seconds. Two solar arrays attached to the sides of the service module. After reaching orbit each deployed to 3.6 meters length and 1.9 meters width, for a total area of about 14 square meters.

This lunar fly-by plan would use equipment that had already been developed, i.e., the same A-2 boosters and launch facilities being developed for the Voskhod and other unmanned programs. The three-man Soyuz, the Soyuz B rocket body, and the Soyuz V tanker spacecraft were the only new equipment that had to be developed. The plan was a simple EOR type, requiring five launches for each lunar mission. But the techniques needed to accomplish the mission were not simple. Automatic rendezvous and docking would be necessary for the Soyuz V tankers to meet and fuel the Soyuz B rocket in Earth orbit. The manned Soyuz would then have to dock with the fueled Soyuz B rocket, which would launch the cosmonauts out of Earth orbit and around the moon (see Figure 2-4).[7]

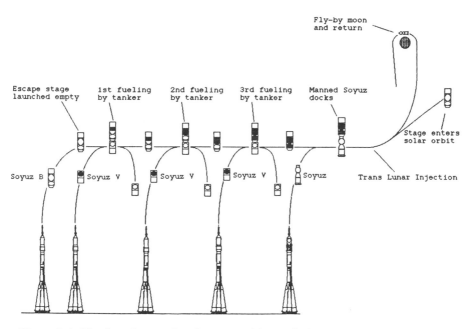

Figure 2-4. The first Soviet plan for manned lunar flights was a classic Earth orbit rendezvous plan, which included multiple refueling flights of a booster and the docking of a manned Soyuz for a flight around the moon.

The Polyot flights could have been tests of the Soyuz B in 1963 and 1964. The Soyuz flights flown from 1967 to 1970 were direct descendants of this first lunar fly-by plan. Since the EOR plan depended on docking in Earth orbit before a lunar flight, the first Soyuz missions tested and finally perfected automatic docking. But, that was not accomplished until late 1967, and a manned docking was not accomplished until 1969, during the Soyuz 4–5 mission. This was far too late to win the race to the moon and by 1964, the Soviets had foreseen that docking could not develop quickly enough to beat the U.S., in light of increasing U.S. confidence and abilities.

By 1964, the Soviets saw that to defeat the U.S. in the race to land on the moon they must make a major effort in new rockets and facilities, just as the U.S. was already doing. The Polyot flights associated with the original Soyuz plan were flown as technology demonstrators (much as the lunar landing equipment would be in the 1970s, after losing the moon landing race). The Soyuz spacecraft was modified for use in the new moon landing effort, and the original circumlunar Soyuz was used to learn the techniques of rendezvous and docking that would be vital to the lunar landing effort.

The Second Lunar Program

With the original lunar fly-by plan canceled, and with the surviving Soyuz's flying research flights to develop the techniques needed to support a lunar landing mission, the Soviets began other work on the lunar landing effort. The lunar landing mission would be a combination of LOR and EOR. A large new booster would be used to launch an Earth orbit escape stage and lunar lander into Earth orbit. Then, a three-man lightweight version of the Soyuz, called a "heavy Zond," would be launched to dock to the lander in orbit and be launched to the moon by the escape stage (see Figure 2-5). From there on the lunar mission would resemble the U.S. Apollo flights (see Chapter 3 for description).

But it was still 1964 when these changes were made, and the Soviets still had not given up the race to reach the moon first. The new lunar landing plan would also provide the equipment needed to perform the lunar fly-by mission by 1969, still beating the planned U.S. missions. A lightweight version of the lunar spacecraft would be used and launched by the large D-1e Proton booster (see Figure 2-6). The Proton booster and Zond (a lightweight Soyuz) spacecraft would be prepared for the lunar fly-by mission and for launching a crew to a lunar lander in Earth orbit for a lunar landing mission. This would also allow the Soviets to test pieces of the lunar landing mission hardware in flight around the moon.

The new lunar landing mission became the Soviets' main program and, at the same time, supplied the means to attempt the first manned lunar fly-by. The lunar landing program included many new pieces of equipment and new facilities on the ground. The Zond spacecraft was a lightweight version of the original Soyuz, and it would perform the same lunar fly-by mission. A heavy Zond, a Zond with extra propellant tanks, would be used to launch and return

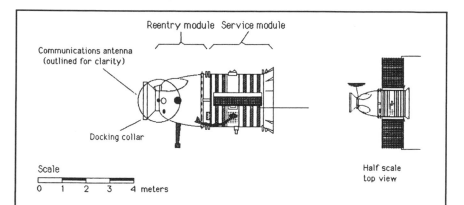

Scale

0   1   2   3   4 meters

Half scale
top view

Figure 2-5. The Zond spacecraft was developed from the Soyuz design to allow a direct launch by a Proton (D-1e) booster on a flight around the moon. The Zond would also be used as a part of the manned lunar landing plan as described in Chapter 3.

Zond Spacecraft

The Zond was a modified Soyuz in which the spacecraft weight was reduced to enable launch by the D-1e booster to lunar distances. The Zond was intended to carry at least one cosmonaut around the moon but was never actually flown manned. Two models of the Zond were apparently developed, one for launch on a solo lunar flight, and another to support a manned lunar landing.

The Zond used the same service module as the early Soyuz, although Zond's did not usually carry torus tanks. The Zond also used a different version of the Soyuz main engine, the KTDU-53 which did not have a reserve engine. The Zond's solar arrays were similar to the Soyuz, measuring 3.1 meters long with antennas on the ends. Overall span of the panels was 9 meters. For the Zond, the Soyuz orbital module was replaced with a docking collar roughly 0.5 meter high and 1.5 meters in diameter, to provide access to the capsule on the launch pad and probably provide internal transfer from the Zond to the Soviet Lunar Module.

cosmonauts to the Earth during the lunar landing missions. To simplify development, both Zond types would be launched by the same Proton booster.

The Soviet Lunar Module (SLM) would be needed to take cosmonauts from a heavy Zond to the lunar surface and back. To get the SLM and heavy Zond

(text continued on page 53)

35273

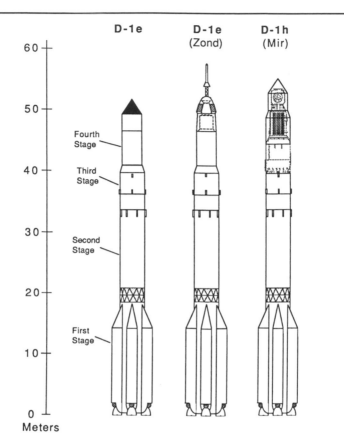

Figure 2-6. The D-type (Proton) booster was to be used to launch the Zond spacecraft around the moon as precursors to manned lunar landing missions. The D-type was also used in its three-stage version to launch space stations and modules into Earth orbit.

D-Type Booster

The type D booster was reportedly developed in 1961 by the Chalomei Design Bureau initially as an ICBM to launch "city buster" 100-megaton hydrogen bombs. The Soviets later announced that the booster would be used for unmanned lunar and planetary missions and to launch improved manned spacecraft (the Zond). Significant features about the design of the Proton successfully remained secret from the general public for 20 years.

(continued on next page)

The D-1 type booster consisted of only first and second stages. The first stage consisted of a center tank for nitrogen tetroxide, and six surrounding tanks (long thought to be strap-on boosters) containing UDMH with six RD-253 staged combustion cycle rocket engines at their base. The RD-253 engine burned the propellant turbine's exhaust in the thrust chamber instead of dumping it overboard like the earlier RD-107. Each engine on the first stage could swivel outward, and the six together provide full steering for the stage. The first stage produced 1,002,000 kg of thrust. The Proton second stage weighed 13,180 kg and had four RD-253 engines for a total thrust of 244,000 kg. During the second stage burn, six minutes after launch, the launch shroud surrounding the payload is jettisoned at an altitude of 145 km.

The three-stage version of the Proton is the D-1h. The same first and second stages were used with a third stage that weighed 5,610 kg. Around 1980, the weight of the third stage was reduced to 4,400 kg. Both versions of the stage had one RD-253 rocket engine and four steerable 30-kg thrust vernier engines, and a total thrust of 61,000 kg. The D-1h booster was 44.3 meters long and 7.4 meters wide at the base of the first stage. The D-1h was used to launch all space station modules into the 1990s.

The D-1e was used to launch probes to high Earth orbit or beyond using a D-1h booster with a fourth stage. The original fourth stage used was referred to as Block-D. The Block-D was 5.5 meters long and 3.1 meters wide. It rested on an interstage that was shaped like a cone on top of a cylinder 4 meters wide at the base and 1.1 meters long. The interstage was jettisoned after separation from the third stage. The weight of the Block-D including the LOX and kerosene propellants was 17,300 kg. The empty or dry structure weighed 1,800 kg. The stage produced about 8,700 kg of thrust using an unidentified engine with a total burn time capability of 600 seconds and that could be restarted twice. After separation from the third stage, the fourth stage drifts in orbit and uses UDMH-nitrogen tetroxide ullage motors to maintain orientation for a maneuver to lift the payload out of low Earth orbit.

(text continued from page 51)

to the moon required an escape stage to launch them out of Earth orbit and a Lunar Braking Module (LBM) to stop them in lunar orbit. All except the manned heavy Zond would be launched by a new super booster, the G-1 type or Enn-Odin (N-I). The lunar landing mission is described in detail in Chapter 3.

During the time period covered in this chapter, the Soviets tested the Zond and Proton booster in Earth orbital flights and then on lunar fly-by missions. Flight testing of the Soyuz spacecraft began first, followed quickly by the Zond. The first manned Soyuz flight ended in disaster, killing Vladimir Komarov, one of the most skilled cosmonauts. The Soyuz program was delayed for over a year.

During the delay, several more unmanned Soyuz spacecraft were tested, and the Zond made its first successful lunar fly-by flight carrying animals. Interestingly, the U.S. Apollo program was similarly delayed by the spacecraft fire that killed the first Apollo crew in 1967. When the Soviets resumed manned flights using the Soyuz, they continued the development of rendezvous and docking needed for the lunar landing mission. The Soyuz was not directly connected to the current lunar program. It was being used similarly to the U.S. Gemini spacecraft for perfection of rendezvous and docking in orbit. The Soyuz also simulated the heavy Zond spacecraft and partially tested its main engine system.

Until 1967, NASA's plans were to land a man on the moon after a series of test flights of the Saturn V booster and Apollo spacecraft. This plan consisted of several intermediate flights of Apollo that would lead to a landing on the moon, but NASA knew that time was running out to land before 1970. Some NASA space center directors and other officials began planning, in secret, to move up a lunar orbit flight to enable a landing before 1970, and, they hoped, to preempt the Soviets of putting a man around the moon first. In August 1968, the secret plan was revealed to and approved by NASA Administrator Webb. The Apollo 8 flight was changed from a conservative high Earth orbit flight to being the first lunar orbit mission. For the first time since the beginning of the space race, the United States was within reach of a space first that would finally rival the well publicized Soviet achievements of the early 1960s, and they were not about to let the chance slip away.

The fears of the NASA officials were well founded, the Soviets were also nearing the point of launching a man around the moon. The next month the Soviets tested a Zond spacecraft sending animals around the moon, which was normal Soviet practice before sending men on the same mission. The Soviets also made statements confirming that the Zond had the ability to carry men around the moon, and that the Zond's Proton booster was intended for manned missions.[8] In October, the first manned launch of Apollo into Earth orbit was successful. Gherman Titov in a statement in Mexico City was sure that the Soviet Union would still be the first to send men to the moon.[9] The next month, the Soviets launched another Zond on a successful trip around the moon. The race to the moon was coming to a very close finish in December 1968.

Monthly launch windows to the moon start several days earlier for the Soviets than for the U.S. After two consecutive successful test flights, the same number as before the first manned launch of Vostok, the Soviets prepared a manned Zond for launch in early December. They attempted the launch for three days before succumbing to technical problems that forced postponement to the next launch window in January.

After that failure, the Soviets could only hope that NASA would encounter similar problems. But, Apollo 8 was successfully launched on December 21, 1968, on a journey that would circle the moon ten times. After Apollo 8, to fly around the moon then would have been admitting second place status, and the moon race was about being first. The Soviets canceled plans for a manned circumlunar mission, but they still had hopes of being first to land on the moon using the G-1 super booster.

If the Soviets had been as confident and prepared as NASA, they clearly could have won at least the race to reach the moon with men. The Soviet moon program delay in 1961 could not have helped the effort, and adding to that delay, Chief Designer Korolev and his top assistant, L.A. Voskreseneky, died before the new generation of Zond and Soyuz manned spacecraft were completed. Deputy Chief Designer Vasily Mishin assumed the top post for finishing the design of the Soyuz.[10]

In the period covered in this chapter (see Figure 2-7), more than 25 percent of the manned missions did not achieve their goals or failed, and the first Soyuz mission resulted in the death of its pilot. Yet, the Soviets almost won the first round in the race to the moon. They would fall further behind the U.S. in the race in the coming years covered in Chapter 3.

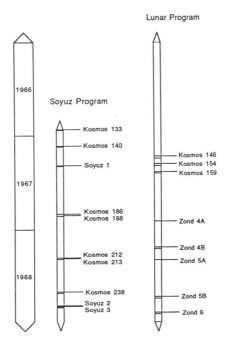

Figure 2-7. The Soyuz test program and Lunar programs were operating from 1966 to 1968. All known flights related to these programs are shown here and are described in this Chapter.

Other related flights not listed in the chapter include possible tests of Soyuz and Zond capsules in sub-orbital reentry tests from December 16, 1965 to June 7, 1966. The Soviets announced that a variant of a landing system for spacecraft would be tested during that time and booster elements could impact the Pacific Ocean around 42°N 177°W, between 12:00 a.m. and 12:00 p.m. local time.[11] Figure 2-8 illustrates a Zond abort test vehicle.

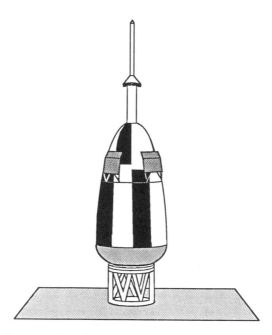

Figure 2-8. This drawing is based on a Soviet photograph and shows a Zond spacecraft abort test vehicle, which indicates that the Soviets conducted abort test flights similar to the NASA Little Joe and Little Joe II series. This also shows that the Zond was a manned spacecraft, because there is no need to recover an unmanned spacecraft from a booster failure.

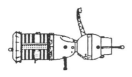

Kosmos 133

First Test of Soyuz—Failed

Launched: November 28, 1966, 2:01 p.m.
Landed: November 30, 1966, 3:00 p.m.
Altitude: 171 × 223 km @ 51.8°

This was the first test flight of the new three-man Soyuz spacecraft originally designed for circumlunar flight. Its orbit was very close to that used by some reconnaissance satellites, but this flight used the same frequencies as Vostok and Voskhod manned flights, indicating its true nature. After testing the new spacecraft's systems for two days in orbit, the spacecraft was commanded to reenter.[12]

During reentry, the capsule's heat shield failed. The module was recovered, but was heavily damaged by a burn through the heat shield, which was the result of a manufacturing flaw.[13,14]

Kosmos 140

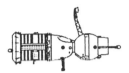

Launched: February 7, 1967, 6:22 a.m.
Landed: February 9, 1967, 6:07 a.m.
Altitude: 165 × 218 km @ 51.7°

This was the second Soyuz test flight, repeating the test flight of Kosmos 133. Kosmos 140 tested the heat shield that had failed during the previous test as well as other Soyuz systems. The capsule was recovered, but there were failures of either the temperature control system, attitude control system or parachutes. The heat shield functioned correctly during the reentry and landing.[15]

Kosmos 146

First Zond Test Flight

Launched: March 10, 1967, 2:32 p.m.
Landed: March 18, 1967, 11:12 p.m.
Altitude: 177 × 296 km @ 51.44°

This was the first test of the Zond launched by a D-1e Proton booster. The weight of the Zond spacecraft at launch was 5,017 kg.[16] The mission was probably intended to be the launch of the Zond into a highly elliptical orbit to simulate lunar return velocities of 11 km/second.

After orbital insertion, the Proton fourth stage apparently failed to restart. On the first day the Zond separated. On March 11, the fourth stage reentered. The Zond maneuvered to 185 × 350 km by March 12. On March 14, the docking collar probably was jettisoned resulting in a slight orbit change. The capsule landed on March 18, and the docking collar reentered the next day.[17] The Soviets later claimed that the mission was a success when it was listed as achieving over 11.2 km per second simulating lunar return velocity.[18] This flight also suffered failures of either the temperature control system, attitude control system, or the parachutes.[19]

Kosmos 154

Failed
Launched: April 8, 1967, 12:01 p.m.
Reentry: April 19, 1967, 11:27 a.m.
Altitude: 183 × 223 km @ 51.3°

This flight was apparently the same as Kosmos 146, and tested either a Zond or heavy Zond. This flight was probably intended to be a high altitude or high velocity reentry test. Again, the fourth stage apparently failed to restart to place the Zond in a higher orbit, from which it could make a high-velocity reentry. The Zond then separated from the fourth stage and jettisoned its docking collar but did not maneuver again. The third stage reentered on the first day, the fourth stage on the second along with another object, probably the docking collar.[20] The Zond decayed after eleven days after the probable failure of the temperature control system or attitude control systems.[21]

Soyuz 1

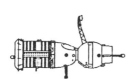

First Spaceflight Fatality—Failed
Launched: April 23, 1967, 3:35 a.m.
Landed: April 24, 1967, 6:23 a.m.
Altitude: 198 × 211 km @ 51.6°
Crew: Vladimir Komarov
Backup: Yuri Gagarin
Call sign: Ruby

In Moscow, rumors were flying before the launch of Soyuz 1.[22] Days before the launch, General Nikolai Kamanin, head of the cosmonaut corp, said that rumors of a six- to eight-man spacecraft were exaggerated, and that resuming the manned flights, meaning Soyuz 1, was not strictly scheduled and the Soviets would not launch until assured of success.[23] This is hard to believe since all previous attempts at launching a Soyuz and its sister ship, the Zond had suffered serious failures. With these problems in mind, the deciding factors to launch a manned Soyuz were most probably political, and related to the need to fly Soyuz to explore flight techniques vitally needed for the manned lunar landing program. James Oberg has reported a rumor that the Soyuz was launched despite design problems because of political pressure. Chief Designer Mishin reportedly refused to approve the launch, but was overruled.[24]

At the time, U.S. intelligence expected two spacecraft to be launched and then dock and exchange up to six cosmonauts. East Berlin radio said, "New

Soviet [manned] flights, after a two-year delay, will accomplish all the elements of the twelve costly Gemini flights in a single step, and penetrate more deeply into space."[25] A day before the launch, United Press International reported from Moscow that within 48 hours two spacecraft would be launched with four to six cosmonauts on board. It went on to say "in-flight hookup between the two ships and the transfer of crews" would occur.[26]

The flight of Soyuz 1 was officially announced the day after its launch. The spacecraft's weight at lift-off was 6,450 kg. The spacecraft separated from the final booster stage in orbit at 185×211 km (see Figure 2-9). Everything appeared normal, according to the news broadcasts. A change from past flights was the announcement of the flight as Soyuz 1, while the first Vostok and Voskhod were not given the "1" designation. (See box for details of the Soyuz launch sequence.)

Late that day, at 10:30 p.m., communications still appeared normal, but as Soyuz 1 flew over Baykonur at midnight, the expected Soyuz 2 flight was not launched. The news announcements now said nothing about the second spacecraft or docking. The Soyuz 1 flight was experiencing difficulties. Although the exact failures remain a Soviet secret, plausible explanations have been developed explaining the flight's disastrous end with the bits of information available. Apparently, one of the Soyuz's solar panels had not deployed properly. This meant that the panel's thermal radiator remained covered and less power was being generated. Also, the television transmitter failed and some antennas did not deploy properly. Because a thermal radiator was covered by the folded solar array, heat generated by the ship's electronics could not be dissipated as required, leading to failure of the controls for the attitude control system. The malfunction was causing the ship to slowly rotate end over end. U.S. tracking facilities later reported that the Soyuz was tumbling during the 15th orbit.[27] NORAD detected no maneuvers during the flight of the Soyuz, suggesting failure of the propulsion system (the attitude control system was needed for pointing during any main engine firing).[28] Despite these failures, after the flight Gagarin said the life support and communication systems had operated well.[29]

Komarov tried the instructions sent up from the ground, but could not correct the problems (see Figure 2-10). As his fuel supply ran low, he got permission to land on the 17th orbit. The Soviets said that with the test program completed, it was suggested that he could finish the flight and land. With the attitude control failure, the ship could not be held steady during the critical retrofire maneuver. Komarov apparently put the ship in a roll to stabilize it during the maneuver. Komarov tried to orient the ship for retroburn on the 17th orbit, but failed. During the 18th orbit, the Soyuz was oriented properly and spinning, and the retro rockets were fired to perform a ballistic reentry, resulting in very high deceleration forces (see Figure 2-11).

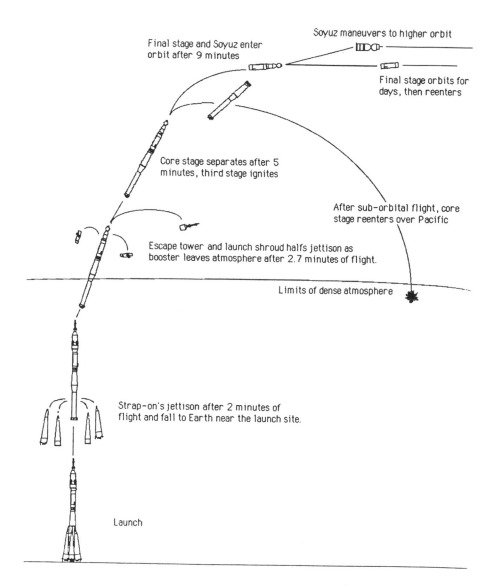

Soyuz maneuvers to higher orbit

Final stage and Soyuz enter
orbit after 9 minutes

Final stage orbits for
days, then reenters

Core stage separates after 5
minutes, third stage ignites

After sub-orbital flight, core
stage reenters over Pacific

Escape tower and launch shroud halfs jettison as
booster leaves atmosphere after 2.7 minutes of flight.

Limits of dense atmosphere

Strap-on's jettison after 2 minutes of
flight and fall to Earth near the launch site.

Launch

Figure 2-9. The Soyuz launch sequence was nearly the same as the Voskhod launch sequence with the exception that abort was possible at any time before or during the launch. The Soyuz abort tractor motor could carry the Soyuz capsule away from the booster while still in the atmosphere. After launch shroud separation, the Soyuz could abort by separating from a malfunctioning booster and reenter before achieving orbit. During the 20-year use of the Soyuz, both types of aborts have occurred saving the cosmonauts' lives.

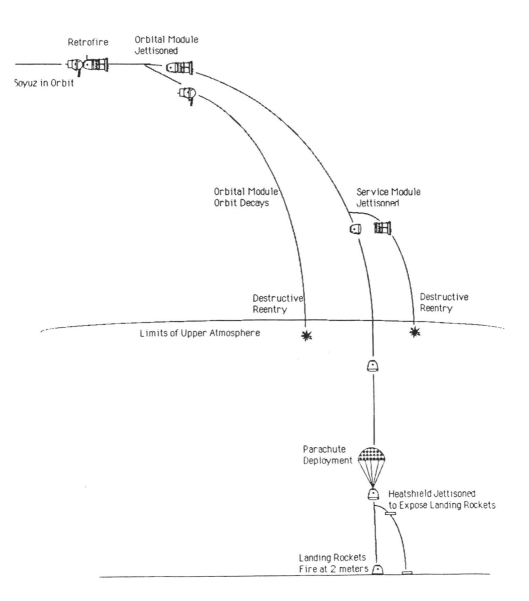

Figure 2-10. The Soyuz landing sequence starts with retrofire, followed by separation of the capsule from the orbital and service modules. After reentry, the capsule deploys a parachute and jettisons its heat shield to uncover four solid rockets and the landing radar antenna. The rockets fire at 2 meters altitude to effectively cushion the landing.

Soyuz Launch Sequence

T − 34:00:00	Booster is prepared for fuel loading.
T − 6:00:00	Batteries are installed in booster.
T − 4:00:00	Loading the booster with liquid oxygen starts.
T − 2:30:00	Crew ingress through orbital module side hatch. While waiting for launch, music is played for the cosmonauts.
T − :45:00	Launch pad service structure halves are lowered.
T − :05:00	Onboard systems switched to onboard control, Commander's controls activated, cosmonauts switch to suit air by closing helmets, launch key inserted in launch bunker, ready for launch.
T − :02:30	Booster propellant tank pressurization starts.
T − :01:00	Vehicle on internal power, automatic launch sequencer on, first umbilical tower separates from booster.
T − :15	Second umbilical tower separates from booster.
T − :10	Engine turbopumps at flight speed.
T − :05	First-stage engines at maximum thrust.
T − :00	Fueling tower separates. Lift-off.
T + :01:10	Booster velocity is 500 meters per second.
T + :01:58	Strap-on boosters jettisoned.
T + :02:00	Booster velocity is 1,500 meters per second.
T + :02:40	Escape tower and launch shroud jettison.
T + :04:58	Core booster separates at 170 km altitude, third stage ignites.
T + :07:30	Velocity is 6,000 meters per second.
T + :09:00	Third stage cut-off, Soyuz separates, antennas and solar panels deploy, flight control switches to Kaliningrad Mission Control.

Source: Ezell, Edward C. and Ezell, Linda N. "The Partnership: A History of the Apollo-Soyuz Test Project," NASA SP-4209, GPO, Washington, D.C., 1978, pp. 542–3.

The retrofire also came earlier than mission rules allowed, which resulted in a landing far short of the normal landing zone. The ballistic path followed during reentry was only possible if the capsule was spinning to even out any lifting effect due to center of gravity of the capsule being off of the center line.[30]

Apparently, he saw the main parachute try to deploy, as the capsule continued rotating and tumbling. The spinning or tumbling caused the parachute straps to twist, while deploying, and stopped the parachute from opening. Komarov then activated the emergency parachute intended for launch aborts,

Figure 2-11. Vladimir Komarov is shown here during training for his Soyuz 1 flight. Unmanned test flights of the Soyuz had not been entirely successful, but he boarded Soyuz 1 even after the chief designer of the Soyuz refused to certify the launch. Komarov's flight was plagued with technical problems, which led to his tragic death while attempting to land. (Source: Sovfoto.)

but it entangled with the main parachute.[31] It may be this last failure that the Soviets refer to when explaining the ultimate cause of the crash.

All communications from the capsule ceased as the ship hit the ground at 500 km/hr about 1,000 km from the normal landing area, near Orenburg in the Urals.[32] The Soviets officially said "During the opening of the main parachute canopy ... the spacecraft descended with high speed ... (as a) result of shroud line twisting."[33]

There were indeed plans to launch Soyuz 2 the day after Soyuz 1, and perform an EVA to transfer two crewmen from Soyuz 2 to Soyuz 1. This was later done during the Soyuz 4-5 mission. Soviet space expert James Oberg uncovered pictures of Komarov training for such a joint flight with the crew of Bykovsky, Khrunov, and Yeliseyev, with Kubasov as Yeliseyev's backup. This joint mission would also have fulfilled the claims of the radio broadcast previously mentioned. After the disaster a board of investigation was set up and investigated all phases of the flight.

Komarov's backup for the first Soyuz flight, Gagarin, was killed along with a cosmonaut instructor in a jet crash on March 27, 1968, during a routine training flight. At the time, Gagarin was reportedly training for one of the first lunar missions. Jet aircraft training flights were used by both the Soviets and NASA to keep astronauts and cosmonauts used to flying high performance vehicles. The two-seat MiG-15 trainer was reportedly flying in violation of flight rules in low clouds and was caught in the wake of another MiG-15. The plane went into a spin and the pilots tried to regain control. Complicating the situation was the fact that the plane's altimeter had been set to a faulty ground instrument reading hundreds of meters higher than actual. The reported altitude of the clouds was also higher than in actuality. The plane exited the clouds at an angle of 70–90° as they tried to recover from the spin. There was no time left to exit the aircraft before hitting the ground, which was much closer than the crew expected.[34]

Kosmos 159

Launched: May 17, 1967, 12:44 a.m.
Reentry: November 11, 1977
Altitude: 350 × 60,637 km @ 51.8°

This was a heavy Zond propulsion module test, launched by a D-1e booster.[35] Spacecraft weight at launch was 4,490 kg, and it was launched 180° away from the moon, like the later flight of Zond 4B. Only the propulsion or service module was launched, there was no reentry capsule. After reaching the final orbit, an object, believed to be the torus propellant tank was jettisoned, identifying the test as a heavy Zond propulsion module.[36] The flight may have tested the heavy Zond engine in a maneuver that simulated a Trans Earth Injection (TEI) from lunar orbit. This cannot be determined from the data publicly available. The last booster stage was left in an orbit at 203 × 431 km. Kosmos 159 showed the Soviet intention to launch a heavy Zond on the Proton despite the fact that it was light enough to be launched by an A-2.

Kosmos 186

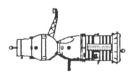

First Automatic Docking
Launched: October 27, 1967, 12:31 p.m.
Landed: October 31, 1967, 11:20 a.m.
Altitude: 172 × 212 km @ 51.7°

This flight and Kosmos 188 were tests of the Soyuz to verify changes made after the failure of Soyuz 1. Pictures released by the Soviets at the time were altered to give the spacecraft a different appearance than the Soyuz. The techniques of automatic rendezvous and docking to be performed by the fight were essential to the manned lunar landing effort then underway. By October 30, Kosmos 186 raised its orbit to 193×258 km in preparation for the Kosmos 188 launch and docking.[37]

Kosmos 188

Launched: October 30, 1967, 11:14 a.m.
Landed: November 2, 1967, 11:53 a.m.
Altitude: 180×247 km @ $51.7°$

Kosmos 188 used a direct ascent launch profile to place it in orbit only 24 km from Kosmos 186, which was the active spacecraft and performed all the docking maneuvers in 30 minutes. Up to 300 meters distance, Kosmos 186 used its main engine for maneuvering. It then switched to large attitude control thrusters for smaller orbital adjustments. During 186's approach, 188 used its attitude control thrusters to keep pointed toward 186 (see box).

Automatic Docking

The docking procedure described was used for the early Soyuz missions, but is similar to the procedure currently used. After rendezvous of two spacecraft is accomplished, the spacecraft should be within a few hundred meters of each other. The docking procedure consists of three parts as described by the Soviets.

During the *capture phase*, radar contact between vehicles is established. Computers detect distance, relative velocity, angular velocity, and relative angle of the spacecraft. Both spacecraft then align themselves to the same axis for docking. The active spacecraft closes on the target, and at 350 meters distance closing velocity is about 2 meters per second.

Next is the *mooring phase* in which the passive spacecraft (or space station) uses attitude control rockets to stay aligned with the active maneuvering ship. The target rolls if the active ship rolls and the active ship extends its probe for soft docking. The extended probe prevents the airtight seals of the two spacecraft docking collars from being damaged if the initial contact is hard, of off center. Soft dock occurs when the mechanical coupling of the probe to the drogue is completed. There are small latches on the top of the probe that catch the center of the drogue. This action completes the soft dock.

(continued on next page)

(continued from page 65)

In the *docking phase*, the active ship reels its probe in and the ship's butt docking collars make an airtight connection. Latches in both docking collars are activated to grip the spacecraft together with about 20,000 kg of pressure. Electrical connections are also made for communications and power distribution between spacecraft. On Progress spacecraft refueling connections are also through the docking collar.

To undock, the ships mechanically disengage probe and docking collar latches, and springs in the docking collars push the ships apart. The active ship usually makes a small maneuver to increase the separation rate.

At 12:20 p.m., October 30, and out of communication range of the ground controllers, the ships docked. Using the television cameras on the ships, ground controllers saw the ships had docked when they regained communications. It is reported that kerosene and nitric acid were transferred between the ships in a test of in-flight refueling.[38] The success of this is not known, but this was a feature of the docking system as planned for Soyuz B and Soyuz V vehicles, which were canceled a few years earlier.

The ships remained docked only 3 hours and 30 minutes, and then undocked during live coverage on Soviet television. The ships separated rapidly as the springs in the docking collars pushed the ships apart. Kosmos 186 remained in orbit until the next day when it made a normal landing. Kosmos 188 spent another three days in orbit before landing.

Zond 4A

Failed
Launched: November 22, 1967
Landed : —
Altitude: ?

The Zond 1 to 3 flights were unmanned planetary probes based on the Venera spacecraft design. These flights were totally unassociated with the manned version Zond's used for flights 4 to 8. Zond 4 to 8 were controlled from the same Crimea Control Center as were the Soyuz flights, and the Soviets admitted the manned capability of the later Zonds in 1968.[39] This was the first flight of the man-related moon ship of the Zond name.

The Zond was launched by a D-1e, which failed during launch. The booster and spacecraft had been preparing for launch since August. The first stage of the Proton failed shortly after launch.[40] This flight would probably have been a high altitude, high-velocity reentry attempt.

Zond 4B
High-Altitude Zond Test
Launched: March 2, 1968, 9:30 p.m.
Landed: March 9, 1968, 9:50 p.m.
Altitude: 192 × 205 km @ 51.5°

Zond 4B was launched by D-1e into a orbit 200 × 400,000 km inclined at 51.5°. The apogee of the orbit was on the opposite side of the Earth as was the moon at the time. This was planned so that its orbit could be computed without the effects of the moon's gravitational field interfering. Spacecraft weight was estimated at 4,425 kg.

Communication or electrical problems early in the flight may account for a lack of standard news reports that occurred later during the Zond 5B flight.[41] The same problems may have contributed to a malfunction that led to the Zond landing off course in China because of a retrofire error or a guidance error during reentry. The booster's third stage reentered March 7.

There was some mystery as to the fate of Zond 4. The NASA Satellite Situation Reports currently list Zond 4 as in a heliocentric orbit around the Sun. Others claimed that the Zond's return to Earth failed, burning up in the atmosphere. Recent reports claim that the capsule is now on display outside the Red Army military museum in Beijing, China, indicating a successful landing within China on March 9.[42] Zond 4 was in orbit for only a short time in any case and there was never official word from the Soviets on its fate. The landing time listed is an estimate.

Kosmos 212
Second Automatic Docking
Launched: April 14, 1968, 12:32 p.m.
Landed: April 19, 1968, 11:15 a.m.
Altitude: 186 × 225 km @ 51.7°

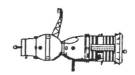

This flight was the same type as Kosmos 186, and served as docking target for Kosmos 213. After achieving orbit, the spacecraft waited until the next day when it passed over the Baykonur Cosmodrome for the launch of another Soyuz type spacecraft.

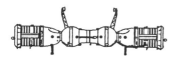

Kosmos 213

Launched: April 15, 1968, 12:32 p.m.
Landed: April 20, 1968, 12:48 p.m.
Altitude: 188 × 254 km @ 51.7°

Kosmos 213 used a direct ascent launch to orbit, initially separated only 5 km from Kosmos 212. The flight used the same procedures as the Kosmos 186 and 188 flight. Kosmos 212 lowered its orbit soon after launch to 184 × 197 km, and then raised it again before docking to 174 × 253 km. The spacecraft docked while out of communication range at 1:21 p.m., April 15. They remained docked for 3 hours and 48 minutes, and undocked at 5:11 p.m.[43] The undocking was observed by ground controllers using the ships docking television cameras. On April 16, Kosmos 213 maneuvered to 195 × 261 km, and then to 193 × 272 km on April 19.

Zond 5A

Failed
Launched: April 22, 1968
Landed: —
Altitude: 40? km

This was another failed Zond flight. The D-1e boosters second stage reportedly failed. This launch matched the Zond 4B launch window, meaning that it would probably have made a high velocity reentry from high altitude to complete Zond 4B's mission.[44,45]

Kosmos 238

Soyuz Test
Launched: August 28, 1968, 1:00 p.m.
Landed: September 1, 1968, 11:53 a.m.
Altitude: 188 × 214 km @ 51.7°

All indications are that Kosmos 238 was a successful test of a Soyuz. But, this flight was probably intended to have been the Soyuz 2 launch that would be a target for a manned Soyuz 3. The manned Soyuz must have encountered difficulty during the countdown, which delayed the launch beyond the few days the target spacecraft could remain in orbit. This forced the target to land and be given a Kosmos name to hide the problems encountered with the manned launch. When the Soyuz 2-3 mission did occur, the unmanned launch of Soyuz 2 was not announced as such until the manned Soyuz 3 was launched.[46]

Zond 5B

First Lunar Fly-by and Return

Launched: September 15, 1968, 12:42 a.m.
Landed: September 21, 1968, 7:08 p.m.
Altitude: 193 × 219 km @ 51.5°

Zond 5B was the first spacecraft to circumnavigate the moon and return to Earth. The capsule carried plants, turtles, flies, and worms to investigate the effects of cosmic radiation on living organisms.[47] After separating from the Proton third stage, the Zond atop its fourth stage escape booster coasted for 67 minutes in low Earth orbit. The third stage reentered September 21 at 11:00 a.m.[48] Before leaving orbit, the Zond jettisoned its docking collar from the capsule to lighten the spacecraft. Then, the escape stage performed the Trans Lunar Injection maneuver and separated leaving the Zond on course to pass by the moon. On September 17, at 6:11 a.m., the Zond made the first in a pair of course corrections. Its closest approach to the moon was 1,950 km on September 18. The spacecraft made a course correction maneuver at the moon to quicken the return and make the course more accurate, rather than relying on the moon's unequal gravity. All future Zond's would use this technique. During the flight, the Zond took pictures of the Earth and moon and transmitted a recorded voice to test the communication system from the 400,000 km distance.[49]

As the spacecraft hurtled back toward Earth, the capsule separated from the service module before it encountered the high reaches of Earth's atmosphere. During reentry, the capsule experienced 10–16 gravities deceleration and exterior heating of 13,000°C. The parachute deployed at 7,000 meters altitude and the Zond made the first water landing of the Soviet space program in the Indian Ocean, at 32° 38'S, 75° 33'E.[50]

Recovery of the capsule was performed by the tracking ship *Borovichiy*. The capsule was taken by an oceanography ship, the *Vasiliy Golovnin*, to Bombay by October 4, where it was flown by an An-12 transport plane to the Soviet Union for examination.[51]

Soyuz 2

Launched Unmanned

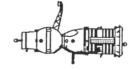

Launched: October 25, 1968, 12:00 p.m.
Landed: October 28, 1968, 11:00 a.m.
Altitude: 170 × 210 km @ 51.7°

Soyuz 2 was launched unmanned to be used as a target for the Soyuz 3, which would follow the next day, if the Soyuz 2 functioned properly on orbit. The

Soyuz 2 launch was not announced until Soyuz 3 was in orbit. If the Soyuz 3 launch had not been successful, the Soyuz 2 flight would have been called a Kosmos flight. This probably happened to Kosmos 238.[52]

Soyuz 3
Aborted

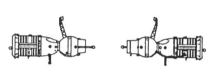

Launched: October 26, 1968, 11:34 a.m.
Landed: October 30, 1968, 10:25 a.m.
Altitude: 177×203 km @ $51°$
Crew: Georgi Beregovoi
Backup: Vladimir Shatalov
Call sign: Argon

This was the first time the Soviets used the U.S. practice of launching the passive target first instead of the active ship, like Kosmos 186 or 212. The spacecraft's weight at launch was 6,575 kg. In addition to Shatalov, Boris Volynov also trained as a backup for the flight.[53]

Soyuz 3 flew a direct ascent to orbit. Beregovoi approached Soyuz 2 during the first orbit, using an automatic system to maneuver within 180 meters. He did not proceed to dock, which was the primary objective of the mission. The ships closed to a few meters, then Soyuz 3 separated to 565 km.[54] Beregovoi then performed a manual approach but did not dock, and again the ships separated. While in orbit, Beregovoi made regular television reports on the ships performance, and spent most of the time observing the ship's systems and photographing Earth.[55]

On October 27, Beregovoi again attempted a docking, and again failed. Soyuz 2 was then commanded into a 197×252 km orbit. Soyuz 2 landed the next day, on October 28, and Soyuz 3's orbit was then lowered to 176×244 km. Later the same day, Soyuz 3 raised its orbit before retrofire to 196×241 km.

Beregovoi manually oriented the ship and used the automatic system for a 145-second retrofire.[56] The capsule landed amid snow drifts, close to the target point, allowing the recovery forces to photograph the decent. There were ships stationed in the Indian Ocean in case an emergency landing there was necessary.[57] Beregovoi retired soon after the flight and assumed the post of head of the Gagarin Training Center.[58]

Zond 6

Launched: November 10, 1968, 10:12 p.m.
Landed: November 17, 1968, 5:10 p.m.
Altitude: 186 × 232 km @ 51.4°

This flight was similar to the Zond 5B mission in that it carried similar animals, experiments, a photo emulsion camera, and a micro-meteoroid detector.[59] The reentry capsule part of the spacecraft weighed 2,046 kg at launch, which was very close to a normal manned Soyuz capsule weight.[60]

The Zond and fourth stage separated from the Proton third stage in low Earth orbit for a 67-minute glide until docking collar jettison and Trans Lunar Injection. The third stage reentered November 12 at 9:58 a.m. A course correction was made at 8:41 a.m. on November 12. The spacecraft passed within 2,420 km of the moon and took pictures of the far side on November 14, at 4:00 and 5:48 a.m. Two course corrections were made during the return at 9:40 a.m., November 16, and at 8:36 a.m., November 17.[61]

Zond 6 used a sophisticated "skip-glide" reentry, which used aero-braking in Earth's atmosphere and then used the capsule's lift to fly back out of the atmosphere in a sub-orbit. The result of the skip-glide technique is that the spacecraft falls back into the atmosphere and reentered at slower velocity (Apollo also used this technique). Maximum deceleration during reentry as a result of the skip-glide was limited to four to seven gravities during the skip portion and the final reentry.[62]

Unlike its predecessor, Zond 6 landed in the Soviet Union and the Soviets announced that Zond 6 and its immediate predecessors were launched to test a manned spacecraft for lunar flight.[63] The Soviets would have only one more chance to win the race to the moon, with the launch of a manned Zond around the moon in December. The flight was attempted, but technical difficulties postponed the flight until January 1969, after the lunar orbit flight of Apollo 8 (see Chapter 3: Zond 7A).

References

1. McDugall, Walter, ...*the Heavens and the Earth, A Political History of the Space Age*, New York: Basic Books Inc., 1985, p. 289.
2. Brooks, Courtney G., Grimwood, James M., and Swenson, Lloyd S., *Chariots for Apollo: A History of Manned Lunar Spacecraft*, NASA SP-4205, Washington D.C.: Government Printing Office, 1979, p. 83.
3. Daniloff, N., *The Kremlin and the Cosmos*, New York: Alfred A. Knopf, 1972, p. 155.
4. Collins, Michael, *Carrying the Fire*, New York: Bantam Books, 1974, p. 282.

5. Gatland, K., "The Soviet Space Program after Soyuz 1," *Spaceflight*, Vol. 9, No. 9, 1967, p. 298.

6. Ertel, Ivan D. and Morse, Mary L., *The Apollo Spacecraft, A Chronology, Vol. I*, NASA SP-4009, Washington D.C.: Government Printing Office, 1969, drawings pp. 88, 90.

7. Clark, Phillip S. and Gibbons, Ralph F., "The Evolution of the Soyuz Program," *Journal of the British Interplanetary Society*, Vol. 36, No. 10, Oct. 1983, p. 436.

8. Sheldon, Charles S., *Review of the Soviet Space Program*, p. 74.

9. Daniloff, op. cit. p. 153.

10. Wachtel, C., "The Chief Designers of the Soviet Space Program," *Journal of the British Interplanetary Society*, Vol. 38, No. 12, Dec. 1985, p. 562.

11. Gatland, op. cit. p. 296.

12. Johnson, Nicholas L., *Handbook of Soviet Manned Spaceflight*, American Astronautical Society: San Diego, 1980, p. 129.

13. Oberg, James E., *Uncovering Soviet Disasters*, Random House: New York, 1988, p. 171.

14. Congressional Research Service, The Library of Congress, *Soviet Space Programs 1976–80, Manned Space Programs and Life Sciences, Part 2*, Washington D.C.: Government Printing Office, 1984, pp. 500–501.

15. Oberg, *Uncovering Soviet Disasters*, op. cit. p. 171.

16. Clark and Gibbons, op. cit. p. 443.

17. Grahn, Sven and Oslender, Dieter, "Cosmos 146 and 154," *Spaceflight*, Vol. 22, No. 3, March 1980, pp. 121–123.

18. Clark and Gibbons, op. cit. p. 443.

19. Oberg, *Uncovering Soviet Disasters*, op. cit. p. 171.

20. Grahn and Oslender, op. cit. pp. 121–123.

21. Oberg, *Uncovering Soviet Disasters*, op. cit. p. 171.

22. Johnson, op. cit. p. 137.

23. Oberg, James, "Soyuz 1 Ten Years After: New Conclusions," *Spaceflight*, Vol. 19, No. 5, May 1977, p. 183.

24. Oberg, *Uncovering Soviet Disasters*, op. cit. p. 172.

25. Johnson, op. cit. p. 139.

26. Oberg, "Soyuz 1 Ten Years After: New Conclusions," op. cit. p. 184.

27. Bond, Peter, *Heroes in Space: From Gagarin to Challenger*, Basil Blackwell: New York, p. 147.

28. Oberg, "Soyuz 1 Ten Years After: New Conclusions," op. cit. p. 184.

29. "Gagarin on Soyuz 1," *Spaceflight*, Vol. 9, No. 9, Sept. 1967, p. 309.

30. Oberg, "Soyuz 1 Ten Years After: New Conclusions," op. cit. p. 187.

31. Pellegrino, Charles R. and Stoff, Joshua, *Chariots for Apollo: The Making of the Lunar Module*, N.Y.: Atheneum, 1985, p. 94.

32. Gatland, op. cit. p. 295.

33. Oberg, "Soyuz 1 Ten Years After: New Conclusions," op. cit. p. 184.
34. News broadcasts, Radio Moscow, Moscow, U.S.S.R., Jan. 1, 1988.
35. Parfitt, John and Bond, Alan, "The Soviet Manned Lunar Landing Program," *Journal of the British Interplanetary Society*, Vol. 40, No. 5, May 1987, p. 232.
36. Congressional Research Service, The Library of Congress, *Soviet Space Programs 1976–80, Manned Space Programs and Life Sciences, Part 2*, Washington D.C.: Government Printing Office, 1984, p. 642.
37. Johnson, op. cit. p. 144.
38. Ibid. pp. 144–145.
39. Clark, Phillip S., "Topics Connected with the Soviet Manned Lunar Landing Program," *Journal of the British Interplanetary Society*, Vol. 40, No. 5, May 1987, p. 235.
40. Vick, Charles P., "The Soviet Super Boosters—1," *Spaceflight*, Vol. 15, No. 8, Aug. 1973, p. 467.
41. Johnson, Nicholas, *Soviet Lunar and Planetary Exploration*, American Astronautical Society: San Diego, 1980, pp. 109–110.
42. Oberg, *Uncovering Soviet Disasters*, op. cit. p. 62.
43. Johnson, *Handbook of Soviet Manned Spaceflight*, op. cit. p. 147.
44. Vick, C. P., "The Soviet G-1-e Manned Lunar Landing Program Booster," *Journal of the British Interplanetary Society*, Vol. 38, No. 1, Jan. 1985, p. 17.
45. Clark, Phillip S., "Soviet Launch Failures," *Journal of the British Interplanetary Society*, Vol. 40, No. 10, Nov. 1987, p. 528.
46. Johnson, *Handbook of Soviet Manned Spaceflight*, op. cit. p. 131.
47. Johnson, *Soviet Lunar and Planetary Exploration*, op. cit. pp. 113–114.
48. Congressional Research Service, The Library of Congress, *Soviet Space Programs 1976–80, Manned Space Programs and Life Sciences, Part 2*, Washington D.C.: Government Printing Office, 1984, p. 641.
49. Dmitriev, A., "Zond 5 Circumnavigates the Moon," *Spaceflight*, Vol. 11, No. 2, Feb. 1969, pp. 64–65.
50. Congressional Research Service, The Library of Congress, *Soviet Space Programs 1976–80, Manned Space Programs and Life Sciences, Part 2*, Washington D.C.: Government Printing Office, 1984, p. 647.
51. Johnson, *Soviet Lunar and Planetary Exploration*, op. cit. p. 113.
52. Johnson, *Handbook of Soviet Manned Spaceflight*, op. cit. p. 148.
53. *Spaceflight*, Vol. 31, No. 2, Feb. 1989, p. 57.
54. Johnson, *Handbook of Soviet Manned Spaceflight*, op. cit. p. 149.
55. Turnill, R., *Spaceflight Directory*, London: Frederick Warne Ltd., 1977, p. 296.
56. Johnson, *Handbook of Soviet Manned Spaceflight*, op. cit. p. 150.
57. Turnill, op. cit. pp. 296–297.

58. Bond, op. cit. p. 290.
59. Johnson, *Soviet Lunar and Planetary Exploration*, op. cit. p. 117.
60. Clark, op. cit. p. 235.
61. Johnson, *Soviet Lunar and Planetary Exploration*, op. cit. p. 117.
62. "Russia's Moon Program," *Spaceflight*, Vol. 11, No. 11, Nov. 1969, p. 382.
63. Clark, op. cit. p. 235.

Chapter 3

Remnants of the Race

The Soviets lost their best chance at winning the race to the moon in December 1968, when technical problems delayed the Zond 7A flight into 1969. After the successful flight of Apollo 8 around the moon later that month, the Soviets apparently ended plans to use the Zond for manned circumlunar missions. To fly one, manned, would have been admitting second place status in space and being number one was what the space race was all about.

The Lunar Landing Plan

While just missing the chance to fly the first manned circumlunar mission, the Soviets had grander plans for manned lunar space flight. A Zond lunar fly-by mission was only a part of what the Soviets were planning. They continued their plans for landing a man on the moon using the G-1 super booster with the hope that NASA would encounter difficulty in meeting their planned 1969 landing date. Cosmonauts were trained in helicopters, presumably to practice vertical lunar landings.[1] The Soviets had even been studying manned lunar bases since 1967, and performed year-long lunar mission simulations in mock-up laboratories on the ground.[2] As the statements in Chapter 2 show, the Soviets intended since at least 1966 to land on the moon in 1971 or later. The Soviets revised their schedule in 1968, after NASA's advance in Apollo's schedule to try to launch a landing mission as early as 1969. Major test flights of lunar mission hardware were attempted in 1969, but failures caused the effort to be pushed back years, and eventually into an indefinite postponement. Some possible cosmonaut assignments for the moon program in the late 1960s are shown in Table 2-1.

The Soviet moon landing plan apparently was to use a combination of Earth Orbit Rendezvous and Lunar Orbit Rendezvous (see Figure 3-1). First, a G-1 booster would launch an Earth escape stage, lunar braking module, and lunar lander into Earth orbit. A manned heavy Zond would then be

75

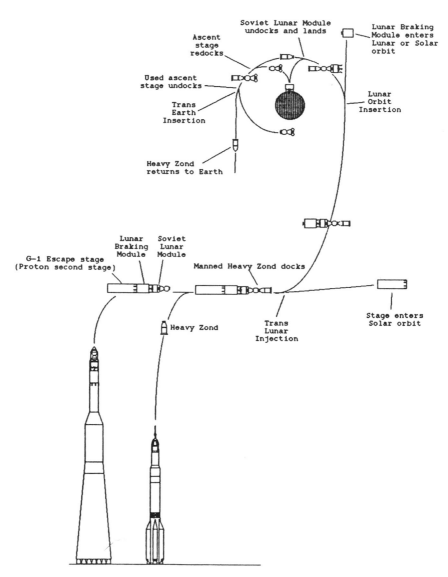

Figure 3-1. The Soviets redirected their manned lunar flight plans after NASA's Apollo program was recognized as a major effort. The Soviets developed a large booster and lunar module to mirror NASA's Apollo program. The replanned program was massive and produced hardware in a very short period, but the years lost to Apollo in the early 1960s could not be made up before the first U.S. lunar landings.

launched and dock to the assembly. This plan avoided the many test flights needed to prepare the new G-1 rocket for manned flight. After the crew docked in the Zond, the assembly would be launched toward the moon by the escape stage. The Soviet Lunar Module and heavy Zond would be stopped in lunar orbit by the braking module, possibly a version of the Proton fourth stage. Two of the crew would probably undock from the Zond and descend in the lander. They would use a Soyuz-type orbital module on the lander ascent stage as an airlock, just as on early Soyuz flights, to egress to the lunar surface. After exploring the lunar surface and collecting rocks and soil, the crew would enter through the airlock and then lift off. The ascent stage would fly into a low lunar orbit and dock to the heavy Zond. After transferring the lunar samples, the heavy Zond would undock from the lander and return to Earth for a landing, probably in the Indian Ocean. This plan is not known as fact, but seems to be the most likely method the Soviets were going to use to land men on the moon judging from the lunar spacecraft test flights in the early 1970s.

Some space analysts have reported that the Soviets may have chosen direct ascent and that a G-1 would launch a highly modified Proton third stage and a fully fueled Soyuz. The Proton stage could have been used to enter lunar orbit, land on the moon and then lift a heavy Soyuz back into lunar orbit for return to Earth. This plan seems very unlikely since when NASA considered the direct ascent method, they were very concerned about how such a large and massive vehicle could be landed safely on the moon even with advanced U.S. spaceflight technology. NASA also determined that direct ascent was not the fastest or cheapest way to land on the moon. It hardly seems likely that the Soviets would ignore the same points, and Feoktistov stated in 1967, years into the Soviet lunar landing effort, that the best way to land on the moon was by LOR. The direct ascent theory also does not match all the available evidence as well as the EOR-LOR method previously described. While the Soviets are slowly releasing details of the lunar landing mission, there is no doubt about the objective of landing on the moon.

Lunar Program Failures

After Apollo 8, the Soviets had moved up their launch schedule to try and match U.S. efforts, but they could not cope with the accelerated NASA program. At this point there was little hope of beating the U.S. to a landing, if there were no serious problems with Apollo. In 1969, Boris Nikolaevich Petrov, chairman of Intercosmos of the Soviet Academy of Sciences, talked about the continuing moon program with some uncertainty: "The major tasks still ahead in the study of the moon will be carried out by automatic means, although that does not exclude the possibility of manned flight."[3]

During 1969, technical problems began multiplying for the Soviets. The lunar lander was not ready in time to be adequately tested, and the first spacecraft was destroyed when its Proton booster failed. The G-1 super booster was tested three times but never successfully. In the 1969 launch attempt, there was reportedly a dual countdown with a Proton booster carrying a heavy Zond with a crew of three (see Zond 7B). A landing attempt would probably not have been tried on the first test. The flight would probably have been an attempt of a lunar orbit mission. The Soviets had also said that they would land animal test subjects (probably dogs) on the moon before landing men. All three of the G-1 boosters launched failed early in the boost phase before second-stage ignition. The booster reportedly suffered from major design faults that were probably a result of inadequate ground testing, which was not completed until 1976[4] (see Figure 3-2).

The slight hesitation by the Soviets in 1961, and then underestimating the U.S. effort until 1964, cost them the time needed to fully test and develop the systems like the G-1 super booster and lunar lander. The Soviets, not wanting to throw away the huge investment in the lunar spacecraft and boosters, made a few more tests in the early 1970s and continued new work on high-energy propellant hydrogen-oxygen engines, an area in which the U.S. had a tremendous lead.[5]

As the program development continued after the U.S. lunar landing of Apollo 11, three test flights, Kosmos 379, 398, and 434, tested the Soviets' lunar lander ascent stage. Kosmos 382B was probably a full test of the lunar lander. The lunar braking module was probably tested on flights of the Proton booster (acting as a different version of the fourth stage). The last two flights of the G-1 were also made. Development work on the giant G-1 booster was the last element of lunar program development easily visible to U.S. intelligence satellites, as huge booster stages were moved from the assembly building to the test stands at the Baykonur Cosmodrome at least into 1976.

At the same time, Soviet officials began down-playing the lunar landing project to save face. Academy of Sciences President M.V. Keldysh had said in 1969, "We no longer have any scheduled plans for manned lunar flights."[6] In what would be repeated many times as the official Soviet line, he later contradicted himself in 1970 by saying that the Soviets had never announced a manned lunar landing program, and there was no such program.

Apparently one of the main reasons for continuing development was the concern about the NASA Apollo Applications Program (AAP), which planned to send space stations around the moon and start large-scale exploration of the moon. Soviet studies of lunar bases had been started in 1967 to match similar AAP plans. Unfortunately, by 1970 it was clear that the Apollo Applications Program would not be funded after the heat of the space race had cooled after the quiet Soviet failures.

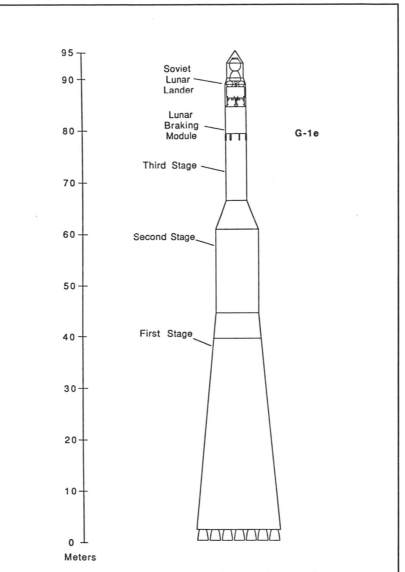

Soviet
Lunar
Lander

Lunar
Braking
Module

G-1e

Third Stage

Second Stage

First Stage

95
90
80
70
60
50
40
30
20
10
0

Meters

Figure 3-2. The G-type (or N-I) booster was developed in a crash program to beat the Apollo lunar landings. The G-type was about the same size as the U.S. Saturn V booster and was designed to launch a lunar lander, which would be met in orbit by a cosmonaut crew launched separately. The G-type failed during launch attempts three times. The booster's facilities were then mothballed for use in the 1980s in the Energia and Soviet shuttle programs.

(continued on next page)

(continued from page 79)

G-Type Booster

The existence of the G-type (or N-I) booster has never been admitted by the Soviets. The G-type was first detected by U.S. intelligence in 1964, and later reported to Congress by then NASA Administrator James Webb. The CIA publicly confirmed the booster's existence in 1976. Recent photos of the Baykonur cosmodrome Energia (K-type) booster launch facilities also show remnants of the outsized G-type facilities that were modified for the Energia booster in the 1980s.

Soviet spaceflight observer C.P. Vick used drawings of the G-1 launch facilities published in a Soviet book to estimate the size of the booster and that the booster had about 20 first-stage engines or nozzles and four second-stage engines or nozzles. The engines may be early versions of the RD-170 engine, which was later used on the J-type booster and Energia. This has not yet been confirmed.

The third stage was probably the Proton second stage. It is assumed that all G-type stages used kerosene and LOX as propellants. The first stage was probably conical shaped with a cylindrical second stage. Based on the Soviet launch pad drawings, the booster was about 100 meters long and 15 meters in diameter at the base. Lift-off thrust was between 4,500,000 to 6,350,000 kg. The super booster was meant to lift a 100- to 135,000-kg payload to low Earth orbit, or place a 45- to 50,000-kg payload in translunar trajectory.

In July 1968, the first flight article booster was rolled to the pad for fueling tests. The booster was repeatedly taken off the pad for modifications and static firings. It was taken to the pad for the second time in late spring 1969 and was being prepared by June 3 for a launch on July 4, 1969. Some reports state that the first booster exploded very shortly after launch on July 4. Other reports state that the explosion occurred before June 14 and without further information it is impossible to determine exactly. The next launch attempt ended in a launch pad abort after some engines failed and were shut down before lift-off. In June, the second booster was launched but failed during the first-stage burn. The third booster reached the launch pad in May or June 1972, and remained there until its launch on November 24. It too failed and was destroyed by range safety command to prevent impact outside of the launch corridor. The major problem of all three launches was reported to be "pogo," which is longitudinal oscillation that produces an up and down acceleration of up to several gravities intensity. The Soviets reportedly tried to synchronize the G-type engines to a frequency that would not couple with the booster's natural resonance, and tried to throttle the engines to compensate for pogo when it developed.

In 1974, the Soviets were observed moving G-1 boosters around the launch complexes and the launch pad destroyed in 1969 was repaired. In 1976, the Soviets built a test stand to enable full upright dynamic testing of the G-1 booster stages stacked in flight configuration. The test stand was about 122 meters tall and was similar to a Saturn V test stand built in the 1960s at NASA's Marshall Spaceflight Center. The G-type program apparently ended after these tests and remained dormant until the redesign that produced the K-type Energia booster.

Soviet Lunar Module

The exact configuration and size of the Soviet Lunar Module (SLM) is still not known, although it probably weighed about 20,000 kg for test flights using the D-1 booster. Since the launch was on a Proton, that limited the diameter of the SLM to about 4.15 meters. The descent stage probably weighed about 11,400 kg, and the ascent stage probably weighed about 7,500 kg. The ascent stage was probably made from modified Soyuz components to serve as the main cabin, engine compartment, and airlock.

Five successful unmanned test flights were made in the 1970s in Earth orbit. The lunar module was to be launched by the G-type booster for manned lunar landing missions. Failures of the G-type booster precluded any lunar test flights of the lunar module.

Space Station Program

Another major factor in continuing lunar hardware development after loosing the moon race was the concern about the military applications and advantages of the boosters and spacecraft developed for the Apollo program. The Soviets were positive that the Apollo-Saturn hardware was going to be used for military purposes in the 1970s. For example, in planning the Apollo-Soyuz Test Project, it was impossible to convince the Soviets that NASA was abandoning Skylab due to a lack of money. They also thought that the backup Skylab, now displayed in the National Air and Space Museum in Washington D.C., was secretly being prepared as a manned military reconnaissance space station.[7]

With these concerns, the Soviets continued development of their advanced lunar hardware like the G-1, and by late 1968, the Soviets had started the Salyut program to counter any future U.S. space stations. Initially, the Salyut space stations were probably developed both to give the manned space effort direction after the lunar program failure, and as a result of military concerns about the planned U.S. Manned Orbiting Laboratory (MOL), which was planned to be launched in the late 1960s, or early 1970s. MOL was planned to be a series of military space stations in which two-man Air Force and possibly Navy crews would perform reconnaissance using powerful cameras and perform other military experiments.

The first of the Salyut related flights was the Proton 4 test flight and the odd triple flight of Soyuz 6, 7 and 8. The Soyuz 9 flight was also a test of some Salyut systems before the first flights to a Salyut station during Soyuz 10 and 11. While the Salyut apparently had military beginnings, the first Soviet Salyut station was mainly a civilian effort and was exploited mainly for its propaganda value, although engineering and scientific research was accomplished. The Salyut program also would enable the first work on researching manned flight to the planets. The Soviets praised the Salyut crews for their work saying that Salyut was a much more rewarding program than the U.S. Apollo flights. The propaganda was mostly for internal consumption in the U.S.S.R. to combat the publicity of the continued U.S. moon landings. While the Salyut program gave a chance for successful manned space missions to offset the lunar program failure, success was still not assured. Failure struck the Soviets when the first crew for Salyut 1 could not open the hatch to enter the station. The next, and last, crew to Salyut 1 completed a record 24-day space flight, only to die during their return, just eleven minutes from the air of Earth. What the Soviets had hoped would draw attention back to the Soviet space program had ended in failure. Unable to boast of success and unwilling to talk of failures meant that the Soviet space program faded from international view in the early 1970s. Thus, the driving factor for the U.S. space program had disappeared.

The Soviet denial of the moon race was well accepted by many in the U.S. After all, many in the United States had been surprised by the Soviet space program from the beginning, so it was easy for them to accept that the Soviets really were not as advanced as they had claimed to be. Certainly the lack of a highly visible Soviet moon program in the late 1960s was the major factor in the decline of the U.S. space program. The Soviets did continue the unmanned Luna program into the early 1970s, with successful sample return missions and robot rovers (see Figure 3-3), continuing well after the last U.S. lunar mission. Still, the lack of Soviet competition caused three Apollo lunar landings to be canceled due to lack of funding. By the early 1970s, the effort to build the Apollo-Saturn equipment was completely abandoned. Lunar program hardware was converted for use in the space shuttle program or scrapped, exactly paralleling Soviet actions in the late 1970s. It was not until at least 1981 that an official Soviet source, Cosmonaut Sevastyanov, again confirmed the existence of the manned lunar landing program of the 1960s.[8] See Figure 3-4 for missions covered in this chapter.

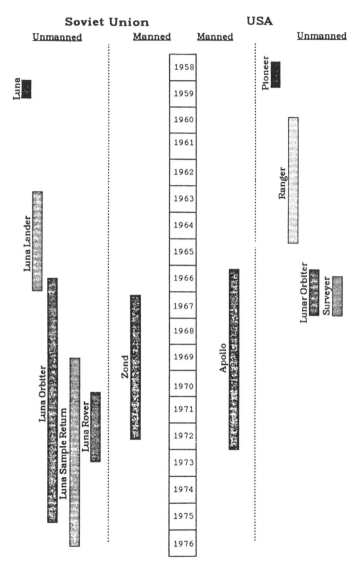

Figure 3-3. This chart shows the Soviet lunar programs and U.S. programs from their start in 1958 to their end in 1976. The Soviet effort was larger and included more flights than the U.S. effort, and extended over a longer period of time. Very successful Soviet lunar rovers and sample return mission were largely ignored in the Western nations after the Apollo landings. The Soviets now plan to continue lunar explorations in the 1990s with more sample returns and rovers.

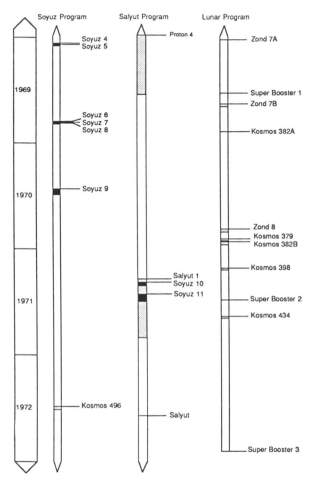

Figure 3-4. The Soyuz test program, Salyut, and Lunar programs were continued from 1969 to 1972. All known flights related to these programs are shown here and are described in this chapter. The gray portions show space stations' lifetimes and black portions show manned mission duration.

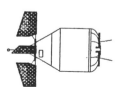

Proton 4

Salyut Test

Launched: November 14, 1968
Reentry: July 22, 1969
Altitude: 248 × 477 km @ 51.55°

Proton 4 was a scientific satellite and test of the Salyut space station structure, weighing 17,000 kg. The flight also tested the D-1h Proton booster that would launch all Salyut space stations and was a derivative of the D-1e, which launched the Zond lunar spacecraft. The Soviets named the flight after three earlier Proton test flights that tested the D-1 booster (first two stages only) in 1965 and 1966, but Proton 4 was completely different, being a test of the Salyut launch vehicle and the basic Salyut structure.

The satellite carried experiments weighing 12,500 kg designed to investigate high-energy particles. The satellite consisted of a cylinder 4.15 meters in diameter (the same as a Salyut) and about 4 meters long with a 5-meter-long cone on one end. It used four small paddle solar arrays for power placed around the cone.[9] The structure was the same as the living section of a Salyut without the service module or forward work and transfer compartments. The satellite was not very complicated and had no maneuvering capability. It reentered the atmosphere within eight months.

Zond 7A

Failed

Launched: January 5, 1969
Landed: —
Altitude: 33–43 km
(Crew: Pavel Belyayev?)
(Backup: Valeriy Bykovsky?)

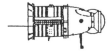

The final decision would have been made to man the Zond around December 1, 1968, two weeks before the launch window, which was December 7 to 9, 1968. Speculation about a possible crew has mentioned Belyayev and Bykovsky. Although there is no direct public evidence that this Zond was to be manned, the Soviets and Belyayev himself said that a lunar mission to beat an Apollo flight would occur. After two successful test flights around the moon it is very likely that this flight would have been the first manned Zond mission, but events would delay the flight and cause a reevaluation of the mission.

The booster and spacecraft were rolled out of the Proton assembly building December 1. The launch was attempted for the three days of the launch window, but was canceled due to problems with the Zond spacecraft. The booster and spacecraft were returned to the assembly building for repairs and rolled out again in late December. Any crew was reassigned after the successful Apollo 8 flight around the moon and the Zond was launched unmanned a month later.

The D-1e booster carrying the Zond was destroyed when the second stage exploded during ignition at 33 to 43 km altitude. The capsule's launch-abort rockets failed to operate and the spacecraft was also destroyed. If there had been a crew on board, they would have died.[10]

Soyuz 4

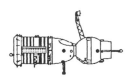

First Soviet Manned Docking

Launched: January 14, 1969, 10:30 a.m.
Landed: January 17, 1969, 9:51 a.m.
Altitude: 161 × 215 km @ 51°
Crew: Vladimir Shatalov
Backup: Anatoliy Filipchenko
Call sign: Amur

The first launch attempt of Soyuz 4 had to be scrubbed once in the final hours of the countdown, and was rescheduled and successfully launched on January 14.[11] The flight would finally accomplish the ill-fated Soyuz 1 mission objectives, which were docking and crew transfer with another Soyuz. The spacecraft's weight at launch was 6,625 kg.

Soyuz 4 maneuvered on the fourth orbit at 4:35 p.m., to 207 × 237 km to prepare to meet with the soon-to-be-launched Soyuz 5.[12] Shatalov then checked out the orbital and reentry modules, and made a television broadcast. As usual for cosmonauts, he slept while out of communication range of ground stations or ships, from 6:16 p.m., to 4:12 a.m., January 15. On Soyuz 4's 16th orbit, Shatalov observed the launch of Soyuz 5 as he flew northeast over the Baykonur Cosmodrome.[13]

Soyuz 5

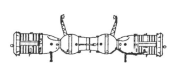

First Soviet Manned Docking

Launched: January 15, 1969, 10:04 a.m.
Landed: January 18, 1969, 10:58 a.m.
Altitude: 210 × 233 km @ 51°
Crew: Boris Volynov, Yevgeniy Khrunov,
and Aleksey Yeliseyev
Backup: Georgiy Shonin, Viktor Grobatko,
and Valeri Kubasov
Call sign: Baikal

After a nearly two-year wait, cosmonauts Khrunov and Yeliseyev were finally launched on their mission to perform an EVA transfer to another Soyuz in orbit as had been planned for Soyuz 1. At the time of launch, the cosmodrome was under snow cover with the temperature about − 30°C. The spacecraft's weight at launch was 6,585 kg. Soyuz 5 flew a normal launch profile taking several orbits to catch up to Soyuz 4 instead of the direct ascent to orbit used by earlier Kosmos docking test flights.[14,15] Soyuz 5 maneuvered on its sixth orbit to 211 × 253 km. On January 16, Soyuz 4 took the active docking role in a 201 × 253

Figure 3-5. Soyuz 4 and 5 cosmonauts Boris Volynov, Aleksey Yeliseyev, Yevgeniy Khrunov, and Vladimir Shatalov (right to left) shown here posing next to a globe of the moon. Their flight proved manned rendezvous, docking, and emergency EVA transfer was possible for the manned lunar missions then planned. (Source: Tass from Sovfoto.)

km orbit at about 9:00 a.m.[16] In preparing for the flight, Shatalov had practiced almost 800 dockings in the simulator at Star City.[17] Shatalov approached manually to within a few kilometers of Soyuz 5. He then activated the automatic docking system at 10:37 a.m., and Soyuz 4 closed to within 100 meters of Soyuz 5.[18] Shatalov took control again and docked to Soyuz 5 at 11:20 a.m., during live Soviet television coverage.[19,20]

After docking, Shatalov manually oriented the two spacecraft so their solar arrays were in sunlight. The crews then tested the docking system communication link and congratulated each other. Khrunov and Yeliseyev then prepared for the EVA, which was to transfer over to Soyuz 4.

The cosmonauts wore newly designed spacesuits during the 37-minute space walk (see Figure 3-5). The spacesuits had life-support packs attached to the front of the upper legs to make passage through the small Soyuz orbital module hatch easier.[21,22] The cosmonauts depressurized the Soyuz 5 orbital module and opened the hatch on the side of the module, which is normally only used when the cosmonauts board the Soyuz on the launch pad. While outside they inspected the docking mechanism, and took pictures of the docked spacecraft.[23] They were also monitored on television using the fixed outside cameras by Shatalov, Volynov, and ground controllers. First, Khrunov moved over to Soyuz 4, using hand rails on the side of the orbital modules. He opened the orbital module's side hatch and Yeliseyev crossed over. They were always attached to one of the ships by tethers. To prove the transfer took place Khrunov and Yeliseyev carried newspapers and letters marked after Shatalov's launch.

The cosmonauts had been advised by Leonov, the Soviets only experienced space walker, on how to move in space. Leonov said to avoid hurried movements or exerting themselves much, since this would build up heat in the spacesuit. He also said to think ten times before moving a finger and twenty times before moving a hand.

After completing the transfer, the spacecraft undocked after 4.5 hours and Soyuz 4 landed near the target point in high winds and $-37°C$ temperatures on January 17, 40 km northwest of Karaganda.[24] Soyuz 5 continued in orbit for the next day, landing on January 18, 200 km southwest of Kustana.

The joint flight finally cleared the way for a lunar mission attempt after proving manned rendezvous and docking, nearly two years behind schedule because of the Soyuz 1 failure.

Super Booster 1

Failed

Launched: July 4, 1969
Reentry: —
Altitude: 0.1 km

With Apollo 11 still on the launch pad, the Soviets tried to launch a test of their lunar module on the Soviet super booster. The flight had been on the launch pad since April 1968, when the launch was scrubbed due to engine plumbing problems. The booster was taken back to the assembly building, repaired, and rolled out to the launch pad again in January 1969.

A Proton booster was reportedly in a simultaneous countdown to launch a manned Zond to meet with the lunar module and G-1 escape stage when they reached orbit, as had been demonstrated by the Soyuz 4-5 flight. The ships of the communications and tracking fleet were on station around the world for the flight.

The G-1 booster lifted off, but never cleared the launch tower, which was 100 meters high. The explosion destroyed the launch pad and heavily damaged the other launch pad, miles away. The explosion was first noticed by a U.S. Air Force Ferret satellite as an electromagnetic disturbance, and was reportedly observed by the NASA Nimbus 3 weather satellite.[25] U.S. ships in the Black Sea reportedly detected the blast. U.S. astronauts also reportedly photographed the blast area from orbit. Despite these reports, some sources also state that the launch attempt was before June 14.[26]

The Soviet tracking ships returned to port after the failure. The cosmonauts linked to the Zond were back at Star City the next day to attend a reception for visiting Apollo astronauts.[27] The launch pad was rebuilt in 18 months starting in late, 1969.

In an unrelated accident in late 1969, there was a major failure during a third-stage fueling test on a static test stand.[28] This accident has often been confused with the July 4 launch attempt. The test stand was also repaired for continued use.

The flight was probably meant to be a high altitude or lunar orbit test of the Soviet lunar module (SLM) after rendezvous with a manned Zond in Earth orbit. A month later Zond 7B was launched. It was only a light Zond and was not equipped with a torus propellant tank needed for lunar orbit. If this spacecraft was originally intended to dock to the SLM, it would have been limited to an Earth orbit test. Of course, a heavy Zond could have been used for the dual countdown to enable a lunar orbit mission. Since the Proton launch waited until the next launch window, it is most likely that a heavy Zond was replaced by a lighter Zond for the unmanned lunar flight test Zond 7B, while the G-1 failure was investigated.

Zond 7B

Lunar Fly-by

Launched: August 8, 1969, 2:49 a.m.
Landed: August 14, 1969, 9:14 p.m.
Altitude: 183 × 191 km @ 51.5°

This flight was nearly the same as Zond 6, but was probably originally intended to be launched to a Soviet lunar module in Earth orbit. The spacecraft may also have been reconfigured or replaced for the lunar fly-by mission.

The estimated spacecraft weight at launch was 5,500 kg. The third stage of the Proton booster reentered on August 8 at 7:48 p.m. after separation from the Zond and escape stage.[29] After the escape stage fired and separated from the Zond, the spacecraft took high-quality color pictures of the Earth on August 8 from a distance of 70,000 km. The photos were developed after return to Earth.

A mid-course correction was made on August 9, and the Zond began taking pictures of the moon on August 11 at 5:28 a.m. at a distance of 10,000 km. The spacecraft passed within 2,000 km of the moon at 7:08 a.m. and began its long fall back to Earth. On August 14, the spacecraft made a skip-glide reentry and landed south of Kustanai.[30]

Soyuz 6

Triple Spacecraft Flight

Launched: October 11, 1969, 2:10 p.m.
Landed: October 16, 1969, 12:53 p.m.
Altitude: 192 × 231 km @ 51.7°
Crew: Georgiy Shonin and Valeri Kubasov
Backup: Vladimir Shatalov and Aleksey Yeliseyev
Call sign: Antaeus

It is rumored that this flight was delayed from an intended launch date in April 1969, due to the concentrated effort to correct pre-Luna 15 launch failures and the effort to launch the G-1 lunar test flight in July.[31]

Soyuz 6's main objective was to test the 50-kg Vulkan welding experiment, which was operated by remote control from the reentry module while the orbital module was depressurized. Kubasov would operate the Vulkan testing electron beam and low-pressure compressible arc welding, and then repressurize the orbital module to collect the samples before landing.[32] Soyuz 6 was the same type of design as the previous Soyuz, but it was not equipped with the docking apparatus. It was also impossible for the crew to make an EVA because they wore no pressure suits when boarding the spacecraft and there were probably no spacesuits on board the Soyuz.[33] The spacecraft weighed 6,577 kg at launch. At 8:08 p.m., October 11, Soyuz 6 maneuvered to a 194 × 230-km orbit and waited for the launches of Soyuz 7 and 8 during the next two days. Soyuz 6 would then observe the joint maneuvers of Soyuz 7 and 8 and performed some rendezvous maneuvers as well.

Soyuz 7
Triple Spacecraft Flight
Launched: October 12, 1969, 1:45 p.m.
Landed: October 17, 1969, 12:25 p.m.
Altitude: 210 × 223 km @ 51.7°
Crew: Anatoliy Filipchenko, Viktor Grobatko, and Vladislav Volkov
Backup: Vladimir Shatalov, Poytr Kolodin, and Aleksey Yeliseyev
Call sign: Snowstorm

As Soyuz 6 passed by the Baykonur Cosmodrome a day after its launch, Soyuz 7 was launched into a matching orbit to conduct joint maneuvers. The Soviets announced that Soyuz 7 would rendezvous with Soyuz 8, which was launched the next day, and investigate navigation of group flight with Soyuz 6.[34] Tass later announced that Soyuz 7 and 8 would dock. Apparently, they carried a new internal docking system designed for the lunar and space station program that would allow internal transfer between spacecraft not requiring spacesuits. Filipchenko also confirmed this at a post-flight news conference in Moscow saying that no spacesuits had been carried on Soyuz 7.[35] The old Soyuz 4–5 docking system would have required spacesuits for a crew to transfer after a docking. The test of the new docking system would validate it for use in the lunar program on the future Salyut space stations. The spacecraft's weight at launch was 6,570 kg. The next day Soyuz 8 joined Soyuz 6 and 7 in orbit.

Soyuz 8

Triple Spacecraft Flight

Launched: October 13, 1969, 1:20 p.m.
Landed: October 18, 1969, 12:11 p.m.
Altitude: 201 × 227 km @ 51.7°
Crew: Vladimir Shatalov and Aleksey Yeliseyev
Backup: Andriyan Nikolayev and Vitaliy Sevastyanov
Call sign: Granite

Shatalov was designated as the group commander of the three Soyuzes in orbit. Soyuz 8, like Soyuz 7, was equipped with docking apparatus but did not dock during the flight.[36] Soyuz 8's weight at launch was 6,645 kg. After entering orbit, Soyuz 8 prepared for close maneuvers with Soyuz 6 and 7 the next day.

On October 15, Soyuz 7 and 8 maneuvered until they were about 460 meters apart.[37] At the time, all three ships were in a 200 × 225-km orbit and Tass claimed that Soyuz 7 and 8 would dock.[38] The ships maintained their distances for 24 hours and tested systems of controlling simultaneous group flight of spacecraft.[39] Soyuz 6 stayed within several kilometers of Soyuz 7 and 8 for several hours during the close maneuvers. Many years later, Shatalov stated in *Flight International* magazine that Soyuz 8 was uncontrollable for a time.[40] This was probably during the final docking maneuvers. Experiments performed during the joint flight included photography and inspection of each other's ships, and using light signals for communication. The crews also concentrated on optical navigation techniques, and manual control of maneuvering engines and the attitude control system. The three spacecraft were supported by eight tracking ships deployed around the world to provide communications with the mission control center in the Crimea.[41]

The crews conducted experiments and photographed Earth before each Soyuz individually landed. On October 16, welding experiments were performed on Soyuz 6 with the electron beam method yielding the best results, but the results were not as good as hoped. After repressurizing the orbital module and retrieving the welded samples, Soyuz 6 landed at 12:52 p.m., 180 km northwest of Karaganda.

Soyuz 7 and 8 remained in orbit for the next day, with Soyuz 8 testing communications through the *Komarov* tracking ship and a Molniya 1 type communications satellite to mission control. On October 7, Soyuz 7 landed at 12:26 p.m., 155 km northwest of Karaganda. The next day Soyuz 8 landed 145 km north of Karaganda ending the first joint flight of three manned spacecraft. During the three flights, 31 orbital changes were made, most under manual control, apparently after the automatic systems failed and while testing several rendezvous maneuvers that would later be used when docking with Salyut stations.[42,43]

Kosmos 382A

Failed
Launched: November ?, 1969
Reentry: —
Altitude: ?

This mission failed to reach orbit. It is rumored to have been a test of the Soviet lunar module on a Proton booster like the one eventually flown on Kosmos 382B a year after this attempt. The flight was to gain experience with the lunar module in orbit after the first spacecraft was lost on the first G-1 test. This test flight would probably have been similar to the Apollo 4 lunar module test flight. The lunar module would probably have fired its decent engine simulating a lunar landing, boosting the spacecraft to a higher orbit. The ascent stage would then separate and fire its engine similar to the Kosmos 379 flight.

Soyuz 9

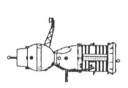

Long Duration: 18 days
Launched: June 1, 1970, 10:00 p.m.
Landed: June 19, 1970, 2:59 p.m.
Altitude: 176 × 227 km @ 51.7°
Crew: Andriyan Nikolayev and Vitaliy Sevastyanov
Backup: Anatoliy Filipchenko and Georgiy Grechko
Call sign: Falcon

A picture of Soyuz 9 shows that it did not carry docking equipment, but it carried a Soyuz 4–5 type housing on top of the orbital module in which a probe or, on which a drogue was normally installed (see Figure 3-6). This shows that Soyuz 9 was a modified Soyuz 4–5 type vehicle and not the first of the Salyut-compatible Soyuzes like Soyuz 7 or 8.[44] Vasili Lazarev, Valery Yazdovsky, and Oleg Makarov were backup cosmonauts for the mission.[45] After separating from the booster stage in a circular orbit at 194 × 198 km, the Soyuz maneuvered to its initial orbit of 176 × 227 km. The orbit was soon changed again to 208 × 254 km, and on June 3, it was changed to 244 × 259 km. The object of the flight was to study the long-term effects of weightlessness, to observe and photograph the Earth, to simulate a lunar duration mission (and break the Gemini 7 space endurance record), and to test prototype Salyut systems. The major piece of Salyut equipment tested was a Salyut-type air regenerator installed in the orbital module for testing. One of the crew's other tasks was to study the oceans, which would lead to better techniques for Soviet fishing fleets. The cosmonauts worked 16-hour days, with 2 exercise periods using exercise

Figure 3-6. Soyuz 9 is shown in the A-type assembly building with its solar arrays wrapped around its service module for launch. Under the Soyuz is the booster interface section to which the launch shroud is attached. The Soyuz is about to be lowered into a horizontal position for insertion into the launch shroud before being taken outside for fueling of its propellant tanks. The booster is shown partially assembled in the background on its rail transporter. (Source: Tass from Sovfoto.)

suits, and 8 hours of sleep.[46] For meals, the cosmonauts ate heated food in aluminum tubes.[47]

On June 4 and 5, the crew conducted various physiological and astrophysical experiments. For the rest of the mission the crew alternated between physiological experiments and astrophysical experiments, working on each for one or two days at a time. On June 5 and 6 the crew tested the so-called sun warping spin stabilized mode. This stabilized the Soyuz so that its solar panels were pointed at the sun continuously. The yaw axis spinning kept the Soyuz from drifting, which would eventually turn the spacecraft to point toward the Earth with its solar panels exposed to only intermittent sunlight, and saved attitude control propellant normally used to prevent drifting. The crew reported that the spinning caused a weak centrifugal gravity, which was unpleasant at times. On June 8 and 9, the crew tested navigational horizon sensors and calculated their orbit using Earth landmarks. On June 11, they had a day off to catch up on work. On June 14, the cosmonauts made joint weather observations with research ships and observed meteors passing into the Earth's atmosphere below them. The next day, they tested the Soyuz main engine in preparation for return to Earth, changing their orbit to 215 × 231 km. On June 16, they took pictures of the U.S.S.R. simultaneously with aircraft, as was standard practice on later Salyut missions. On June 19, at 1:52 p.m., Nikolayev oriented the ship for retroburn. The Soyuz was programmed to ensure no more than three gravities deceleration during reentry, which started at 2:17 p.m. Soviet television carried live coverage of the final parachute descent to a landing 75 km west of Karaganda.[48]

The cosmonauts had set a new space endurance record of almost 18 days, breaking the Gemini 7 record of 14 days. After landing, the crew had difficulty getting out of the capsule.[49] The long flight in the small Soyuz capsule and with too little exercise had weakened the cosmonauts. They said the Earth's normal gravity felt like the several gravities they normally experienced in the centrifuge simulators at Star City. They were put into isolation at a new quarantine facility at Baykonur, similar to the NASA Lunar Receiving Laboratory in Houston, Texas, to prevent illness in their weakened state.[50] They had difficulty sleeping for four days and did not fully recover for ten days.[51] Along with the usual awards at the end of a successful space flight, Nikolayev was also promoted to Major General.[52]

After the mission, the usual Soviet news reports mentioned the crew's re-adaptation period was unexpectedly long. The Western wire services misinterpreted the statement as to mean that something had gone wrong with the mission, and that a report about routine tests of the cosmonauts' color perception during the flight meant that their eye sight had deteriorated. When U.S. Senators Walter Mondale and William Proxmire heard the wire service reports, they authored an amendment to NASA's fiscal 1971 budget request that would cut 20% overall and terminate the shuttle program on the basis that long duration spaceflight was dangerous to humans, even though the shuttle was not designed to stay in space for long durations. The amendment failed.[53]

Zond 8

Last Zond Flight
Launched: October 20, 1970, 10:55 p.m.
Landed: October 27, 1970, 4:55 p.m.
Altitude: 202 × 223 km @ 51.5°

Zond 8 was probably the back-up spacecraft to the Zond 7B. Since the spacecraft was already prepared, the Soviets probably flew it as a purely research flight. This flight differed from the other Zond flights in that Zond 8 reentered over the Northern hemisphere to a landing in the Indian Ocean. The Soviets also announced the launch and landing times for the first time ever. The spacecraft's weight was estimated at 5,500 kg. After launch, Zond 8 and its escape stage separated from the third stage, which reentered October 26 at 10:55 p.m.[54] Zond 8 was then propelled on a trajectory that would loop around the moon and return it to Earth. The escape stage then separated and Zond 8's solar panels and antennas deployed for the lunar journey.

On October 21, Earth was photographed while on route to the moon at 65,000 km. On the next day, there was a course correction, and on October 24, Zond 8 made its closest approach to the moon, at 1,120 km altitude. It took black and white, and color photographs of the moon and Earth, and provided television pictures as well.[55]

On October 27, the capsule reentered Earth's atmosphere and made a ballistic reentry, flying over the North Pole, which enabled the reentry to be monitored by ground stations in the U.S.S.R. By flying a reentry over the Soviet Union the capsule could not land within Soviet territory but further south in the Indian Ocean where Soviet naval forces waited to recover the capsule with its cargo of film and experiment results. Going to this trouble, instead of a normal and relatively routine landing in the area around Baykonur, indicates that monitoring the capsule's reentry was a major objective of the mission. The capsule landed 730 km southeast of Chagos Archipelego, in the Indian Ocean, at 4:55 p.m. It was recovered quickly, put into a shipping canister, taken to Bombay, and loaded onto a plane for shipment to Moscow.[56-58]

Kosmos 379

First Soviet Lunar Module Flight
Launched: November 24, 1970, 8:15 a.m.
Reentry: September 21, 1981
Altitude: 192 × 232 km @ 51.62°

This was the first flight of a Soviet lunar module (SLM) ascent stage on an A-2 booster.[59] The spacecraft's weight at launch was estimated at about 7,500 kg.[60] The final booster stage was left in orbit at 189 × 214 km.

The spacecraft maneuvered to 190 × 1,210 km, by November 25, and then separated from its orbital module. The ascent stage then maneuvered to 175 × 14,035 km @ 51.69° by November 30. It has been reported that taped voice messages were transmitted by the spacecraft to test the communications system as was normal practice on previous Soviet manned spacecraft test flights.[61]

Kosmos 382B

Full Test of Lunar Module

Launched: December 2, 1970, 8:00 p.m.
Reentry:
Altitude: 305 × 5,045 km @ 51.54°

This flight was probably the first full test of Soviet lunar module on a D-1 booster.[62] The exact configuration of the lander it still not known. Spacecraft weight would have been about 19,000 kg for launch on a Proton booster, and the spacecraft itself was used to achieve orbit instead of additional Proton upper stages.

After separating from the Proton second stage, the SLM descent stage was used to put the spacecraft into a 305 × 5,045 km orbit. By the next day, the spacecraft had fired again to increase its orbit to 409 × 5,045 km. It later boosted its orbit again to 1,590 × 5,081 km by January 1.

The descent stage fired by December 7 putting the spacecraft into a 1,615 × 5,072 km @ 51.55° orbit where the descent stage and orbital modules were jettisoned. The ascent stage then maneuvered to a 2,577 × 5,082 km @ 55.87° orbit.[63] The test appears to have been successful and the major components are still in orbit.

Kosmos 398

Launched: February 26, 1971, 8:06 a.m.
Reentry: September 21, 1983
Altitude: 189 × 252 km @ 51.61°

This was another lunar module ascent stage test, similar to Kosmos 379.[64] The booster's final stage was left in orbit at 186 × 239 km and it decayed on March 1. The spacecraft maneuvered by February 28 to 186 × 1,188 km @ 51.60° and separated the orbital module. It then maneuvered to 203 × 10,903 km @ 51.59°. At least two of the four tests of this type transmitted recordings of voices to test the communications systems.[65]

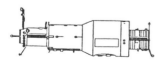

Salyut 1

First Space Station

Launched: April 19, 1971, 3:40 a.m.
Reentry: October 11, 1971

Altitude: 200 × 210 km @ 51.6°

Salyut 1 was Earth's first manned space station. The design of Salyut 1 began in 1969, and fabrication began in 1970. It has been reported that the

Salyut 1 space station experienced design and production problems that delayed the program and the first flight (see Figure 3-7). Salyut 1's main task was to serve as an engineering test for future military and civilian space stations, and to perform general scientific research (see Figure 3-8).

Figure 3-7. The world's first space station, Salyut 1, is shown here before being attached to its Proton (D-1h) booster. The Salyut's solar panels are wrapped around the transfer compartment and the docking drogue is visible to the left. The large circular opening to the scientific instruments is on the right side of the picture. (Source: Tass from Sovfoto.)

Salyut 1 was launched into low Earth orbit by a Proton D-1h booster. After passing out of the atmosphere, the station's launch shroud fell away in halves and the Orion telescope covering was jettisoned. The Proton third stage shut down in an orbit at 176×211 km and the Salyut separated to be propelled into an initial orbit by its Soyuz-type engine module. Salyut 1's low orbit was characteristic of the military stations that would follow in the years ahead. Because of the low orbit it was necessary for the Salyut to be boosted in its orbit periodically, to prevent it from reentering due to atmospheric drag.[66] When the station was not being used for photography or telescope observations, it was put into a three-degree-per-second yaw spin to save propellant and orient the solar arrays toward the sun much like was done on solo Soyuz flights.

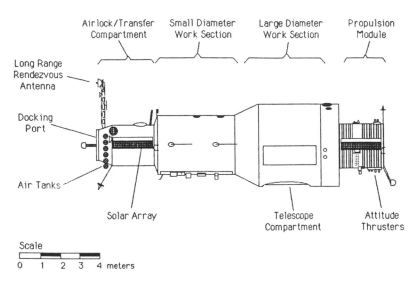

Figure 3-8. This drawing shows the major components of the Salyut 1 station. The Soyuz manned transport docked to the transfer compartment, which could also be used as an airlock for EVA. The crew used the small and large diameter sections as living space and laboratory. A modified Soyuz service module provided propulsion for the station. The space station and its general design carried on through the stations of 1970s and 1980s.

Salyut 1

The first Salyut space station was made of three main components—a transfer compartment, a two-part work compartment, and a service module. Its total weight was about 19,000 kg and its length was 15.8 meters. The transfer compartment, or docking adapter, was 3 meters long and 2 meters in diameter. The forward end had a docking drogue to accept a Soyuz docking probe. It also served as an airlock with a hatch for EVA, but it was never used for that purpose during the short Soyuz 11 mission.

The work compartment was made of two cylinders, the first 3.8 meters long and 2.9 meters in diameter. This was connected by a 1.2-meter-long cone to a cylinder 4.1 meters long and 4.15 meters in diameter. In the forward part of the work compartment were the dining area, recreation area, and control panels. Four of the seven station control panels were at the forward work compartment and included the navigation system, motion control gyroscope system, propulsion controls, life support, thermal regulation and communication controls, warning indicators, the Earth globe, a stellar globe, a television monitor and checklist, chronometers, light controls, automatic sequencer controls, experiment controls, and duplicate station orientation and propulsion controls. The aft work compartment area contained food and water storage, toilet, exercise equipment, a telescope and camera housing, and its control panel. The compartment also had two spherical scientific airlocks consisting of two spheres, one inside the other. The inner sphere had a hole that matched the opening on the outer hull of the station. To open the airlock, the outer sphere was closed and the inner one rotated until its opening matched the outer opening. The same type airlocks were used on all future space stations including Mir.

Attached to the aft end of the station was the service module, which contained the engine systems needed to orient and maintain the station's orbit. It was a modified Soyuz service module, 2.17 meters long and 2.2 meters in diameter. The KTDU-66 main engine was modified from the Soyuz version to provide up to 1,000 seconds of service.

Soyuz 10

First Mission to a Space Station—Aborted

Launched: April 23, 1971, 2:54 a.m.
Landed: April 25, 1971, 2:40 a.m.
Altitude: 209 × 258 km @ 51.6°
Crew: Vladimir Shatalov, Aleksey Yeliseyev, and Nikolay Rukavishnikov
Backup: Aleksei Leonov, Valeri Kubasov, and Pyotr Kolodin
Call sign: Granite

This was the first mission to the Salyut 1 space station. The crew was apparently chosen because Shatalov and Yeliseyev were experienced with Soyuz rendezvous maneuvers from previous flights, and Rukavishnikov was a Salyut systems expert.

The pre-dawn launch occurred three days after Salyut 1 was launched. After separating from the final booster stage, the Soyuz made three orbit changes enabling the crew to sight the station at a distance of 15 km.[67] The Salyut performed the same role as the passive Soyuz targets, keeping its docking port pointed at the approaching Soyuz. The Salyut also made four orbit changes to match orbit with the approaching Soyuz. The automatic docking system brought the Soyuz to 180 meters away from the Salyut.[68] Shatalov completed the docking to the station, under manual control, from 180 meters distance at 4:47 a.m., April 24.[69]

No crew transfer was made into the Salyut. The Soyuz carried a new docking system that had not been tested in flight before. The Soyuz 7–8 flight carried the new docking system, but they failed to dock to test the new system. After the flight, Salyut designer Feoktistov said "the docking of this type is a more difficult task as compared to the docking of Soyuz or Kosmos spaceships—craft of roughly the same mass."[70] The probe and hatch of the Soyuz docking system probably refused to open because the Soviets later successfully sent Soyuz 11 to Salyut 1, which meant the Salyut was probably without fault.

Soyuz 10 remained docked to the Salyut for five and a half hours, which means they were probably hard docked, but could not remove the probe assembly to move into the Salyut or there was no seal between the ships. The cosmonauts carried no spacesuits for the mission and an EVA to board the station would have been impractical and probably impossible. The mission was probably planned to last about 23 to 33 days.[71]

After waiting to analyze the situation, the decision was made to end the mission and let the next planned Salyut crew (Soyuz 11) cope with the problem. Soyuz 10 undocked and circled the Salyut for an hour taking pictures of the station.[72] The orbit was lowered before retrofire to 190×231 km in the usual test of the Soyuz engine before return to Earth. Soyuz 10 reentered 16 hours later in darkness, which was not normal procedure, and landed 120 km northwest of Karaganda.[73] The capsule landed 50 meters from a lake. The cosmonauts said they were "saved by a gust of wind" from landing in the lake.[74]

On April 28, Salyut 1 raised its orbit to 251×271 km to await the next Soyuz. Before the Soyuz 11 launch, the Salyut's orbit was circularized to 200 km.

Soyuz 11

Long Duration—24 days

Crew Lost

Launched: June 6, 1971,
      7:55 a.m.

Landed: June 30, 1971, 2:17 a.m.
Altitude: 189×209 km @ 51.6° (4th orbit)
Crew: Georgi Dobrovolsky, Vladislav Volkov, and Victor Patsayev
Backup: Aleksei Leonov, Valeri Kubasov, and Pyotr Kolodin
Call sign: Amber

This was the second, and last flight to Salyut 1. The Soyuz 10 crew had failed to enter the space station after docking. It was judged best to return the Soyuz

10 and prepare a new mission to deal with the hatch problem. The original Soyuz 11 crew was Leonov, Kubasov, and Kolodin. They had been preparing to occupy the Salyut after the month-long mission of Soyuz 10. After the Soyuz 10 failure, they prepared to perform the Salyut activation for human occupation, and to correct the hatch problem that had kept the Soyuz 10 crew from entering the Salyut.

Two days before the launch, a medical problem was discovered with Kubasov.[75] Under the prevailing mission rules, if a crew member was disqualified from flight status, the entire crew might be removed from the flight due to contamination or compatibility problems. A well trained but sick crew could easily be less productive in orbit than a less trained but healthy crew. The back-up crew of Dobrovolsky, Volkov, and Patsayev was somewhat less experienced but was forced to replace the primary crew (see Figure 3-9). The choice to fly the back-up crew was also dictated by the condition of the space station in orbit. The Salyut would slowly run low on unreplaceable supplies of propellant if a new crew of veteran cosmonauts was trained. On the back-up

Figure 3-9. The Soyuz 11 crew is shown here training in a Soyuz simulator at the Gagarin Training Center at Star City. The Soyuz's simple control panel contains indicators and only a few controls, because most of the flight is controlled automatically. The crew is wearing normal flight suits, which were standard because the use of pressure suits was discontinued after Vostok (except for Voskhod 2). After a stay on Salyut 1, the crew was killed during reentry when the air escaped from their capsule. (Source: Novosti from Sovfoto.)

crew, only Volkov had experience in flight, and the others did not even have acknowledged experience as back-ups for any flight. On entering orbit, Soyuz 11 was 3,000 km behind Salyut.[76] While out of communication range the crew slept, Dobrovolsky and Volkov in the orbital module and Patsayev in the reentry module. The next day, as the Soyuz approached the station, the automatic docking system was activated at 6 km distance.[77] On the 16th orbit, a second orbital correction was made by the automatic system to put the Soyuz 100 meters from the station in orbit at 177×209 km. Dobrovolsky then manually docked on June 7, at 7:55 a.m. When docked to the station, the Soyuz made communications and electrical connections so the Soyuz's solar panels could supply the station with extra power. The crew entered the station at 10:45 a.m. (see Figure 3-10).[78]

For the next two days the crew prepared the station for operation and shut down their Soyuz. The cosmonauts then activated scientific and technical

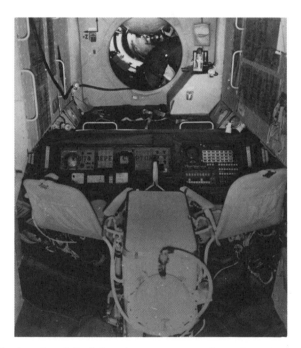

Figure 3-10. The interior of the Salyut 1 space station is shown here during preparations for launch at Baykonur. The photo was taken from the telescope housing near the rear of the station. The station's control panels are in the center of the photo, the one on the left is a modified Soyuz control panel. The seats at the controls are virtually useless in weightlessness, but it took years to finally remove them from space station designs made by Earth-bound designers. A technician is visible working in the transfer compartment/airlock section. (Source: Novosti from Sovfoto.)

experiments and complained of being overworked. Duties included astronomical observations, live television shows, growing tadpoles and plants, and Earth photography.

On June 9, the Salyut performed a 73-second maneuver to raise the orbit from 239×265 km to 256×264 km. The cosmonauts made repeated checks of the station's atmosphere and tested the elastic penguin exercise suit (so named because on Earth subjects wearing the suit waddled like penguins). On June 10th, they tested the treadmill, performed medical tests and made a television broadcast. The next day, the Anna III telescope was used, and communications through a Molniya 1 communications satellite was tested.

Until June 15, the crew did routine work including checking dosometers, medical checks, Earth photography, tending Oasis, and making television broadcasts. The next day, they tested the station's attitude controls and their accuracy. On the 17th, they tested a high-frequency electron resonance transmitting antenna design. On June 18, they began Orion 1 observations. Until June 27, the crew continued normal duties, television broadcasts, and medical checks.

On June 27, they activated and checked Soyuz systems to prepare for return to Earth. On June 29, the crew packed their experiment results, exposed film, and log books in the Soyuz reentry capsule for return to Earth. They closed the hatches to the station at 9:15 p.m. and undocked 13 minutes later, reporting all was well. On the ground, the Soviet Union was preparing for the triumphant return of the "Heroes of the Soviet Union" (a title bestowed to nearly all returning cosmonauts). The mission had been given extensive coverage by the Soviet media, while saying that Salyut was a much better and rewarding program than the U.S. Apollo moon landings. Despite the glowing reports from the Soviet press, there are rumors that several problems including a serious electrical fire on June 27, caused a premature end of the mission.[79] Estimates of the landing windows of the Soyuz indicated a normal landing opportunity would have occurred between 33 and 45 days into the Soyuz 11 mission.[80]

At 1:35 a.m., June 30, the crew fired the Soyuz retrorockets to deorbit and 12 minutes later separated from the orbital and service modules. At this time, the orbital module was normally separated by 12 pyrotechnic devices, which were supposed to fire sequentially, but they incorrectly fired simultaneously. This caused a mechanism to release a seal on the capsule's pressure equalization valve, which normally opens at low altitude to equalize cabin air pressure to the outside air pressure. This caused the cabin to lose all its atmosphere in about 30 seconds, while still at a height of 168 km. This killed the crew while they were trying to manually shut off the valve, a procedure that would take over 60 seconds. They managed to half close the valve before losing consciousness and dying. Fifteen and a half minutes after retrofire, the pressure reached zero in the capsule and remained that for eleven and a half minutes, at which point the cabin started to fill with air from the upper atmosphere.[81] The rest of the descent was normal and the capsule landed at 2:17 a.m. The recovery forces located the capsule and opened the hatch only to find the cosmonauts motionless in their seats. On first glance they appeared to be asleep, but closer examination showed why there was no normal communication from the capsule during the descent.

The Soviets had to give a detailed report on the accident to NASA in preparation for the Apollo-Soyuz Test Project. They said that the amount of tissue damage to the cosmonauts' bodies caused by the boiling of their blood during the 11.5 minutes of exposure to vacuum could at first have been misinterpreted as being the result of a catastrophic and instantaneous decompression. The cause of death was pulmonary embolism. Only through analysis of the telemetry records could the failure of the valve be determined. The telemetry showed that the attitude control thrusters had automatically fired to counteract the force of the air escaping from the valve. Examination also found traces of pyrotechnic residue in the valve, which also proved that the valve had opened when the pyrotechnics were fired. NASA verified these conclusions after viewing the telemetry data.[82]

As a result of the accident, the head of the cosmonaut corp, General Nikolai Kamanin was replaced by Cosmonaut Shatalov. The Soviets had hoped to use the Salyut 1 again to support another manned mission, but when a redesign of the Soyuz was found to be needed, the time this would take was beyond the possible lifetime of the Salyut. Temperature, pressure in the station, and orbit altitude had been maintained to collect engineering data until attitude control propellant supplies ran low. Only when enough was left for reentry, on October 1, 1971, was Salyut 1 deorbited to prevent uncontrolled decay of the orbit.

Super Booster 2

Failed

Launched: June 24, 1971
Reentry: —
Altitude: 12 km

After almost two years of redesign, the G-1 booster was ready to be tested again. Even though the moon race had been lost, the Soviets still saw a use in the G-1 to launch large space stations to continue the work begun on Salyut. They might also have intended to continue the lunar landings begun by the U.S. Apollo program, and then portrayed the United States as foolish and only interested in a space race, and not true scientific discovery after ending all lunar exploration in 1972.

The booster was rolled out in March or April to the secondary launch pad that was damaged in the first launch attempt in 1969. It is also reported that the launch was attempted once before June 1971, but was aborted while still on the launch pad due to engine failure during the engine start sequence.[83] When it was eventually launched, the G-1 test flight ended when the first stage failed due to instrumentation and engine failure, possibly caused by excessive pogo. This caused the vehicle to break up at 12 km altitude.[84]

Kosmos 434

Last Soviet Lunar Module Test

Launched: August 12, 1971, 8:30 a.m.
Reentry: August 22, 1981
Altitude: 188 × 267 km @ 51.6°

This was the last lunar module ascent stage test. The flight was the same as Kosmos 398.[85] Kosmos 434's booster stage separated in an orbit at 194 × 261 km. The spacecraft maneuvered by August 16 to 189 × 1,328 km @ 51.6°. The orbital module then separated, and the ascent stage maneuvered to 186 × 11,804 km @ 51.6°. The spacecraft completed its tests and drifted in orbit until its orbit decayed due to atmospheric drag.

After reentry in 1981, and to calm fears that Kosmos 434 might carry a nuclear power source (after the Kosmos 954 satellite spread radioactive parts across Canada), the Soviets announced that Kosmos 434 was a test of an experimental lunar cabin and did not carry nuclear materials.[86] The "lunar cabin" statement confirmed the long-held belief among Western space analysts that the Kosmos flight was a part of the Soviet manned lunar landing program, because the term "lunar cabin" was also used by the Soviets to refer to the NASA lunar lander.

Kosmos 496

Launched: June 26, 1972, 5:53 p.m.
Landed: July 2, 1972, 5:24 p.m.
Altitude: 187 × 321 km @ 51.6°

This was the first Soyuz test since the failed flight of Soyuz 11. The Soyuz now was configured to carry a crew of two in pressure suits, for protection against pressure loss in the cabin, during launch, and landing. The equipment needed to support the suits (air supplies, filters, and cooling units) took up the same space as a person and weighed about 100 kg, thus causing the reduction in crew size. The Soyuz was still equipped with solar panels, indicating that the

Soyuz was still intended to be used with the Salyut 1 type station, which required the power from the Soyuz arrays. Later Salyut stations had larger solar panels, which eliminated the need for the Soyuz to have solar panels.

This flight was probably flown in preparation for a manned Soyuz flight in July or August 1972, to a new Salyut 1 type station, probably Salyut 1's back-up. The new flight and space station would finally fulfill the mission of the failed Soyuz 10 and 11 flights.

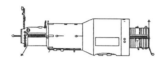

Salyut

Failed
Launched: July 29, 1972
Reentry: July 29, 1972
Altitude: 75 km?

The Soviets had again entered a space race without telling anyone. They desperately wanted to perform the first successful space station mission after the terrible failure of the Soyuz 11 crew, and beat the U.S.'s Skylab and its first crew. This desire may have caused them to rush the next two Salyut's into being destroyed during flight.

This Salyut launch was scrubbed in May 1972. Since the Soyuz was still at this time equipped with solar panels, this Salyut was probably of the Salyut 1 type and probably Salyut 1's back-up.

During the launch, at least one of the Proton's second-stage engines shut down early, but the stage continued to fire the programmed time, and shut down with unspent propellant remaining. Clearly, state-of-the-art technology was lacking in the Proton booster to compensate for the loss of thrust, which had also occurred during Apollo launches. The Proton third stage and Salyut station never reached orbit and reentered over the Pacific.[87,88]

Super Booster 3

Failed
Launched: November 23, 1972
Reentry: —
Altitude: 40 km

The results of the second G-1 test flight must have looked promising in comparison to the first flight, and they planned one more test. The Soviets still probably were intending to use the G-1 in the large space station program and to start a new round of lunar exploration after the last Apollo landing. The booster lifted off successfully and flew through the first minutes, but encountered serious problems at the end of the first-stage burn. The booster was destroyed at 40 km altitude after suffering instrumentation and engine failure during the last part of the first-stage burn, possibly caused by excessive pogo.[89,90]

The manned lunar and large space station programs were then stopped while a new dynamic test stand was built at Baykonur to finally test the G-1 in 1976.[91] After the three unsuccessful tests, and dynamic testing in the new test stand in the middle of 1976, the G-1 facilities were mothballed for future use with the Energia booster and Soviet space shuttle in the 1980s. The G-1 launch pads and assembly buildings would be converted for Energia and Shuttle use. When photos of the facilities were finally made public in 1988, they matched the plans for launch pads of the giant G-1 booster in many details.

References

1. Collins, Michael, *Carrying the Fire*, New York: Bantam Books, 1974, p. 282.
2. Foreign Broadcast Information Service, JPRS-USP-86-006, November 12, 1986, Joint Publications Research Service, pp. 106–115.
3. Ezell, Edward C. and Ezell, Linda N., *The Partnership: A History of the Apollo-Soyuz Test Project*, NASA SP-4209, Washington D.C.: Government Printing Office, 1978, p. 96.
4. Turnill, R., *Spaceflight Directory*, London: Frederick Warne Ltd., 1977, p. 361.
5. Vick, C. P., "The Soviet G-1-e Manned Lunar Landing Program Booster," *Journal of the British Interplanetary Society*, Vol. 38, No. 1, Jan. 1985, p. 17.
6. Daniloff, N., *The Kremlin and the Cosmos*, New York: Alfred A. Knopf, 1972, pp. 152, 170.
7. Oberg, James E., *Red Start in Orbit*, Random House, New York, p. 143.
8. Mills, Phil, "The Soviet Manned Lunar Landing Program and Soyuz Missions 1–5," *Journal of the British Interplanetary Society*, Vol. 41, p. 129.
9. Congressional Research Service, The Library of Congress, *Soviet Space Programs 1976–80, Unmanned Space Activities, Part 3*, Washington: Government Printing Office, 1985, p. 812.
10. Vick, Charles P., "The Soviet Super Boosters—1," *Spaceflight*, Vol. 15, No. 8, Aug. 1973, pp. 467–468.
11. Oberg, James E., *Uncovering Soviet Disasters*, Random House: New York, 1988, p. 174.
12. Turnill, op. cit. p. 297.
13. Johnson, Nicholas L., *Handbook of Soviet Manned Spaceflight*, American Astronautical Society: San Diego, 1980, pp. 151-2.

14. Borisenko, I. and Romanov, A., *Where All Roads to Space Begin*, Moscow: Progress Publishers, 1982, p. 55.
15. Bond, Peter, *Heros in Space: From Gagarin to Challenger*, Basil Blackwell: New York, p. 291.
16. Johnson, op. cit. p. 152.
17. Foreign Broadcast Information Service, USSR, Space, JPRS-USP-86-005, Sept. 1986, Joint Publications Research Service, p. 66.
18. Turnill, op. cit. p. 297.
19. Johnson, op. cit. p. 156.
20. Ertel, Ivan D. and Newkirk, Roland W., *The Apollo Spacecraft, A Chronology, Vol. IV*, NASA SP-4009, Washington: Government Printing Office, 1978, p. 387.
21. Turnill, op. cit. p. 297.
22. Oberg, James E., "Soyuz 1 Ten Years After: New Conclusions," *Spaceflight*, Vol. 19, No. 5, May 1977, p. 189.
23. Johnson, op. cit. p. 157.
24. Turnill, op. cit. p. 297.
25. Vick, Charles P., "The Soviet Super Boosters—2," *Spaceflight*, Vol. 16, No. 3, March 1974, p. 96.
26. Baker, David, *The Rocket*, London: New Cavendish Books, 1978, p. 225.
27. Anderman, D., "Soviet Type G Booster—A Skeptical View," *Journal of the British Interplanetary Society*, Vol. 40, No. 5, May 1987, p. 227.
28. Vick, "The Soviet G-1-e Manned Lunar Landing Program Booster," op. cit. p. 17.
29. Congressional Research Service, The Library of Congress, *Soviet Space Programs 1976–80, Manned Space Programs and Life Sciences, Part 2*, Washington: Government Printing Office, 1984, p. 641.
30. Johnson, Nicholas L., *Soviet Lunar and Planetary Exploration*, American Astronautical Society: San Diego, 1979, p. 119.
31. Clark, Phillip, *The Soviet Manned Space Program*, Orion Books: New York, 1988, p. 52.
32. Johnson, *Handbook of Soviet Manned Spaceflight*, op. cit. p. 162.
33. Bond, op. cit. p. 294.
34. Johnson, *Handbook of Soviet Manned Spaceflight*, op. cit. p. 160.
35. Turnill, op. cit. p. 298.
36. Clark, Phillip S. and Gibbons, Ralph F., "The Evolution of the Soyuz Program," *Journal of the British Interplanetary Society*, Vol. 36, No. 10, Oct. 1983, p. 440.
37. Ertel and Newkirk, op. cit. p. 390.
38. Johnson, *Handbook of Soviet Manned Spaceflight*, op. cit. pp. 161, 163.
39. Turnill, op. cit. p. 298.

40. Furniss, Tim, "Soviets Open 1988 Space Account," *Flight International*, Jan. 26, 1988, p. 26.
41. Bond, op. cit. p. 295.
42. Johnson, *Handbook of Soviet Manned Spaceflight*, op. cit. pp. 161–163.
43. Turnill, op. cit. p. 298.
44. Von Braun, Wernher, Ordway, Frederick, and Dooling, David, *Space Travel: A History*, 1975, photograph p. 206.
45. *Spaceflight*, Vol. 31, No. 2, Feb. 1989, p. 57.
46. Johnson, *Handbook of Soviet Manned Spaceflight*, op. cit. p. 165.
47. Foreign Broadcast Information Service, USSR, Space, JPRS-USP-84-003, June 1984, Joint Publications Research Service, p. 44.
48. Johnson, *Handbook of Soviet Manned Spaceflight*, op. cit. pp. 166–168.
49. Bond, op. cit. p. 297.
50. Congressional Research Service, The Library of Congress, *Soviet Space Programs 1976–80, Manned Space Programs and Life Sciences, Part 2*, Washington: Government Printing Office, 1984, p. 516.
51. Turnill, op. cit. p. 299.
52. Congressional Research Service, The Library of Congress, *Soviet Space Programs 1976–80, Manned Space Programs and Life Sciences, Part 2*, Washington: Government Printing Office, 1984, p. 516.
53. *Aviation Week and Space Technology*, July 6, 1970, p. 13.
54. Congressional Research Service, The Library of Congress, *Soviet Space Programs 1976–80, Manned Space Programs and Life Sciences, Part 2*, Washington: Government Printing Office, 1984, p. 641.
55. Johnson, *Soviet Lunar and Planetary Exploration*, op. cit. p. 120.
56. Ibid.
57. Turnill, op. cit. p. 334.
58. Congressional Research Service, The Library of Congress, *Soviet Space Programs 1976–80, Manned Space Programs and Life Sciences, Part 2*, Washington: Government Printing Office, 1984, p. 649.
59. Vick, "The Soviet G-1-e Manned Lunar Landing Program Booster," op. cit. p. 17.
60. Woods, D. R., "Lunar Mission Cosmos Satellites," *Spaceflight*, Vol. 19, No. 11, Nov. 1977, p. 383.
61. Parfitt, John and Bond, Alan, "The Soviet Manned Lunar Landing Program," *Journal of the British Interplanetary Society*, Vol. 40, No. 5, May 1987, p. 231.
62. Vick, "The Soviet G-1-e Manned Lunar Landing Program Booster," op. cit. p. 17.
63. King-Hele, D. B., Walker, D. M. C., Pilkington, J. A., Winterbottom, A. N., Hiller, H., and Perry, G. E., *The R.A.E. Table of Earth Satellites 1957–1986*, New York: Stockton Press, 1987, p. 243.

64. Vick, "The Soviet G-1-e Manned Lunar Landing Program Booster," op. cit. p. 17.
65. Vick, "The Soviet Super Boosters—1," op. cit. p. 465.
66. Johnson, *Handbook of Soviet Manned Spaceflight*, op. cit. pp. 169, 232.
67. Bond, op. cit. p. 229.
68. Turnill, op. cit. p. 299.
69. Johnson, *Handbook of Soviet Manned Spaceflight*, op. cit. p. 284.
70. Ezell and Ezell, op. cit. p. 139.
71. Clark, Phillip S., "Soyuz Missions to Salyut Stations," *Spaceflight*, Vol. 21, No. 6, June 1979, p. 262.
72. Ertel and Newkirk, op. cit. p. 395.
73. Turnill, op. cit. p. 300.
74. Johnson, *Handbook of Soviet Manned Spaceflight*, op. cit. p. 284.
75. *Spaceflight*, Vol. 30, Feb. 1988, p. 72.
76. Turnill, op. cit. p. 300.
77. Bond, op. cit. p. 301.
78. Johnson, *Handbook of Soviet Manned Spaceflight*, op. cit. pp. 291–292.
79. Oberg, James E. and Oberg, Alcestis R., *Pioneering Space*, New York: McGraw-Hill, 1986, p. 201.
80. Clark, "Soyuz Missions to Salyut Stations," op. cit. p. 262.
81. Ezell and Ezell, op. cit. p. 230.
82. Ibid. p. 231.
83. Vick, "The Soviet Super Boosters—2," op. cit. p. 98.
84. Vick, "The Soviet G-1-e Manned Lunar Landing Program Booster," op. cit. p. 17.
85. Ibid.
86. Parfitt and Bond, op. cit. p. 231.
87. Johnson, *Handbook of Soviet Manned Spaceflight*, op. cit. p. 234.
88. Clark, Phillip S., "Soviet Launch Failures," *Journal of the British Interplanetary Society*, Vol. 40, No. 10, Nov. 1987, p. 529.
89. Vick, "The Soviet G-1-e Manned Lunar Landing Program Booster," op. cit. p. 17.
90. Vick, "The Soviet Super Boosters—2," op. cit. p. 102.
91. Turnill, op. cit. p. 361.

Chapter 4

The First Space Stations

The Salyut space station, both civilian and military, became the only manned Soviet space program after the lunar landing program was canceled in the early 1970s. The investigation into the failure of the lunar program's booster continued until 1976, and the lack of the booster stopped any plans for lunar exploration or large space stations in the 1970s. For that time, small space stations were the main manned space activities. The Soviets, of course, wanted to have the first successful space station (Salyut 1 didn't count since the missions to it failed) before the U.S. after losing the prestigious moon race. This unannounced space race lead to many failures for the Soviets in the early 1970s.

During the four years covered in this chapter (April 13, 1973—February 25, 1977), the Soviets pushed forward as fast as possible with operations of space stations. From 1972 to 1977, half of the Salyut stations would fail during or shortly after launch. In an attempt to upstage the U.S. Skylab space station, the Soviets launched three Salyuts only to have each fail, during or shortly after launch. In 1973, Skylab was launched and hosted three crews before the first successful Soviet space station mission. The U.S. program was not intended as a permanent space station, and due to budget limitations, extended missions or resupply were never attempted. Well after the last Skylab mission had returned to Earth after setting a record 84 days in space, the Soviet military station, Salyut 3, was successfully launched and occupied.

The existence of the military space station program had been fortunate for the civilian program, which needed purpose after the manned lunar effort had been canceled. It was the civilians that would make the most of the Salyut stations, and the military would have to use civilian stations by the end of the 1970s. The dominant theme in the Salyut program, after losing the race with Skylab, was to increase mission duration. This is a very important factor in future missions to the planets. Manned flight to the nearest planets, especially Mars, would require flights of many months duration and leaps in technology. Testing men and equipment for those periods would be a major part the Salyut program. The hardest part, in extending duration, would be proving that people

111

could withstand the long periods of weightlessness and return with no great ill effects. Weightlessness can cause muscles and bones to deteriorate dangerously if not countered by strict exercise and diet.

Military Space Stations

There is much information on the civilian Salyuts, but information on the military stations is very sparse. Training of the cosmonauts for the first military station was begun in 1972 under the direction of Major General Gherman Titov and Colonel Yevgeniy Khrunov. The military cosmonaut group consisted of:*

Commanders	Flight Engineers
Popovich	Demin
Volynov	Artyukhin
Gorbatko	Zholobov
Sarafanov	Glazkov
Zudov	Rozhdestvenski
Berezovoi	Lisun
Kozelsky	Preobrazhensky
	Illarionov

*Sources: Peebles, Curtis. Guardians: Strategic Reconnaissance Satellites, Novato, CA: Presidio Press, 1987, p. 260. Spaceflight, Vol. 31, February 1989, p. 57.

The first mission to Salyut 3 failed to reach the old Soviet 18-day record set during Soyuz 9. The next mission to Salyut 3 failed to dock, and the station was destroyed on command a few weeks later, after running out of attitude propellant. Salyut 4 functioned for more than two years and hosted a crew for 29 and 63 days. The Soyuz 18A crew was to have occupied Salyut 4, but they instead flew the first aborted manned launch in history. Salyut 5 was the last military space station, and a malfunction forced its first crew to land after only 50 days. The next crew failed to dock and the last did not even attempt a long flight. It would take the next generation of Salyut stations to finally break the Skylab records.

After the experiments with manned military space stations, a new generation of unmanned military reconnaissance satellites was introduced using the Soyuz design. They complemented and eventually replaced the old Vostok- and Voskhod-based first- and second-generation reconnaissance satellites. The Soviet experiments into manned military spacecraft apparently showed that the most cost effective means of reconnaissance was unmanned satellites. This does not mean that people in space have no military value. Some military work was done by cosmonauts on civilian Salyuts in the 1970s and 1980s. Currently, there are many different type missions flown by several types of unmanned reconnaissance spacecraft based on the Soyuz (see box).

Reconnaissance Versions of Soyuz

Fourth Generation†: These were Soyuz derivative and long-duration types launched by the A-2 only. Types without solar panels flew from Kosmos 758 to the present. Types with solar panels and a two-capsule carrier in place of the reentry module flew between Kosmos 1611 to the present. Altitude ranged from 150 to 400 km. Flights usually lasted about 60 days. When launched from Baykonur, they went into 64.7° orbits; when launched from Plesetsk they went into 67.1° orbits. Later versions may use both film and digital imagery.

Fifth Generation: This series was a modification of the fourth-generation spacecraft and was launched from Baykonur into 64.8° orbits and from Plesetsk into 67.1° orbits; altitude was usually from 210 to 310 km. Maximum duration in orbit was about 259 days. This generation was capable of near real-time digital data transmission and used charge-coupled detectors; it did not carry any return capsules. Data were received by geostationary satellites for relay to ground stations. The first flight was Kosmos 1426.

---

*Source: Peebles, Curtis, *Guardians: Strategic Reconnaissance Satellites*, Novato, CA: Presidio Press, 1987, pp. 164, 388.
†For first-, second-, and third-generation reconnaissance satellites, see Chapter 1.

Space Shuttles

During the development and use of the Salyut space stations, the Soviets also began the first serious efforts at building a space shuttle to service their planned second-generation space stations during the middle 1970s. Rumors had existed about a Soviet shuttle since even before Sputnik's launch! The first was in the middle 1950s when Soviets were reportedly studying the possibility of a space plane bomber. They had obtained three copies of the 1940's Sanger-Bredt Antipodal Bomber report, captured from the Germans after World War II. The reported Soviet shuttle's design and dimensions were very similar and the methods identical to the German proposal. Stalin apparently was having the project studied and tried to have Sanger kidnapped in France, but the Soviets soon discovered the high complexity of such a spacecraft and ended their investigations. In the middle 1950s, the rumor was still circulating that they hoped to develop a space plane for launch around 1967, to counter the then-planned U.S. Air Force X-20 "Dynasoar." The X-20 space plane was eventually canceled in 1963, a year before its first atmospheric drop tests. The Soviets apparently did little serious work on building a shuttle until the 1970s.

At the end of the period covered by this chapter, the first test flight associated with the Soviet efforts to build a reusable space shuttle was launched. The goal was to develop a small reusable shuttle or space plane. The shuttle's mission would be to supply second-generation Salyut space stations, perform military reconnaissance missions, and also beat the large U.S. shuttle program. The Soviets, being very careful not to be observed, conducted landing tests of the advanced spacecraft, using a Tu-95 "Bear" bomber as a carrier aircraft at a remote site in the middle 1970s. They also began talking publicly of a reusable spacecraft to be flown in the 1980s.[1] In 1974, at the IAF conference,[2] Leonov said that development of a recoverable space plane was underway. Radio Moscow reported on the development of a space shuttle with a wing span of 8.5 meters in October 1978.[3] This initial shuttle development should not be confused with the Soviet shuttle launched in 1988. The shuttle of the 1980s was a completely different program from the one of the 1970s. Both of the Soviet shuttle programs and second-generation space stations are discussed in Chapter 5. Figure 4-1 shows all missions and programs covered in this chapter.

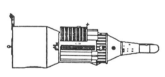

Salyut 2

First Military Space Station—Failed
Launched: April 3, 1973, 12:00 p.m.
Reentry: May 28, 1973, 2:46 p.m.
Altitude: 207×248 km @ 51.6°

This was the first military Salyut and its exact configuration is unknown except that its solar arrays were the same type as used on the Salyut 4 space station, and the docking port was in the rear with the engines and propellant tanks moved to the outer diameter of the station[4] (see Figure 4-2). The military space station also transmitted on a different radio frequency from previous Soviet manned spacecraft. The frequencies were used with previous military reconnaissance spacecraft. The frequency change was characteristic of military Salyut flights along with extensive use of code words and a lack of the usual information released about civilian manned flights. The launch of the Salyut 2 station was first scheduled for August or September of 1972. A launch attempt of the station's Proton booster failed 10 minutes before launch in September 1972. The booster and Salyut were taken back to the assembly building for repairs and refurbishment for a successful April 1973 launch.[5]

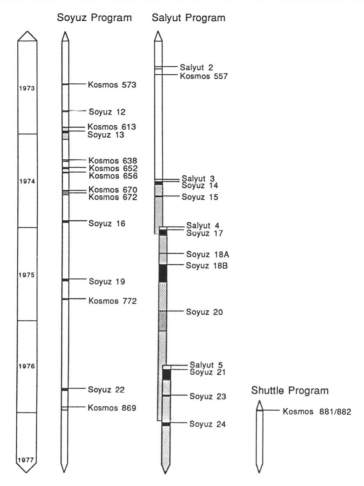

Figure 4-1. The Soyuz test program, Salyut, and Shuttle programs were conducted from 1973 to 1977. All known flights related to these programs are shown here and are described in this chapter. The gray portions show space stations' lifetimes and black portions show manned mission duration.

The day after being launched, the station raised its orbit to 239×260 km.[6] Twenty-three fragments appeared in orbit at this time, along with the station, although they may have been debris from booster separation. Concern about solar flare activity may have delayed the Soyuz mission to the station on April 4 or 5. On April 6, a Soyuz was reported to be on the launch pad ready for launch. After two normal Soyuz launch windows passed, still with no launch, Salyut 2 raised its orbit on April 8 to 257×278 km.

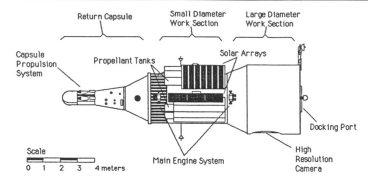

Figure 4-2. This drawing is a conceptual design based on Soviet statements of the design of the Salyut 3 and 5 military space stations. The main elements in the design are two large cylinders connected by a cone-like Salyut 1. The Salyut also carried a return capsule and was about the same length as Salyut 1. The military Salyut's propulsion system was also distributed around the outer hull of the station according to rumor.

On April 10, a manned mission to the Salyut was still being prepared.[7] Four days later, the Salyut maneuvered to 261×296 km. Soon after this, the station developed a severe tumbling motion and broke up into 25 pieces, which decayed on May 28.[8] The delay in sending a crew to the Salyut may have indicated problems with the Salyut or Soyuz before the breakup on the 14th.

On April 18, the Soviets denied the station would have been manned. Most sources agree that it is most likely that an attitude control thruster stuck on and caused the station to tumble until it broke up (a similar problem caused Gemini 8 to make an emergency landing in 1966). There is some speculation that the 25 pieces of debris detected around Salyut 2 may have been its own booster that exploded and in some way damaged the station. This is unlikely since the station operated for days after that, and debris after launch is normal because the port hole covers and engine covers are jettisoned.

After the break-up, tracking ships around the world returned to port to resupply in anticipation of the next Salyut launch attempt. Possible crew members for Salyut 2 missions were Popovich, Bykovsky, Volynov, and Shonin.

Kosmos 557

Civilian Salyut Station—Failed

Launched: May 11, 1973, 3:20 a.m.
Reentry: May 22, 1973, 6:07 a.m.
Altitude: 214×243 km @ $51.6°$

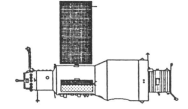

Before the launch of the Salyut named Kosmos 557, another Salyut was once believed to have been launched on April 25, 1973. On that day a launch did occur from the Baykonur Cosmodrome, but failed. The booster's second stage failed and the satellite payload reentered and fell into the Pacific. That satellite was once thought to be a Salyut station, but the satellite that was destroyed that day was almost certainly a Radar Oceanographic Reconnaissance satellite (RORSAT).

The reasons for this conclusion are that the tracking ships deployed for the Salyut 2 launch returned to ports in Newfoundland and the Dutch West Indies to refuel before returning to stations for the Kosmos 557 launch, which was definitely a Salyut station. Also, U.S. Air Force planes were flying over the Pacific searching for high radiation levels after the failure.[9] This indicates that U.S. intelligence expected a radioactive source to be on the satellite. At that time, the only Soviet spacecraft to carry significant amounts of radioactive material were Luna probes and RORSATs. The failed launch did not occur during a normal lunar launch window, thus the failed satellite was most probably a RORSAT and not a Salyut or a Luna.[10]

The launch of this Salyut was the last attempt to launch a space station to beat the U.S. Skylab, which was launched on May 14. Skylab almost failed too when a micrometeoroid shield and solar array were ripped off during launch. The new Salyut was launched into an orbit that widely separated it from the derelict Salyut 2 which was still in orbit. The Proton boosters third stage separated from the station in orbit at 209×226 km. The third stage decayed from its orbit by May 17. Usually, a space station would make a separation maneuver to enter a higher orbit away from the booster stage, however, the station made no maneuvers. During the launch phase or shortly after, the Salyut suffered a major failure and the Soviets lost control of the station. It was given the generic Kosmos 557 label in an attempt to hide the failure. The launch was announced by the Soviets at 6:00 p.m. the day of the launch.

Radio signals received by the Kettering group in England, on May 12, indicated that the space station may have been tumbling because the telemetry signals faded 2.5 times a second. The radio signals were typical of a civilian type Salyut station, and Kosmos 557 was visually brighter than Salyut 1 or 2, indicating the use of larger solar panels as on the Salyut 4 civilian type station.[11]

Soyuz 12 was scheduled to dock with the Kosmos 557 station. Soyuz 12 was a new version of the Soyuz that had no solar panels, because the new civilian Salyut had enlarged solar panels and did not require extra power once needed by the Salyut 1 station. These facts indicate that the Kosmos 557 was the first of the new civilian space stations like Salyut 4. A mockup of the new type Salyut was shown at Star City in the summer of 1973.[12]

Soviet statements about the Kosmos 557 flight were typical of the standard reactions to space flight failures in the 1960s and 1970s. The Soviets insist that Kosmos 557 successfully completed its mission, and that the flight was not related to the manned space program.[13] Tracking ships that were deployed to support the expected Salyut 2 or Kosmos 557 manned missions again returned to port after the Kosmos 557 failure. Some analysts speculate that the Soviets intended that at least two of the three failed Salyut stations be in orbit and manned when the U.S. launched Skylab in a propaganda effort similar to the Soyuz 6-7-8 triple flight.

Kosmos 573

First Soyuz Ferry Test Flight

Launched: June 15, 1973, 9:00 a.m.
Landed: June 17, 1973, 10:12 p.m.
Altitude: 192 × 309 km @ 51.6°

This was the first flight of the new Soyuz Ferry, which had no solar panels and instead carried a two-day battery power supply to save weight[14] (see Figure 4-3). Both the military Salyuts and the new civilian stations had larger solar

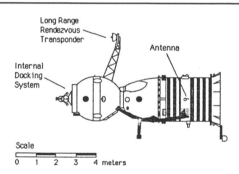

Figure 4-3. The Soyuz ferry was an evolutionary development based on the original Soyuz design. The cosmonauts wore lightweight pressure suits for protection during launch and landing after the Soyuz 11 tragedy. The ferry also eliminated solar arrays to allow for more cargo capacity but limited the Soyuz to two to three days of battery power for independent flight.

Soyuz Ferry Spacecraft

The Soyuz Ferry was a modified version of the original Soyuz that was used to transport cosmonauts to and from Salyut space stations. The most

(continued on next page)

visible change was the removal of the old Soyuz's solar panels to save weight to enable greater payloads. The Soyuz 11 tragedy produced changes inside from a three-person crew to a crew of two and support equipment for their pressure suits. The orbital module was changed from a work area into a cargo hold for ferrying supplies and equipment to a space station. The orbital module still could be used as an airlock if necessary. The weight of the ship was still around 6,800 kg.

The ferry version had the Salyut type docking system that was flown on the old type Soyuz 7, 8, 10, and 11 spacecraft. The internal docking system made the spacecraft significantly shorter than the first Soyuz's. The orbital module was shortened to 2.65 meters with the deletion of the docking housing carried by the original Soyuz. Internal volume remained the same at 10 cubic meters. The Soyuz ferry was about 8.6 meters long including the antennas, 7.5 meters excluding them. The Soyuz Ferry used the Igla (needle) automatic rendezvous and docking system. The space station docking procedure was the same as earlier Soyuz docking missions with both spacecraft maneuvering actively. The loss of the solar panels limited the Soyuz to about two days flight time using battery power, which was enough for a trip to a space station and back.

---

panels than Salyut 1, and did not require extra power from the Soyuz like Salyut 1. The elimination of the arrays from the Soyuz allowed for a little more payload to be carried to the space stations.

Soyuz 12

First Manned Soyuz Ferry Flight
Launched: September 27, 1973, 3:18 p.m.
Landed: September 29, 1973, 2:34 p.m.
Altitude: 181 × 229 km @ 51.6°
Crew: Vasili Lazarev and Oleg Makarov
Backup: Georgly Grechko and Alesksei Gubarev
Call sign: Urals

As the first manned test of the new Ferry version Soyuz, this flight was to have flown to one of the two Salyut stations that failed in the last year, but was finally flown to test the new systems, in the absence of usable Salyuts. The Soyuz service module for the ferry version had no solar panels and instead carried batteries, which limited the Soyuz to about two days flight time. Two days were just enough to fly to and return from a Salyut. Another change for the Soyuz Ferry was pressure suits for the crew to wear during launch, docking, and landing to prevent the loss of a crew, as during the Soyuz 11 flight (see Figure 4-4). This allowed only enough room for two cosmonauts and their suits' environmental control equipment. The Soyuz also had improved attitude control system sensors. The main objective of this flight was to test and observe the Soyuz's systems flying a simulated flight to a space station.[15,16]

The Soyuz was maneuvered on the second day to 326 × 344 km, which was later a standard Salyut 4 orbit. To gain some use from the short test flight, a multispectral camera was carried in the orbital module. The cosmonauts took pictures of Earth in coordination with aircraft, which took close-up photographs of the same areas. The camera had a 100-meter resolution in nine visible and infrared wavelengths.[17] Possible application for the photographs was reported as surveying crop and forest conditions. They also tested communicating to

Figure 4-4. After the loss of the Soyuz 11 crew, pressure suits were worn during launch and landing. Here, Leonov and Kubasov are shown in their pressure suits used for the Apollo-Soyuz mission. The lightweight suits were designed for easy donning and as much comfort as possible. They were not intended for EVA use, although they could be used for short emergency EVAs if necessary. (Source: National Aeronautics and Space Administration.)

mission control through a Molniya 1 satellite when out of range with ground stations.

In preparing for retrofire, a large object, possibly the orbital module or experiment equipment on the orbital module was jettisoned. The object remained in orbit for 116 days. After retrofire, the capsule landed 400 km southwest of Karaganda.[18] They were the first cosmonauts to use pressure suits for reentry since Voskhod 2. The flight was said to have been flawless.[19]

Kosmos 613

Soyuz Long-Duration Storage Test

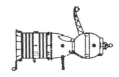

Launched: November 30, 1973, 8:20 a.m.
Landed: January 29, 1974, 10:12 a.m.
Altitude: 188 × 273 km @ 51.6°

This flight tested the Soyuz's systems over a long period in preparation for missions to future Salyut stations. While docked to a space station, a Soyuz is normally shut down to conserve its electrical and environmental systems, but spacecraft systems like propellant lines and engine valves can degrade over long periods of time when they are exposed to the toxic propellants used in the engine systems. The systems must function for cosmonauts on the return to Earth from a Salyut, and the Soviets wanted to test them for the length of a long mission.

After six days, it was maneuvered to a higher orbit at 269 × 385 km. It was powered down from December 9 until January 23, 1974,[20] and then it was maneuvered to 239 × 363 km and shut down again for six days. The spacecraft was then activated and jettisoned an object simulating undocking from a Salyut station. This object may also have consisted of a small solar panel to keep the Soyuz batteries charged during the long period in orbit (normally, the batteries only had two or three days capacity with a crew on board).[21] After separating from the object, the Soyuz performed retrofire and successfully landed.

Soyuz 13

Short-Duration Scientific Flight

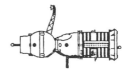

Launched: December 18, 1973, 2:55 p.m.
Landed: December 26, 1973, 11:51 a.m.
Altitude: 188 × 247 km @ 51.6°
Crew: Pyotr Klimuk and Valentin Lebedev
Backup: Lev Vorobyov and Valery Yazdovsky
Call sign: Caucasus

This flight was to perform some of the experiments that were to be done on the failed Salyuts in 1972 and 1973. This flight was the first to be controlled by the new Kaliningrad Mission Control Center near Star City and Moscow (see Figure 4-5). The Soyuz 13 crew also joined the Skylab 4 crew in orbit, but neither crew reported sighting the other even though they were in similar orbits. The Soyuz maneuvered to 225×273 km on its fifth orbit.

Figure 4-5. The Kaliningrad control center was completed in 1973. Earlier missions were controlled from a control center in the Crimea. This photo shows the center during the Soyuz 16 mission. A third control room was added in the early 1980s to control Kosmos and Progress space station elements and by 1988, a fourth control room was added solely dedicated to Soviet space shuttle operations. (Source: Tass from Sovfoto.)

The Soyuz was an old type with solar panels and a specially outfitted orbital module without docking equipment. In place of this was the Orion-2 ultra-violet telescope, which was similar to the Orion on Salyut 1, an X-ray camera for observing the sun and Earth, and the KSS2 spectrograph, which used a special NASA film for stellar studies. The instruments on the outside of the orbital module were covered by a housing that was opened for 1 to 20 minutes for observations. To operate the experiments one cosmonaut in the reentry module aimed the ship using attitude control thrusters and the other operated the experiments in the orbital module. Using the Orion-2 telescope, the crew made observations of stars, comet Kohoutek, and Earth. The crew was trained extensively before launch at the Byurakan observatory in Armenia.[22] They also used a multispectral camera like the one on Soyuz 12 for measuring the atmosphere and pollution.

They also tested the Oasis-2 closed ecology from which the cosmonauts harvested protein after two days of operation. During the flight, the bio-mass increased to 30 times its original amount. One of the medical experiments performed was Levka, which measured blood distribution to the brain in weightlessness and its effect on space sickness.[23]

Shatalov said that there was some concern about the crew landing in a heavy snowstorm, but the cosmonauts were recovered only minutes after landing, at 11:50 a.m., 200 km southwest of Karaganda.[24,25]

Kosmos 638

ASTP Test Flight
Launched: April 3, 1974, 10:30 a.m.
Landed: April 13, 1974, 7:48 a.m.
Altitude: 187 × 309 km @ 51.8°

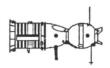

This was the first test flight of the Apollo Soyuz Test Project (ASTP).[26] The ASTP Soyuz was equipped with the universal docking system, had modified life-support systems to support four crewmen, and was equipped with newly designed solar panels to provide longer orbit time (see Figure 4-6). The Soyuz was also launched on an improved version of the A-2 booster, which was planned to be used for the ASTP flight.

The spacecraft maneuvered to planned ASTP orbital parameters, which was a circular orbit at 225 km. By August 7, the Soyuz maneuvered to 258 × 274 km, and later to 268 × 390 km.

Kosmos 652

ASTP Test Flight
Launched: May 15, 1974, 11:47 a.m.
Landed: May 23, 1974, 10:12 a.m.
Altitude: 173 × 343 km @ 51.8°

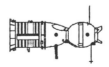

This was the second ASTP Soyuz test flight and was similar to the first. The Soyuz was equipped with the universal docking system, had modified life-support systems to support four crewmen, and was equipped with newly designed solar panels to provide longer orbit time. The Soyuz was also launched on an

Figure 4-6. The photo shows a mockup of the modified Soyuz ferry used for the Apollo-Soyuz Test Project. The Soyuz had new solar panels to extend orbital stay time and a universal docking system designed jointly by NASA and Soviet engineers. The new docking system had a short life after ASTP, because the Soviets continued using their old docking system design on space stations, and NASA generated new designs for future space stations.

improved version of the A-2 booster, which was planned to be used for the ASTP flight. The Soyuz flew a simulated ASTP mission as did the Kosmos 638 flight. A docking ring simulating the docking system of the U.S. docking module was jettisoned before retrofire.[27,28]

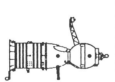

Kosmos 656

Retest of Soyuz Ferry
Launched: May 27, 1974, 10:30 a.m.
Landed: May 29, 1974, 10:12 a.m.
Altitude: 195 × 364 km @ 51.6°

This was the third test flight of the Soyuz Ferry in anticipation of the launch of Salyut 3 in June. The two-day flight may have tested a new automatic rendezvous and docking system.[29]

Salyut 3

First Successful Military Station

Launched: June 25, 1974, 1:38 a.m.
Capsule Landed: September 23, 1974
Module Reentry: January 24, 1975
Altitude: 213 × 253 km @ 51.6°

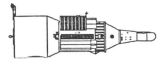

This was the first successful military Salyut station, and was designed very differently from the civilian stations. The Soyuz docked at the aft end instead of the forward end as on civilian Salyuts. Salyut 3 also had a redesigned propulsion system that was built onto the outer hull of the station and eliminated the Soyuz type propulsion module that Salyut 1 and 4 carried. The Soviets referred to the docking section as the front of the station, which was called the rear on the civilian stations (see box, Figure 4-2).

Salyut 3 Space Station

Salyut 3 was the first successfully launched military Salyut station. The military Salyuts were designed by Academician Chalomei, who also later designed the Star module spacecraft that was used with Salyut 6, 7, and Mir.

The design of the military stations differed greatly from the civilian stations. Salyut 3 consisted of two cylinders of different diameters. At the forward end was a capsule that may have been a Voskhod-type capsule with a retro package or a Star module-type capsule for the return of experiments and film. The station's dimensions were 14 meters long, 2.9 and 4.15 meters in diameter. Salyut 3 had the same 100 cubic meters habitable volume as Salyut 1. Two large solar arrays that could rotate to track the sun were mounted on the middle of the space station. The station used magnetically suspended spherical motor flywheels for attitude control. A high-resolution camera with 10-meter focal length replaced Salyut 1's telescope housing. It probably had a resolution of about 50 cm.

Inside, the floor was also covered with velcro to aid movement and hold things in place. Soviet descriptions state that a corridor ran the length of the large diameter section on the left side of the station to connect the control, working, and living compartments. The living section was next to the control section and had four portholes and two beds, one was fixed to a wall and the other folded against a wall to save space. It also had a table for medical experiments, a small library, a chess set, tape recorder, water tank, a shower, and toilet. Salyut 3 also tested a system for water recovery from the station's atmosphere that was used later on Salyut 4.

Operations of the Salyut were also different from civilian stations. Salyut 3 was the first station to be constantly oriented toward Earth (Skylab was constantly pointed to the sun). This was helpful for keeping its cameras always pointed at Earth. The designed lifetime for supporting a crew was three months, although this was not achieved due to failure of the second and last mission to the station. During its mission, the station would receive 8,000 commands, make 500,000 thruster firings, and have experiments performed a total of 400 times.[30] By June 28, the station maneuvered to 266 × 269 km in preparation for the Soyuz 14 launch.

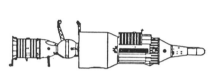

Soyuz 14

Only Salyut 3 Crew—16 Days

Launched: July 3, 1974, 9:51 p.m.
Landed: July 19, 1974, 3:21 p.m.
Altitude: 195 × 217 km @ 51.5°
Crew: Pavel Popovich and
      Yurity Artyukhin
Backup: Gennadi Sarafanov and
      Lev Demin
Call sign: Golden Eagle

This was the first mission to Salyut 3 and the first successful space station mission for the Soviets. The Soyuz was a standard ferry version and was launched eight days after Salyut 3. The main objective was to evaluate the Salyut's potential as a manned military reconnaissance platform.[31] The flight's backup crew also had two backup military Salyut crews, which would later fly military missions including Volynov, Zholobov, Zudov, and Rozhdestvenski.[32]

After separating from the booster's upper stage, Soyuz 14 maneuvered into a transfer orbit 3,500 km behind the Salyut 3 in orbit at 255 × 277 km. At 1,000 meters distance the automatic rendezvous was used to bring the Soyuz to within 100 meters, from which Popovich docked manually at 11:51 p.m., July 4, in orbit at 268 × 271 km. The crew entered the station at 4:30 a.m., July 5, and began activating the station's systems.

The next day some Soviet physicists thought the crew should land when there was major solar flare activity. The cosmonauts checked their radiation gauges and determined there was no significant increase, because the station's low orbit was somewhat protected from solar radiation.[33,34] The crew continued their work activating the Salyut's systems and learning how the new systems installed in Salyut 3 performed in flight, like the large new solar panels.[35]

The crew's main duties were to make extensive Earth observations of the Soviet Union and military points of interest. Special targets were deployed at the Baykonur Cosmodrome to test the ability of the cosmonauts to see detail on the ground. Communication with the station was nearly continuous via tracking ships and Molniya satellites for the entire flight. The crew often used code words in their conversations with mission control, in addition to using the traditional military radio frequencies.

The cosmonauts' daily routine consisted of two hours of exercise a day to prevent the adverse effects of weightlessness. They also used the penguin elastic exercise suit, ran on the treadmill, carried out the normal medical checks, and performed what would later be known as resonance tests to measure any dangerous resonances that built up during crew movements, such as exercising. They observed polarization of light reflected from Earth and the atmosphere's optic quality during day, night, and twilight.

On July 11, the Soviets announced that the mission was half over. In the following days they tested new systems for station thermal regulation and water recovery from the air in the station. They checked the station's manual controls and spent most of the time taking pictures of Earth. They also tested navigation using horizon sensors and made observations for the Tropex-74 Earth resources program. The cosmonauts tested a treadmill and the Polynom-2M blood flow experiment, and made several TV broadcasts during the mission showing various experiments. All during the flight, there was a group of specialists working to emulate the cosmonauts' activities in the Salyut 3 simulator at Star City to aid in the resolution of any problems encountered in space.

On July 13, they tested Soyuz 14's improved manual control systems, preparing for the upcoming landing. On the 17th, the Soyuz was tested and the crew started loading film and experiment results in the capsule. They undocked at 12:03 p.m., on July 19. Retrofire followed two orbits later, at 2:34 p.m. The Soyuz capsule landed less than an hour later, at 3:21 p.m. only 2 km from the target, 140 km southeast of Dzhezkazgan. The cosmonauts climbed out of the capsule without waiting for the recovery forces. They were fully recovered from the 16 days in weightlessness within a few days. The crew left Salyut 3 with enough supplies to last the next crew for at least six months.[36-39]

Kosmos 670

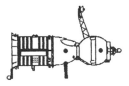

Launched: August 6, 1974, 3:14 a.m.
Landed: August 9, 1974, 3:00 a.m.
Altitude: 211×294 km @ $50.6°$

This flight was first thought to be a reconnaissance satellite, but was later identified as a Soyuz type spacecraft with solar panels.[40] The flight was probably a Soyuz T or a Progress related test, but its purpose is still unknown. The mission used a 50.6° inclination orbit never before used in the manned program. The Soyuz capsule landed three days after it was launched.

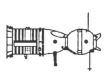

Kosmos 672

ASTP Test Flight
Launched: August 12, 1974, 9:25 a.m.
Landed: August 18, 1974, 7:48 a.m.
Altitude: 195 × 221 km @ 51.8°

This was the third test of the ASTP Soyuz and was launched by the improved A-2 (see Figure 4-7). The new version of the A-2 booster was also used to launch ten other missions in addition to the ASTP test flights and the ASTP mission.[41] The spacecraft maneuvered to 227 × 238 km before landing six days after being launched.[42,43]

Soyuz 15

Aborted
Launched: August 26, 1974,
         10:58 p.m.
Landed: August 28, 1974,
         11:10 p.m.
Altitude: 173 × 236 km @ 51.6°
Crew: Gennadi Sarafanov and
         Lev Demin
Backup: Boris Volynov and
         Vitaliy Zholobov
Call sign: Danube

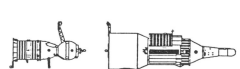

Salyut 3 raised its orbit to 255 × 275 km in preparation for the Soyuz 15 launch. The Soyuz was to have docked on the 16th orbit with Salyut 3, but there was no docking. On orbit 16, the Soyuz was in orbit at 249 × 259 km, but 120 km behind the station.[44] During the second day, they tried to dock several times, but the automatic system malfunctioned twice, pushing the ship out of control with excessive engine burns while only 30 to 50 meters from the station.[45] Sarafanov attempted to manually dock several times but ran low on propellant before successfully docking.

The crew was then forced to return to Earth as their propellant supply ran low. The retrofire of the Soyuz was monitored by the tracking ship *Morzhovets* in the South Atlantic. The Soyuz capsule landed 48 km southwest of Tselinograd at 11:10 p.m. in rain, and was located within a minute.[46] Analysis of the launch

Figure 4-7. This A-2 type booster was used to launch all Soyuz missions. The booster has just been placed on the launch pad and the service towers are being raised to surround the booster. The launch pad arms and service towers are built onto a large turntable, which rotates to place the booster on the correct launch azimuth, eliminating the need for the booster to roll during launch. This feature is a direct descendant from the German V-2 design. This particular photo is of an ASTP test flight. (Source: National Aeronautics and Space Administration.)

windows for the flight indicates a mission of 19 to 29 days' duration was planned.[47]

It was later revealed that Soyuz 15 carried a new automatic docking system that would be used on future Progress transports.[48] The Soviets said the new system was being observed by the cosmonauts, and they were not expected to manually dock, however, this crew never flew again after their failure to correct the error. Shatalov officially said after the flight that the objective was only to dock testing the new systems, redock in further tests, and then return to Earth.[49] However, before the flight the mission was said to be a continuation of the Soyuz 14 crew's work on Salyut 3. After the Soyuz 15 docking abort, the Salyut's sample return capsule separated, performed retrofire, and landed on September 23 at the end of a normal Soyuz landing window.[50]

On December 25, the Soviets issued a statement saying that Salyut 3 had completed its mission. The next day the civilian station Salyut 4 was launched. Salyut 3 had run low on supplies of attitude control propellant before a new mission could be launched and performed. On January 24, 1975, Salyut 3 was maneuvered into a destructive reentry over the Pacific Ocean by mission control command.[51]

Soyuz 16

ASTP Test Flight
Launched: December 2, 1974, 12:40 p.m.
Landed: December 8, 1974, 11:04 a.m.
Altitude: 184 × 291 km @ 51.8°
Crew: Anatoliy Filipchenko and Nikolay Rukavishnikov
Backup: Yuri Romanenko and Aleksandr Ivanchenkov
Call sign: Snowstorm

The Soviets had told NASA that they would test fly a manned ASTP spacecraft, but did not give the exact time until one hour after launch.[52] They had offered to tell NASA the exact time if NASA would keep it secret, but NASA refused to keep the secret. A second backup crew consisting of Dzhanibekov and Andreyev was also officially assigned to the flight.[53] They would ultimately serve as the prime crew for the backup ASTP spacecraft.

This flight was preceded by three unmanned tests of the ASTP Soyuz. Soyuz 16, like the unmanned tests before it, was also launched by the improved A-2 booster.[54] After reaching orbit, the spacecraft was maneuvered to the correct ASTP 225-km circular orbit. On the fifth orbit, they maneuvered to 177 × 223 km, and then to 190 × 240 km. A rough circularization maneuver then brought the orbit to 221 × 224 km, which was then corrected to 225 × 225 km. The crew also tested modified environmental systems, new solar panels, and improved control systems. The Soyuz also had new docking radar equipment and light beacons for docking with the Apollo. Soyuz 16 also carried an ASTP universal docking system and a simulated 20-kg U.S. docking module docking ring, attached to the Soyuz docking ring. Also during the first day, they lowered air pressure from 760 mm, 20% oxygen, to 540 mm, 40% oxygen, in a test to reduce the planned transfer time between the Soyuz and the Apollo from two to one hour.[55]

On December 4, they made television broadcasts, continued testing spacecraft systems, and performed medical and biological experiments. On December 5, they practiced docking procedures by donning pressure suits and testing the capture and latch engaging several times, on orbits 32 to 38. The next day, they tested the mechanism again on orbit 48, and made more television broadcasts. On December 7, they jettisoned the docking ring with explosive bolts. This was to ensure separation of two docked spacecraft if problems arose with the capture latches of the docking system. They also simulated flying maneuvers and equipment required for a solar eclipse experiment to be performed during the ASTP flight and tested the Soyuz engines as normal prelanding procedure. While testing the solar eclipse equipment, the cosmonauts observed a cloud of debris surrounding the Soyuz that would affect the planned test.[56] Spacecraft give off many gases and debris like paint chips in normal operations, and the particles trail along with the spacecraft in the same orbit. The phenomenon can affect many types of measurements and is currently being studied by both the Soviets and NASA. During the flight, the cosmonauts also monitored some simple biological experiments like studying fungus growth in weightlessness. On December 8, after a retroburn lasting 166.5 seconds, the Soyuz returned to Earth. The capsule was first spotted at 5,000 meters altitude by recovery forces. The capsule landed at 11:04 a.m., on frozen ground 30 km north of Arkalyk. The test flight was a complete success.[57-59]

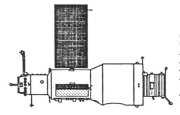

Salyut 4

Launched: December 26, 1974, 7:15 a.m.
Reentry: February 3, 1977, 2:31 a.m.
Altitude: 212 × 251 km @ 51.6°

This Salyut was probably the backup to Kosmos 557, which failed in 1973. Salyut 4 was to continue the civilian program begun by Salyut 1, and was planned to support crews for up to a total of six months. Salyut 4 was basically a modified Salyut 1 with three new rotatable solar arrays. Salyut 4 still used the Soyuz-type propulsion module at the rear of the station, even though the military Salyut station's engine system was completely different (see Figure 4-8).

The day before Salyut 4 was launched, the Soviets announced that Salyut 3 had completed its mission. This statement usually meant that the spacecraft was about to be de-orbited and Salyut 3 was commanded to reenter within a

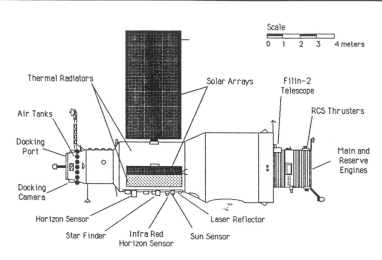

Figure 4-8. The Salyut 4 space station was a modified version of the original Salyut 1. Changes included new solar arrays, new environmental control systems for air regeneration and water recovery, and new telescopes to study the sun.

Salyut 4 Space Station

Salyut 4 was a civilian-type space station and was basically the same design and dimensions as Salyut 1, with the exception that it was equipped with 60-square-meter solar arrays mounted on its midsection to provide 4 kilowatts of power. One of the main experiments on the station was the OST-1, which was a 25-cm-diameter solar telescope that took the place of the reconnaissance cameras of the military Salyuts.

At the forward end of the work compartment was the navigation post. A new control system called Kaskad was installed for experimental use for future stations. The Delta navigation unit was the standard navigation instrument. Aft of that were the main control panels; on the walls next to the controls were the air regeneration units. Forward of the regeneration units on the left was the water recovery system and to the right was the life support system controls. The Salyut's air regeneration filters were about the size of golf bags. Air entering the system was dehumidified by a cooling process. When the dry air was circulated over sheets of potassium superoxide, along with the separated and reheated water, the carbon dioxide reacted with the sheets forming potassium carbonate and oxygen, which was returned to the cabin.

(continued on next page)

On Salyut 4 and later stations, water was scavenged from the filter and reused after treatment. Carbon dioxide remaining in the air is removed by lithium hydroxide canisters, which are also used on the U.S. spacecraft.

Aft of the main controls was the large diameter work section including a medical post with a vestibular chair and exercise equipment, telescope housing, and on the aft end of the compartment was the sanitary equipment. A shower was to have been installed but was delayed too long and was flown on Salyut 6. Showers had also flown on at least Salyut 3 and this may indicate the wide division between the military and civilian elements of Soviet space programs.

Salyut 4 Experiments and Equipment

OST-1: solar telescope, covers 800–1,300 Å with 1 Å resolution
Kaskad: star sensor navigation system (experimental)
Stroka: teletype for personal communication and data
Zentis: mirror resurfacing by aluminum spray, for OST-1
Polynom-2M: cosmonaut cardiovascular monitor
Tonus-2: muscle micro-electrical stimulator
Filin-2: X-ray spectrometric telescope, 1–60 Å
ITS-K: infrared spectrometric telescope, 300-mm mirror, 1–7 micron
MMMS: micrometeor monitor system
MMK-1: micrometeoroid detector
KDS-3: diffraction spectrometer
BA-3K: star camera
Delta: automatic navigation system
Treadmill: 40×90-cm belt type
Spektru: upper atmosphere analyzer
Oasis: plant growth experiment
KATE-140: 140-mm multispectral camera
Raketa: vacuum cleaner
KATE-500: 500-mm multispectral camera
KM: microbe cultivator
Vshk-2 & LV-1: optical orientation sensors
Fakel: photometer
VPA-1: visual polarimetric analyzer
FKT: cell division experiments
Freon: liquid behavior experiment
S-2: sextant
RT-4: 200-mm X-ray telescope, 40–60 Å
KKS-2: solar spectrometer
Silya-4: spectrometer   *(continued on next page)*

(continued from page 133)

Emissiya: interferometer
Chibis: low-pressure pants
Neytral: velocity vector sensor
Amak-3: blood analyzer
Vektor: ionic position sensor
Plotnost: bone tissue density meter
Ion: mass spectrometer
Bioterm-2M: plant growth experiment
Ergometer: exercise machine
Bioterm-3 & 4: plant growth experiments
Ya-2: luminescence meter

*Sources: Johnson, Nicholas, L. *Handbook of Soviet Manned Spaceflight*, American Astronautical Society: San Diego, 1980, pp. 309–311.
Hooper, Gordon R. "Missions to Salyut 4," *Spaceflight*, Vol. 18, No. 1, Jan. 1976, p. 14.

month on January 24. Meanwhile, by December 30, Salyut 4's orbit had been raised to 276 × 341 km. On January 6, Salyut 4 was placed into a 343 × 355-km orbit in preparation for the launch of Soyuz 17. While ground controllers were making routine checks of the station's systems, the OST solar telescope was damaged when it was exposed to the sun accidentally. This resulted in a burned-out sensor making automatic operation of the telescope impossible.

Soyuz 17

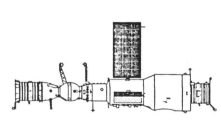

Long Duration—29 Days
Launched: January 11, 1975, 12:43 a.m.
Landed: February 9, 1975, 2:03 p.m.
Altitude: 185 × 249 km @ 51.6°
Crew: Alesksei Gubarev and Georgiy Grechko
Backup: Vasili Lazarev and Oleg Makarov
Call sign: Zenith

The second backup crew for this mission was Pyotr Klimuk and Vitaliy Sevastyanov.[60] They would also serve as the backup for the next Soyuz flight. Soyuz 17 was launched 16 days after Salyut 4. After making orbit corrections, the Soyuz was in a 274 × 347-km transfer orbit. The automatic docking system was engaged at 4 km and the cosmonauts began manual control at 100 m.[61] Gubarev docked manually with the Salyut, which was in a rough 350-km circular orbit at 4:00 a.m., January 12.[62] On January 13, the orbit of the space station complex was 342 × 355 km.

The crew quickly found the "wipe your feet" sign left next to the hatch by the technicians at Baykonur when the station was launched. The crew started activating the station and checking the communication systems.[63] On January 14, they set up a ventilation hose from the Salyut environmental control system to the Soyuz through the transfer compartment to ventilate the Soyuz while its systems were shut down until landing.

The crew reportedly did not fully adapt to weightlessness until five to eight days into the mission. The crew worked very hard during the first weeks despite instructions to take one day a week to rest. They did agree to rest for one hour every evening instead. The crew normally worked 15 to 20 hours a day, including their exercise period, which was 2.5 hours a day.[64] One of their activities was testing communications to tracking ships and to mission control through a Molniya satellite. During these tests, the Academy of Sciences communications ship *Korolev* was stationed in the Atlantic along with the *Ristna* and *Nevel*.[65]

The Salyut's OST solar telescope was damaged when the sun accidentally burned out a sensor during systems checks before the Soyuz 17 launch. As a result, a reflecting mirror could not position itself for automatic observations. The cosmonauts tried to manually operate the telescope, but were unable to see the mirror to position it correctly. They could, however, listen to the mirror as it moved using the station's medical kit stethoscope. They judged from the sound when the mirror had traveled enough to be in correct position for observations. The telescope would have been a complete failure without the cosmonauts on board. Unfortunately, the telescope had no solar event alarms, as did Skylab, to alert the crew to important photo opportunities.[66]

On January 17, the Salyut raised its orbit to 336 × 349 km, its operational altitude. The crew tested the water recovery system, which could recover one liter per man per day; they also tended the garden, and performed experiments with the X-ray and the infrared telescopes. Other activities included resurfacing the mirror on the solar telescope to counter dust build-up using a hot tungsten wire to vaporize aluminum that was deposited on the mirror.[67] They also tested a water flash evaporator temperature control system with some success.[68]

On February 7, they activated and began packing the Soyuz with the results of their experiments and cleaned and shut down the Salyut. They undocked in the Soyuz at 9:08 a.m., February 9, in orbit at 334 × 361 km. They landed 110 km northeast of Tselinograd, at 2:03 p.m., in a snowstorm with clouds at 250 meters, visibility of 500 meters, and winds blowing at 72 km/hr. The cosmonauts wore gravity suits, which restricted blood flow into their legs for landing to ease the effects of readaptation.[69,70]

The cosmonauts had lost 3.6 kg weight average during the flight, which they recovered by the first week after landing.[71] During their recovery from the effects of weightlessness, they reportedly still tired easily two weeks after landing.[72] Salyut 4 raised its orbit to 343 × 356 km in March in preparation for the Soyuz 18A mission.

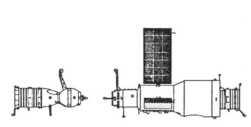

Soyuz 18A

Aborted

Launched: April 5, 1975,
       2:00 p.m.
Landed: April 5, 1975, 2:20 p.m.
Altitude: 180 km
Crew: Vasili Lazarev and
       Oleg Makarov
Backup: Pyotr Klimuk and
       Vitaliy Sevastyanov

Call sign: Urals

This flight was intended to dock with the Salyut 4 space station. The Soviets were using an older model of the A-2 booster for this flight. A newer version had been developed and had flown several times for the ASTP mission, but the Soviets were using all the old models, apparently to save money.[73]

The flight was going normally until a malfunction of the A-2 booster. A sequencer relay prematurely fired two of the four pyrotechnic latches that hold the core stage to the upper stage. Later, when the core stage shut down, the remaining latches had been disabled by the premature firing of the others, and they did not release, as the upper stage engines began to fire normally. With the core still attached and the upper stage firing, the rocket began to deviate from the planned attitude. When the flight path deviation reached 10°, the abort system automatically shut down the upper stage, while the spacecraft separated from the booster.[74]

The Soyuz separated at 180 km altitude and at 5.5 km-per-second velocity. The cosmonauts had no control over the abort system and urgently asked mission control if it was working properly. After separating and falling away from the booster, the Soyuz turned around, with the rear of the ship pointing in the direction of flight, and fired its engines in a retroburn that put the ship into a ballistic trajectory ending in the western U.S.S.R. This subjected the cosmonauts to up to 20 gravities deceleration during reentry.[75] During the descent, the cosmonauts asked mission control several times if they were going to land in China.

The Soyuz landed, in darkness and waist deep snow, north of where the Soviet, Mongolian and Chinese borders meet, in the Altai mountains, southwest of Gorno-Altaisk, 320 km short of the border. The cosmonauts quickly left the capsule and built a fire, while waiting for the rescue forces, which arrived after several hours. Analysis of the landing windows for the launch indicated mission duration of around 51 to 63 days.[76]

Soyuz 18B

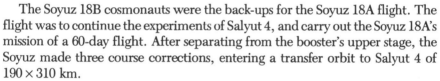

Long Duration—63 Days

Launched: May 24, 1975, 5:58 p.m.
Landed: July 26, 1975, 5:18 p.m.
Altitude: 186 × 230 km @ 51.6°
Crew: Pyotr Klimuk and
        Vitaliy Sevastyanov
Backup: Vladimir Kovalyonok and Y. Ponomaryov
Call sign: Caucasus

The Soyuz 18B cosmonauts were the back-ups for the Soyuz 18A flight. The flight was to continue the experiments of Salyut 4, and carry out the Soyuz 18A's mission of a 60-day flight. After separating from the booster's upper stage, the Soyuz made three course corrections, entering a transfer orbit to Salyut 4 of 190 × 310 km.

On May 26, 1,500 meters from Salyut 4, the automatic rendezvous system fired the Soyuz engine, reducing the closing velocity with the station. At 10:21 p.m., the Soyuz was 800 meters from the Salyut. The spacecraft then passed into Earth's shadow, causing Klimuk to turn on the Soyuz spotlight to see the station. After some time searching for the station, Klimuk docked manually from 100 meters distance with the station at 338 × 349 km.

The crew entered the station about two hours later, at 12:30 a.m., May 26, and within 48 hours, began experiments and repairing or replacing components. The Silya spectrometer was repaired, one gas analyzer was replaced, and one of six pumping condensers in the water regeneration system was replaced by a hand pump.[77] The station's orbit was raised to 344 × 356 km and checked by laser tracking from the ground.[78]

On May 29 and 30, the crew performed medical and biological experiments, and started the Oasis garden. On June 2 and 3, they began studies of the stars, planets, Earth, and its atmosphere. The station's Kaskad orientation system aided in these observations by orienting the station automatically. Over a period of 60 orbits, 2,000 photographs were taken.

By June 7, the station's orbit was 356×341 km. The next few days were devoted to extensive medical checks and Earth resources photography using different modes of space station orientation testing the Kaskad system. The crew also studied their reactions to weightlessness and said that the body remembers and knows ways to adapt to the environment. The crew also used a Penguin elastic exercise suit during normal work hours to help condition their muscles.[79]

The crew attempted to grow plants, including peas, in the Oasis unit. The plants grew for four weeks and then died. They then tried growing onions. They also performed experiments with insects including flies and beetles. The cosmonauts also tried an experimental daily schedule that advanced their wake up times every day. This led to increased nervous tension by opposing the body's natural tendency to sleep later each day. The crew generally worked a schedule that allocated eight hours' sleep, eight hours' work, three hours' personal time, 2.5 hours' exercise, and 2.5 hours for meals and meal preparation. This schedule resulted in an average of 4.5 hours' experiment operation per man per day. In comparison, the three U.S. Skylab crews averaged about 11 hours per day operating experiments. The backup crew for the cosmonauts carried out the same experiments in the Salyut 4 trainer at Star City to aid in any problems that arose in orbit (see Figure 4-9).

June 23 was another day of extensive medical examinations as the cosmonauts broke the Soviet record of 29 days set by Soyuz 17. This was still far less than the record of 84 days set in 1974 by the Skylab 4 crew. But the Soviets had no intention of trying to break the U.S. record during the Soyuz 18B mission; Leonov had said that the mission was to last only two months.

On June 29, the station was in an orbit at 336×362 km. On July 3, the mission was given the official okay to last beyond the ASTP flight, but the exact length was not yet officially announced despite Leonov's recent statement. On

Figure 4-9. This is the Salyut 4 trainer at the Gagarin Training Center. The Salyut's large solar panels are attached to the middle section of the station and thermal radiators are visible at the base of the arrays. A Soyuz docking probe is visible next to the Salyut on the right and a Soyuz capsule is also visible on the left. (Source: Tass from Sovfoto.)

July 14, Radio Moscow announced that the flight would end in the last 10 days of July. To avoid any conflicts of resources during the ASTP and Soyuz 18B flights, the Soviets controlled the Soyuz 18B mission from the old Crimean Control Center, at Yevpatoria, which was used until the Soyuz 12 flight, and controlled the ASTP Soyuz 19 from the Kaliningrad Control Center. The crew exchanged greetings with the Soyuz 19 on two occasions, but the communications were very short in duration.[80,81]

Soyuz 19

Apollo-Soyuz Test Project

Launched: July 15, 1975, 3:20 p.m.
Landed: July 21, 1975, 1:51 p.m.
Altitude: 186 × 220 km @ 51.8°
Crew: Aleksei Leonov and Valeri Kubasov
Backup: Anatoliy Filipchenko and Nikolay Rukavishnikov
Backup Soyuz 19: (see Soyuz 22)
  Crew: Vladimir Dzhanibekov and Boris Andreyev
  Backup: Yuri Romanenko and Aleksandr Ivanchenkov
Call sign: Union

This was the joint flight between the U.S. and the U.S.S.R. that was called for in a treaty signed by both nations in 1972. The Soviets were the first to mention using a universal docking adapter for ASTP, during early negotiations. The U.S. negotiating teams were considering a Soyuz-Skylab or an Apollo-Salyut docking using one of the existing systems. The Soyuz-Skylab option quickly vanished since NASA was abandoning Skylab missions, due to a lack of money. Because the Soviets were continuing the Salyut series, an Apollo-Salyut mission became the obvious alternative. But, the universal system eliminated the possibility of an Apollo-Salyut mission, thus saving the Soviets from publicly explaining any complications with military Salyut missions and possible embarrassment with inefficient space station operations. The Soviets also didn't want to offer their space station when they thought that the backup U.S. Skylab was being prepared secretly as a manned military reconnaissance platform. The Soviets thought that NASA was lying about a Skylab 2, and that did not help the possibility of an Apollo-Salyut mission.[82]

On July 11, fuel loading preparations began at Baykonur and on July 13, the backup booster and Soyuz were taken to a secondary launch pad 20 km away from the primary launch pad, where the ASTP prime spacecraft was being prepared. The back-up booster and spacecraft were readied for use, until five hours before the scheduled launch. Two tracking and communications ships, the *Gagarin* and the *Korolev*, also took up stations to support the ASTP mission to supplement ground stations and the NASA communications network.

The ASTP launch was the first Soviet manned launch ever whose time was announced in advance and was the first to be televised live. It was launched into clear skies with a light wind and hot temperatures on July 16. The spacecraft's weight was 6,690 kg at launch. The U.S. Apollo would not be launched until the next day. On the second orbit, the cosmonauts removed their pressure suits and found that the television camera control unit had failed. After four orbits they maneuvered to 192 × 232 km with a seven-second burn. After 17 orbits, another maneuver changed the orbit to 222 × 225 km, with a 21-second burn. This was not quite as good an orbit as planned. The Soyuz 16 rehearsal did better, but the difference was not really significant considering the Apollo spacecraft's abilities to match orbits with the Soyuz (see Figure 4-10).

On July 16, the crew rewired the portable camera, using medical tape, allowing for interior pictures (but all the other cameras were still out), and they spoke to the Salyut 4 crew briefly. Salyut 4's orbit crossed that of Soyuz

Figure 4-10. The Soyuz carrying Leonov and Kubasov is seen in orbit from the Apollo. The Soyuz reentry and orbital modules are covered with dark (green) thermal blanket insulation, and the white bands around the Service module are thermal radiators. (Source: National Aeronautics and Space Administration.)

19 twice during the mission allowing for 30 seconds and 90 seconds of communication between the crews.[83] The crew also lowered the cabin pressure from Earth normal pressure of 760 mm, 20% oxygen to 540 mm, 40% oxygen. This allowed for faster cycling of the airlock once docked to the Apollo. The Soyuz was in a 218 × 231-km orbit at the time. The same day, the Apollo was launched at 10:50 p.m., Moscow time, into a 173 × 155-km orbit. After transposition and docking with the docking module carried in the adapter section of the booster upper stage, the Apollo maneuvered to 172 × 172 km. The Apollo then maneuvered to 173 × 234 km, and finally 229 × 229 km the next day. In preparation for the docking, the Soyuz crew donned their pressure suits again. At 7:09 p.m., July 17, the Apollo docked to the Soyuz with Stafford in control and with the Apollo docking system in active mode. The only action of the Soyuz during the process was to keep pointed at the Apollo and roll to match Apollo maneuvers. Three hours later, Stafford and Slayton opened the Apollo docking module hatch and were greeted by Leonov and Kubasov, as they opened the orbital module hatch. Four crew exchanges took place in all, with each crewman visiting the other's ship at least once, and performing some experiments ranging from materials processing to taking Earth photographs (see Figure 4-11).

Figure 4-11. Leonov is entering the orbital module from the reentry module while Kubasov works nearby. The orbital module was used as cargo space and living quarters, and contained experiments and equipment needed for the mission. On Salyut missions, the orbital module is used to store cargo being delivered to stations. Garbage is stored in the module when returning to Earth, because the module is jettisoned and is destroyed on reentry. (Source: National Aeronautics and Space Administration.)

On July 19, at 3:02 p.m. the ships undocked and then the Apollo flew extensive precise maneuvers around the Soyuz to perform the artificial solar eclipse experiment, where the Apollo's diameter blocked sunlight to the Soyuz and allowed pictures of the solar corona to be taken. Slayton then redocked the Apollo to the Soyuz with the Soyuz docking system in active mode at 3:20 p.m. With the sun reflecting brightly off the Soyuz, the docking was harder then the first and consumed more propellant, but the docking system performed well. The ships undocked for the last time after only a few hours, at 6:26 p.m. and drifted apart. The Apollo would stay in orbit until July 26, performing the last manned space experiments until the U.S. space shuttle was launched more than five years later (see Figure 4-12).

On July 20, the Soyuz tested its engines as usual in preparation for landing. The Soyuz capsule's landing point was changed a few hours before retrofire to avoid clouds and ensure a good television picture. On July 21, at 1:51 p.m., the capsule landed 87 km northeast of Arkalyk. The descent was televised live from the recovery helicopters for the first time in the U.S.S.R.[84,85]

The Apollo crew of Stafford, Slayton and Brand splashed down 12:18 a.m., Moscow time, July 26. The crew encountered difficulty when the capsule was flooded with highly dangerous fumes from the attitude control thrusters before landing, when the pressure equalization valve opened. The attitude control thrusters had been accidentally left on. Slayton put oxygen flow onto high and the capsule hit the water and turned over. Stafford dropped out of his seat and got oxygen masks out and put one on Brand, who was by then unconscious. The astronauts were not wearing spacesuits for the landing, which would have prevented any emergency. The crew spent two weeks under observation and in recovery, and later went to a reunion in the Soviet Union in October.

After the flight, Leonov was promoted from colonel to general, Stafford was promoted to major general (and went on to command Vandenburg Air Force Base), and Slayton went on to head Shuttle flight testing.

Figure 4-12. The ASTP Apollo and docking module are viewed here by the Soyuz. This was the last Apollo flight and the start of a long lull in the U.S. manned space program as Apollo facilities were scrapped and rebuilt to support the shuttle. (Source: National Aeronautics and Space Administration.)

Meanwhile, the Soyuz 18B crew was still orbiting in the Salyut 4 space station finishing their two-month mission. Living conditions had degraded in the station over the months. The environmental control system was failing and the windows were fogged over. The crew had requested to return complaining that there was a green mold growing on the walls of the station.[86] The Soyuz also had problems when it was exposed to the sun for unusually long periods due to the nature of the station's orbit. Extra cooling fans were used to move air from the Soyuz to the Salyut.[87]

During the last ten days of the mission, the cosmonauts wore exercise suits during their work and increased their exercise period to more than two hours per day. On July 18, the crew began preparing the Salyut for unmanned flight and the crew increased water and salt intake to rehydrate their bodies. They also continued with final Earth resources photography sessions work.

On July 24, the Soyuz was activated and the crew tested its engine boosting the Salyut's orbit to 349 × 369 km. The capsule was packed with 50 kg of experiment results and film over the next two days. Two days later, the crew began the return to Earth. The Soyuz 18B undocked from Salyut 4 at 1:56 p.m., July 26, and landed at 3:18 p.m., 56 km northeast of Arkalyk. During the landing, the crew tested a new orientation system for the Soyuz capsule that would enable precise night landings. Continuing with the policy of the ASTP flight, the Soviets televised the landing.[88]

Doctors had requested that the cosmonauts be taken out of the capsule with stretchers, but the cosmonauts refused. However, it was two days before Klimuk would take a ten-minute walk again and a week to make a full recovery. They both lost an average of 2.9 kg of weight. They started writing their mission reports on August 2, and they returned the Star City on August 5.[89–91]

The crew had taken 2,000 pictures of Earth and 600 of the sun. During the mission, they spent a total of thirteen days on geophysical experiments, thirteen days doing astrophysical work, ten days doing medical experiments, two days on station photography, two days on atmospheric experiments, ten days for relaxation, and seven days for Salyut activation and deactivation.

Kosmos 772
Progress Test Flight
Launched: September 29, 1975, 7:19 a.m.
Landed: October 1, 1975, 7:48 a.m.
Altitude: 195 × 299 km @ 51.8°

This flight was similar to the Kosmos 670 flight, and was a rehearsal for the Soyuz 20 unmanned test of Progress automated transport systems.[92] The spacecraft carried no solar panels, but had battery power for about three days of independent flight.[93] The Progress automated spacecraft was being developed to support the second- and third-generation space stations of the late 1970s and 1980s.

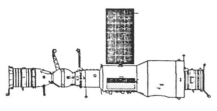

Soyuz 20

Unmanned—Progress Test Flight

Launched: November 17, 1975,
5:37 p.m.
Landed: February 16, 1976, 5:24 a.m.
Altitude: 177 × 251 km @ 51.6°
Crew: none

Salyut designer Feoktistov said this flight was to test the Progress automatic transport systems, although the ship was a Soyuz type, and had a reentry module. The spacecraft also tested the in-orbit storage mode for a record duration. A biological payload of turtles, flies, seeds, and plant specimens gave the flight a secondary scientific mission much the same as the Kosmos 782 Bio-satellite.[94]

The Soyuz used a transfer orbit of 199 × 263 km, then 247 × 282 km, and finally 328 × 355 km. The Soyuz docked with Salyut 4 on November 19, at 350 × 359 km. After docking, the orbit was 343 × 367 km. In the first week after docking, the station complex maneuvered changing the orbital inclination from 51.8° to 51.5°.[95]

By the end of the mission some serious degradation of critical systems on Soyuz 20 was noticeable.[96] This led the Soviets to limit the Soyuz to around 91 days in orbit. The ship tested its engines on February 15, slightly changing the station's orbit before undocking on February 19. The capsule landed about three hours later.[97]

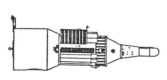

Salyut 5

Last Dedicated Military Station

Launched: June 22, 1976, 9:04 p.m.
Capsule Landed: February 26, 1977, 12:36 p.m.
Module Reentry: August 8, 1977
Altitude: 208 × 233 km @ 51.6°

There was some speculation in the West shortly after the launch that Salyut 5 was equipped with two docking ports, but it was not.[98] Salyut 5 was the last dedicated military Salyut station. As with Salyut 3, the reports from Salyut 5 were not as descriptive as from Salyut 4, or later space stations, and little is known about the military stations. Salyut 5 was never described exactly, but it was probably very similar to Salyut 3 (see box, Figure 4-2).

Salyut 5 Space Station

Salyut 5 was the last dedicated military Salyut station. It was never described exactly, but it was probably very similar to Salyut 3. Salyut 5's estimated weight was about 19,500 kg to allow launch by its D-1h booster. Like Salyut 3, Salyut 5 also carried a high-resolution camera with 10-meter focal length in the telescope housing. A gyroscope attitude control system was again used as on Salyut 3.

Few experiments were described as most of them were of a secret nature. The ones publicly announced were assorted biological experiments on plants, insects, and fish. Sfera was a materials processing furnace for melting bismuth, lead, tin, and cadmium. RSS-2M was a spectrograph for atmospheric observations. Kristall was a crystal growth unit. Diffuzyia was an experiment to produce dibenzyl and toluene alloys. Reaktsia was an enclosure with stainless steel pipes that could be joined with nickel manganese solder. Potok was a capillary action pumping experiment investigating resupplying future space stations.

The entire interior of the work compartment was covered with a blue fabric cushion to prevent injuries on corners of equipment. This was done at the request of previous Salyut crews. A treadmill replaced Salyut 3's ergometer as exercise equipment. A Stroka teletype was also installed as on Salyut 4 to relay instructions to the crew. As on Salyut 3, the station carried a small library and slide projector for leisure use.

Soviet tracking ships *Korolev*, *Gagarin*, *Morzhovets*, and *Bezhitsa* left their ports before the launch indicating the upcoming mission. By June 25, the station's orbit was raised to 212×244 km, and raised again by June 27 to 214×257. On July 4, a maneuver boosted the orbit to 262×285 km, and the next day the station's orbit was circularized at 274×274 km, ready for the Soyuz 21 mission.[99]

Soyuz 21

Long Duration—49 Days

Launched: July 6, 1976, 3:09 p.m.
Landed: August 24, 1976, 9:34 p.m.
Altitude: 193×253 km @ $51.6°$
Crew: Boris Volynov and Vitaliy Zholobov
Backup: Vyacheslav Zudov and Nikolay Rukavishnikov
Call sign: Baikal

As on earlier military Salyut missions, the Soviets officially assigned multiple backup crews to this flight. The second and third backup crews consisted of Gorbatko and Glazkov, and Berezovoi and Lisun.[100] Soyuz 21 was launched 14 days after Salyut 5, in 40°C temperatures during a 25-day-long second launch window. The Soyuz used a final transfer orbit of 246 × 274 km before Volynov docked at 4:40 p.m., July 7, at 254 × 280 km. The crew entered the station five hours later and began activating the station's systems. After completing the activation, on July 8, the cosmonauts gave a televised tour of the station. The orbit was raised on July 10 to 269 × 281 km.[101,102]

A main objective of the mission was apparently the observation of Operation Sevier, which was a massive Soviet military exercise in Siberia. Scientific work was kept to a minimum; however, the crew operated the Sfera, Kristall, Potek, Diffusia, and Reaction experiments. Medical experiments Levka, Rezeda, Impulse, and Palma were also performed, and they tested a new mass meter.[103] Many operations of controlling the station were performed under ground control to relieve the cosmonauts from the routine tasks. Communication with the crew was also kept to a minimum. Reportedly, only an emergency situation would be reason enough to require informing the cosmonauts!

The crew also conducted experiments observing Danio fish to research vestibular systems. Smelting experiments melting bismuth, lead, tin, and cadmium were also performed. The crew also spent time studying Earth's atmosphere, testing a mock-up propellant transfer system, and performing medical tests. The crew's main job was making extensive observations of Earth and using the high-resolution camera Salyut 5 carried. The crew exercised for a couple of hours a day like previous crews to keep their muscles strong for the eventual return to Earth.

July 11 was a rest day for the crew, but they continued light work by calibrating a spectrograph and making Earth observations. On July 14, they started the crystal growth experiments. The next day was devoted to routine medical checks. During medical checks, the crew made use of low-pressure suits to exercise their cardiovascular systems. On July 16, the crew conducted the Potek capillary liquid flow experiment. On July 28, the crew performed the Reaction soldering experiment in which 15-mm diameter tubes were soldered together with a nickel manganese solder heated by a flame.[104] The pipes soldered could withstand 500 atmospheres of pressure. They continued Earth observations over the next few days. On August 3, the station's orbit was 261 × 268 km, and on August 7, a second crystal growth experiment was started.

Two days later, they finished another crystal growth experiment using the Kristal unit in which a crystal had been growing for 26 days.

The crew then began several days of observations of Earth's atmosphere using an improved version of the Salyut 4 infrared telescope spectrometer. On August

15, the crew held a question-and-answer session with children on a Young Pioneers tour of the mission control center. On August 17, the crew spent most of the day controlling the attitude of the space station manually.

On August 23, they made observations with the infrared telescope spectrometer of the sun and Earth. The next day, the Soviets announced that the flight would end in only ten hours.[105] The quick notice of the landing even took the reporters of Radio Moscow by surprise, making the normally major news story only a one-line addition to the news broadcast. They crew undocked at 6:12 p.m., and landed at 9:33 p.m., 200 km southwest of Kokchetav on the Karl Marx collective farm, far from the normal landing area.[106,107]

The cause of the quick end of the mission was reported to be an acrid odor that developed in the environmental control system. The cosmonauts endured it while they tried to find the cause, but they could not correct the problem and were forced to quickly deactivate the Salyut, pack and activate the Soyuz.[108] The problem may have begun as early as August 17. Analysis of the landing opportunities indicates the mission was to have lasted for 54 to 66 days.

After landing, the cosmonauts were in worse condition than earlier crews, but managed to climb out of the capsule unaided. Both had lost about 1.5 kg in weight, and they spent the next days completing their flight logs while recovering.[109] The recovery at Baykonur took about a week and they were flown back to Star City, September 2. The next day they received the usual awards of Hero of the Soviet Union, indicating no fault on their part for the early end to the mission.

Soyuz 22

Short-Duration Earth Observation Flight

Launched: September 15, 1976, 12:48 p.m.
Landed: September 23, 1976, 10:40 a.m.
Altitude: 185 × 296 km @ 64.75°
Crew: Valeriy Bykovsky and Vladimir Aksyonov
Backup: Yuri Malyshev and Gennadiy Strekalov
Call sign: Hawk

Three months before the launch of Soyuz 22, Klimuk had said that no more manned solo Soyuz missions would be flown, making this solo Soyuz flight an unexpected development.[110] Perhaps Klimuk was not informed about the flight because of its partial military mission. The flight's cosmonauts were not however, from the military group, implying civilian control of the mission. The second backup crew for the flight was Leonid Popov and Boris Andreyev.[111] An objective of the mission was probably to observe the NATO Exercise Teamwork that was taking place in Norway. Some of the NATO exercises were well above 51° latitude and therefore out of good view of the Salyut's orbit. The Soyuz used an orbit similar to that used by some Soviet reconnaissance satellites enabling it to better view the exercises. To accomplish any reconnaissance and to test equipment to be used on the next Salyut space station, the Soyuz was equipped with a new MKF-6 camera.

The Soyuz used was the modified ASTP backup ship. The ASTP docking mechanism was replaced with the 204-kg East German multispectral camera (MKF-6M). The camera, which was protected by a cover on the outside of the ship when not in use, recorded four visible wavelengths and two infrared wavelengths simultaneously. Resolution of the camera was 20 meters and in ten minutes 800,000 km^2 could be photographed. The resolution was very poor compared to the usual military reconnaissance satellite, making the cosmonauts' own manual observations more important to their mission.

The launch was observed by a delegation from East Germany, where the camera was built. On the fourth orbit, the Soyuz maneuvered to 250×280 km and to 251×257 km on the 16th orbit. During the mission, the crew photographed the NATO exercises, and parts of the Soviet Union and East Germany using the camera. The photographic sessions required both cosmonauts' attention. Bykovsky controlled by pointing the Soyuz, to which the camera was attached, while Aksenov operated the camera controls in the orbital module.

Aksenov was getting to fly after only three years as a cosmonaut. This was much less time than the average. Aksenov also had a specialized Air Force background, which probably gave him the knowledge to select the best military targets for observation.[112]

The MKF-6M camera tests began on orbit 15. Targets were Earth's horizon, the moon, parts of Asia, and the Baikal-Amur railway, which was mentioned many times on future flights as a target for the MKF-6M. For the Raduga experiment, pictures were taken in parallel with an AN-30 aircraft, flying at high altitude with an MKF-6M. Later, investigators using both photos could see areas on the space photos and then check the aircraft photos with greater resolution. The cosmonauts took 2,400 pictures in all. Other experiments on board included the Biogravistat plant growth centrifuge, which spun to create artificial gravity to aid plant growth.[113]

On September 22, the film canisters were transferred to the reentry module and the cosmonauts tested the Soyuz main engine, changing the orbit to 239×253 km. The next day, they landed 150 km northwest of Tselinograd.[114]

Soyuz 23
Aborted

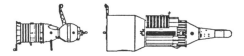

Launched: October 14, 1976, 8:40 p.m.
Landed: October 16, 1976, 8:47 p.m.
Altitude: 188 × 224 km @ 51.6°
Crew: Vyacheslav Zudov and Valeri Rozhdestvenski
Backup: Viktor Grobatko and Yuri Glazkov
Call Sign: Radon

Soyuz 23's mission was to dock with the Salyut 5 station for a 73- to 85-day mission. The second backup crew for the mission was Anatoliy Berezovoi and Michail Lisun. By the 16th orbit, the Soyuz had maneuvered close to the Salyut in a 246 × 269-km orbit. Salyut 5 was nearby in a 253 × 268-km orbit.[115] On the next orbit, at 9:58 p.m., October 15, during the automatic approach phase, the automatic docking system malfunctioned before the Soyuz was within 100 meters of the station.[116,117] Normally, the crew manually docks from 100 meters distance, but the Soyuz 23 crew had not been trained for a manual approach, and was ordered to land at soonest opportunity.

The Soyuz transport had less than two days' battery power left, so the cosmonauts were told to shut off all unnecessary electrical systems, including the radio to conserve power.[118] The Soyuz had already passed the landing opportunity for the day and would have to wait another day for the next pass over the Baykonur Cosmodrome to land. The orbit was raised slightly to 246 × 266 km during the usual test of the Soyuz engine before retrofire. On October 16, as the Soyuz flew over the South Pacific, the retroburn started at 8:02 p.m., lowering the trajectory of the spacecraft so it entered the atmosphere over North Africa. This was the normal landing corridor for the Soyuz, but weather in the landing area near the Baykonur Cosmodrome was not favorable. Shatalov told the cosmonauts to stay in their seats after landing, because of squall force winds causing blizzard conditions in the landing area. There was no choice about landing because of the limited battery power for the Soyuz, and there was no thought of an alternate landing site since the capsule could land in any weather. Only finding the capsule and crew would be more difficult in bad weather.

The capsule drifted down under its parachute into squall force winds and fog, with temperatures at − 22°C, and splashed down in the 32-km-wide Lake Tengiz, at 8:46 p.m. Lake Tengiz was in the middle of the spacecraft recovery zone, about 140 km west of Arkalyk, and has a surface area of 1,590 km². The capsule landed about 8 km from the northern shore in the partially frozen lake. During landing, Zudov suffered an unspecified injury which was probably not serious since he jettisoned the parachute as part of his normal after-landing duties.[119] The capsule cooled rapidly in the freezing water, and the cosmonauts removed their pressure suits and put on their normal flight suits, expecting a quick recovery. Helicopters began searching for the capsule, but the capsule's light beacon became obscured as the helicopters descended to the capsule, in 50- to 70-meter-thick fog.

The cosmonauts were exhausted after removing their pressure suits in the small capsule and decided to eat some of the capsule's rations while waiting to be recovered. Recovery teams also tried using rubber rafts to reach the capsule but were blocked by blocks of ice and icy sludge on the surface of the freezing lake shore, making it impossible for them to reach the capsule. Helicopters air-lifted amphibious vehicles to the lake, but they could not reach the capsule either, because of the bogs surrounding the lake. After being blocked from reaching the capsule by air, water and land, the recovery forces were ordered to wait until dawn for the helicopters to take in frogmen.

There was no immediate threat facing the cosmonauts, because the capsule was sea worthy, but there was a concern over power. Normally, the capsule's batteries were only needed for the short landing sequence of 40 minutes. This forced the cosmonauts to shut down everything except a small interior light. Some food rations were available for just such an emergency in the capsule and air to breath was available through the pressure equalization vent, which was above the water line.

The next day the helicopters brought in frogmen, and the cosmonauts put on their emergency water survival suits, in case they had to exit the capsule through the top hatch. The cosmonauts also turned on the exterior light beacons again so the helicopters could find the capsule.

This time they managed to find the capsule and frogmen attached flotation aids to it, as the helicopters hovered overhead. An Mi-8 helicopter attached a line to the capsule but it was too heavy to lift out of the water. Instead, the helicopter pulled the capsule the 8 km to the shore in about 45 minutes.[120]

The recovery was finally over after 9 hours, and return of the crew was announced at 7:00 a.m. Moscow time, October 17. The crew was flown back to Baykonur the same day and received the usual mission awards. Shatalov later praised the heroism and courage of the recovery forces. The mission was probably intended to last from 73 to 85 days.[121]

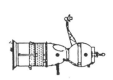

Kosmos 869
First Soyuz T Test Flight
Launched: November 29, 1976, 7:04 p.m.
Landed: December 17, 1976, 12:36 p.m.
Altitude: 198 × 293 km @ 51.8°

This was the first test of a new version of the Soyuz called the Soyuz T—the first of a family of spacecraft being developed for use with the second generation of Salyut stations to be flown in 1977. The spacecraft was equipped with solar arrays enabling independent, longer-than-two-to-three-day flight possible in the current Soyuz ferry. The Soyuz T also was capable of jettisoning the orbital module to save weight and propellant needed for the retrofire maneuver. This would become standard procedure for the Soyuz T, but apparently was not tested on this flight.[122] The spacecraft maneuvered to an orbit at 268×390 km and then to 299×309 km. It then drifted in storage mode for 18 days before landing.[123]

Kosmos 881/882

First Shuttle Reentry Test Flight

Launched: December 15, 1976, 4:34 a.m.
Landed: December 15, 1976, 6:05? a.m.
Altitude (Kosmos 881): 198×233 km @ $51.6°$
Altitude (Kosmos 882): 189×213 km @ $51.6°$

Kosmos 881 and 882 were two 9,090-kg lifting bodies or winged space planes that were launched on the same D-1h booster from Baykonur. The booster's third stage separated from the payload in orbit at 189×213 km. One of the spacecraft entered a higher orbit after separating from the booster stage. The craft were probably recovered around $44°N\ 73°E$ in a desert area of the U.S.S.R. Analysis found that the flight seems to have required dawn for visibility during the gliding portion of the flight, prior to landing. Two other double flights were launched during the next few years, each around midnight to have the proper lighting conditions in the landing area, considering the seasonal variations.[124]

These flights were probably tests associated with the small shuttle in development in the late 1970s. Each of the spacecraft weighed about twice as much as the old U.S. X-20 shuttle design, which would have weighed 5,200 kg. The only comment from U.S. government observers was that the flight was definitely man related.[125]

Soyuz 24

Short Duration—18 Days

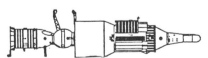

Launched: February 7, 1977, 7:12 p.m.
Landed: February 25, 1977, 12:36 p.m.
Altitude: 173×323 km @ $51.6°$
Crew: Viktor Grobatko and Yuri Glazkov
Backup: Anatoliy Berezovoi and Michail Lisun
Call sign: Terek

This crew was the back-up for the aborted Soyuz 23 flight. The second backup team for Soyuz 23 also moved up into the backup position for Soyuz

24, being replaced by the second backup team of Vladimir Kozelsky and Vladimir Perobrazhensky.[126] Grobatko had also been the backup for Voskhod 2, Soyuz 5 which involved an EVA. Glazkov wrote a thesis on EVA,[127] which indicates that EVA was possibly a planned element of Soyuz 24. Shatalov had said, in 1976, that an EVA had been planned for Salyuts 3, 4 and 5, but was always canceled, due to time constraints, as it apparently was on this flight.[128] Soyuz 24's transfer orbit to Salyut 5 was 217 × 264 km. Grobatko docked from 80 meters distance, at 10:30 p.m., February 8, in orbit at 222 × 287 km.

On February 12, the orbit was circularized to 251 × 258 km. The cosmonauts had an unusual sleep period before entering and reactivating the Salyut station, at 8:45 a.m., February 9.[129] The problem encountered during the Soyuz 21 mission with Salyut 5's environmental control system must have been resolved or the cosmonauts were prepared to repair the problem. The 11-hour delay before entering the station may have had something to do with this. On February 11, they completed reactivating the station's systems, and they replaced components of the station's computer.[130]

The crew carried out studies like the Soyuz 21 crew, but this flight seemed to have a specific objective and was not meant to be a long-duration attempt. Salyut 5's supplies of maneuvering propellant were too low to attempt the long-duration flight that had been planned for Soyuz 23. It is possible that the Soyuz 24 crew was to load the Salyut's sample return capsule with experiment results and film generated by the last crew, which was forced to return unexpectedly.

The Terek crew did perform some Earth resources work, biological, and materials experiments. Areas observed included the Caucasus and the Sakhalin coast area. On February 15, more observations were made with the infrared spectroscopic telescope while the station's orbit was 253 × 274 km. On February 16, the Soviets announced that the mission had reached the half-way point. The next day, the cosmonauts developed slight colds, but they took no medication.[131]

On February 21, the crew carried out an air changing experiment, which was shown on TV, during which they slowly vented air from one end of the station, while releasing 100 kg of air from tanks in the Soyuz orbital module at the other end. This was a test of a method of air replenishment that the Progress transports to the future Salyut stations would use. Unfortunately, the poor quality TV pictures helped little in the further description of the secret military station.

On February 23, they began to activate and pack the Soyuz, and deactivate the Salyut for the last time. They undocked at 9:21 a.m., February 25, and landed at 12:36 p.m., 36 km northeast of Arkalyk in strong winds, snow, and a temperature below − 17°C. There was a report that the Soyuz made a ballistic reentry, but this is not certain. The crew received the usual mission awards.[132-134]

On February 26, the Salyut ejected a capsule that was recovered on Soviet territory, during a normal Soyuz landing window.[135] The capsule probably contained exposed film and experiment results. On March 5, Salyut 5's orbit was circularized to 251×265 km, possibly indicating another mission was planned, but it never materialized. On March 22, the orbit was raised again to 250×273 km. The Soviets were just observing the Salyut system's performance as temperature and atmosphere were maintained. The station was only intended to have a six-month lifetime.

References

1. Johnson, Nicholas L., *Handbook of Soviet Manned Spaceflight*, American Astronautical Society: San Diego, 1980, p. 381.
2. Lawton, Tony, "The Soviet Shuttle," *Spaceflight*, Vol. 29, July 1987, p. 4.
3. Congressional Research Service, The Library of Congress, *Soviet Space Programs 1976-80, Manned Space Programs and Life Sciences, Part 2*, Washington: Government Printing Office, 1984, p. 655.
4. Grahn, Sven, "Future Salyut Missions," *Spaceflight*, Vol. 16, No. 10, Oct. 1974, p. 393.
5. Vick, Charles P., "The Soviet Super Boosters—2," *Spaceflight*, Vol. 16, No. 3, March 1974, p. 102.
6. Johnson, Nicholas L., "The Military and Civilian Salyut Space Programs," *Spaceflight*, Vol. 21, No. 8-9, Aug.-Sept. 1979, pp. 364-366.
7. Johnson, op. cit. p. 235.
8. *Spaceflight*, Vol. 15, No. 8, Aug. 1973, p. 293.
9. O'Toole, Thomas, "2nd Russian Space Shot Fails," *Washington D.C. Post*, May 4, 1973, p. 1, 14.
10 "Russian Ocean Surveillance Satellites," *The Royal Air Force's Quarterly*, Spring 1978, Vol. 18, No. 1, pp. 60-63.
11. The Kettering Group, "Reception of Radio Signals from Cosmos 557," *Spaceflight*, Vol. 16, No. 1, Jan. 1974, p. 40.
12. Johnson, op. cit. p. 237.
13. Ezell, Edward C. and Ezell, Linda N., *The Partnership: A History of the Apollo-Soyuz Test Project*, NASA SP-4209, Washington D.C.: Government Printing Office, 1978, p. 232.
14. Johnson, op. cit. p. 133.
15. "Latest Soyuz Flights," *Spaceflight*, Vol. 16, No. 2, Feb. 1974, p. 110.
16. Johnson, op. cit. pp. 169-171.
17. "Latest Soyuz Flights," *Spaceflight*, Vol. 16, No. 2, Feb. 1974, p. 110.

18. Turnill, R., *Spaceflight Directory*, London: Frederick Warne Ltd., 1977, p. 302.
19. Johnson, op. cit. pp. 169–171.
20. Grahn, op. cit. p. 393.
21. Christy, Robert D., "Orbits of Soviet Spacecraft at 51.6° Inclination," *Spaceflight*, Vol. 22, No. 2, Feb. 1980, p. 80.
22. Congressional Research Service, The Library of Congress, *Soviet Space Programs 1976–80, Manned Space Programs and Life Sciences, Part 2*, Government Printing Office: Washington, 1984, p. 518.
23. Johnson, op. cit. pp. 172–4.
24. Turnill, op. cit. p. 302.
25. "Latest Soyuz Flights," *Spaceflight*, Vol. 16, No. 2, Feb. 1974, p. 110.
26. Congressional Research Service, The Library of Congress, *Soviet Space Programs 1976–80, Manned Space Programs and Life Sciences, Part 2*, Government Printing Office: Washington, 1984, p. 523.
27. Ezell and Ezell, op. cit. p. 302.
28. "Soyuz 16 Mission," *Spaceflight*, Vol. 17, No. 3, March 1975, p. 112.
29. Johnson, op. cit. p. 133.
30. Ibid. pp. 242–244.
31. Ibid. p. 302.
32. *Spaceflight*, Vol. 31, No. 2, Feb. 1989, p. 57.
33. Johnson, op. cit. pp. 301–304.
34. "Soyuz 14 Mission Report," *Spaceflight*, Vol. 16, No. 11, Nov. 1974, p. 430.
35. Grahn, op. cit. p. 392.
36. "Soyuz 14 Mission Report," op. cit. p. 430.
37. Turnill, op. cit. p. 303.
38. Grahn, op. cit. p. 392.
39. Johnson, op. cit. p. 304.
40. Ibid. p. 134.
41. Ezell and Ezell, op. cit. p. 302.
42. "Soyuz 16 Mission," op. cit. p. 112.
43. Congressional Research Service, The Library of Congress, *Soviet Space Programs 1976–80, Manned Space Programs and Life Sciences, Part 2*, Government Printing Office: Washington, 1984, pp. 522–523.
44. Turnill, op. cit. p. 304.
45. "Russia's 'Universal Spacecraft'," *Spaceflight*, Vol. 18, No. 3, March 1976, p. 95.
46. "The Soyuz 15 Mission," *Spaceflight*, Vol. 16, No. 12, Dec. 1974, p. 470.
47. Clark, Phillip S., "Soyuz Missions to Salyut Stations," *Spaceflight*, Vol. 21, No. 6, June 1979, p. 263.

48. Hooper, Gordon R., "Missions to Salyut 4," *Spaceflight*, Vol. 19, No. 2, Feb. 1977, p. 64.
49. Johnson, op. cit. p. 306.
50. Clark, op. cit. p. 263.
51. Johnson, op. cit. pp. 244, 307.
52. Turnill, op. cit. p. 305.
53. *Spaceflight*, Vol. 31, No. 2, Feb. 1989, p. 57.
54. Ezell and Ezell, op. cit. p. 302.
55. "Soyuz 16 Mission," op. cit. p. 111.
56. Johnson, op. cit. pp. 177–179.
57. Turnill, op. cit. p. 305.
58. "Soyuz 16 Mission," op. cit. p. 112.
59. Congressional Research Service, The Library of Congress, *Soviet Space Programs 1976–80, Manned Space Programs and Life Sciences, Part 2,* Government Printing Office: Washington, 1984, p. 523.
60. *Spaceflight*, Vol. 31, No. 2, Feb. 1989, p. 57.
61. Turnill, op. cit. p. 306.
62. "Missions to Salyut 4," *Spaceflight*, Vol. 17, No. 6, June 1975, p. 219.
63. Johnson, op. cit. pp. 307–308.
64. Turnill, op. cit. p. 306.
65. "Missions to Salyut 4," op. cit. p. 219.
66. Hooper, Gordon R., "Missions to Salyut 4," *Spaceflight*, Vol. 18, No. 1, Jan. 1976, p. 16.
67. Kidger, Neville, "Salyut 6 Mission Report—Part 8," *Spaceflight*, Vol. 23, No. 8, Oct. 1981, p. 267.
68. "Missions to Salyut 4," op. cit. p. 225.
69. Bond, op. cit. p. 344.
70. "Missions to Salyut 4," op. cit. p. 225.
71. Johnson, op. cit. p. 313.
72. Turnill, op. cit. p. 307.
73. Ezell and Ezell, op. cit. p. 302.
74. Johnson, op. cit. p. 314.
75. Oberg, James E., *Uncovering Soviet Disasters*, Random House: New York, 1988, p. 173.
76. Clark, op. cit. p. 263.
77. Johnson, op. cit. p. 320.
78. Hooper, op. cit. p. 14.
79. Turnill, op. cit. pp. 308–309.
80 Hooper, op. cit. pp. 16–17.
81. Ezell and Ezell, op. cit. p. 308.
82. Oberg, James E., *Red Star in Orbit*, Random House, New York, p. 143.
83. Ezell and Ezell, op. cit. p. 309.

84. Turnill, op. cit. p. 81.
85. Hooper, op. cit. p. 18.
86. Oberg, *Red Star in Orbit*, op. cit. p. 138.
87. Turnill, op. cit. p. 309.
88. Johnson, op. cit. p. 321.
89. Bond, op. cit. p. 347.
90. Turnill, op. cit. p. 310.
91. Hooper, op. cit. pp. 17–18.
92. Johnson, op. cit. p. 134.
93. Hooper, Gordon R., "Missions to Salyut 4: Part 3," *Spaceflight*, Vol. 19, No. 2, Feb. 1977, p. 64.
94. Johnson, op. cit. p. 323.
95. Hooper, "Missions to Salyut 4: Part 3," op. cit. pp. 64, 80.
96. Kidger, Neville, "Salyut 6 Mission Report," *Spaceflight*, Vol. 22, No. 2, Feb. 1980, p. 60.
97. Turnill, op. cit. p. 311.
98. Johnson, op. cit. p. 324.
99. Hooper, Gordon R., "Mission to Salyut 5," *Spaceflight*, Vol. 19, No. 4, April 1977, p. 138.
100. *Spaceflight*, Vol. 31, No. 2, Feb. 1989, p. 58.
101. Hooper, "Mission to Salyut 5," op. cit. p. 139.
102. Turnill, op. cit. p. 311.
103. Johnson, op. cit. pp. 325–327.
104. Hooper, "Mission to Salyut 5," op. cit. pp. 140–142.
105. Turnill, op. cit. p. 312.
106. Johnson, op. cit. p. 327.
107. Hooper, "Mission to Salyut 5," op. cit. p. 144.
108. Oberg, *Red Star in Orbit*, op. cit. p. 146.
109. Turnill, op. cit. p. 312.
110. Johnson, op. cit. p. 194.
111. *Spaceflight*, Vol. 31, No. 2, Feb. 1989, p. 58.
112. Turnill, op. cit. p. 314.
113. Johnson, op. cit. pp. 196–197.
114. Turnill, op. cit. p. 314.
115. *Spaceflight*, Vol. 31, No. 2, Feb. 1989, p. 58.
116. Foreign Broadcast Information Service, U.S.S.R., Space, JPRS-USP-84-003, June 1984, Joint Publications Research Service, p. 49.
117. Hooper, "Mission to Salyut 5," op. cit. p. 145.
118. Johnson, op. cit. p. 329.
119. Furniss, Tim, "Soviets Open 1988 Space Account," *Flight International*, Jan. 26, 1988, p. 26.

120. Foreign Broadcast Information Service, U.S.S.R., Space, JPRS-USP-84-003, June 1984, Joint Publications Research Service, pp. 50–55.
121. Clark, op. cit. p. 263.
122. King-Hele, D. B., Walker, D. M. C., Pilkington, J. A., Winterbottom, A. N., Hiller, H., and Perry, G. E., *The R.A.E. Table of Earth Satellites 1957–1986*, New York: Stockton Press, 1987, p. 470.
123. Johnson, op. cit. p. 134.
124. Williams, Trevor, "Soviet Re-entry Tests: A Winged Vehicle?," *Spaceflight*, Vol. 22, No. 5, May 1980, pp. 213–214.
125. Congressional Research Service, The Library of Congress, *Soviet Space Programs 1976–80, Manned Space Programs and Life Sciences, Part 2,* Government Printing Office: Washington, 1984, p. 654.
126. Spaceflight, Vol. 31, No. 2, Feb. 1989, p. 58.
127. Johnson, op. cit. p. 331.
128. Hooper, Gordon R., "Missions to Salyut 6," *Spaceflight*, Vol. 20, No. 6, June 1978, p. 232.
129. Hooper, Gordon R., "Mission to Salyut 5: Part 2," *Spaceflight*, Vol. 19, No. 7–8, July–Aug. 1977, p. 266.
130. Foreign Broadcast Information Service, U.S.S.R., Space, JPRS-USP-USP-85-001, Feb. 4, 1985, Joint Publications Research Service, p. 5.
131. Johnson, op. cit. p. 332.
132. Turnill, op. cit. p. 315.
133. Hooper, "Mission to Salyut 5: Part 2," op. cit. p. 268.
134. Kidger, op. cit. p. 60.
135. Clark, op. cit. p. 263.

Chapter 5

Second-Generation Space Stations

The success of the civilian space stations Salyut 1 and 4, and the fact that the military Salyut's were less successful, led to the next series of stations being developed and run mainly by civilian elements of the Soviet space program. While Salyut 4 and 5 were in use, preparations for the second-generation space stations were already well underway in the middle 1970s in the form of new buildings, increased personnel, and development of three new spacecraft types. The development of the three new spacecraft and the second-generation Salyut demonstrated a very large effort by the Soviets during this period. American astronauts observed the effort in 1975, while preparing for the ASTP flight. They reported large amounts of construction at Star City. Of course, at the time, the Soviets did not reveal the purpose of the expansion.

Space Station Program

The effort put into the second-generation Salyut program should not be ignored. During the missions of the space stations, Salyut 6 and Salyut 7, 64 boosters and spacecraft were needed to support the space stations. In 1980, Shatalov said that one months' data from a Salyut could take up to two years to analyze.[1] The cost of the facilities, personnel, and resources needed to build and operate the stations over a period of several years was substantial. The cost of a single Salyut or Star module has been estimated at between $1 to $2 billion. Six of the large spacecraft were built and flown over the nine years covered in this chapter. Each Progress flight also cost about $40 million, making the total cost for automatic resupply mission from 1978 to 1986 near $1 billion.[2]

The second-generation Salyut stations differed from their predecessors in that they could be resupplied in orbit, while the crew stayed on board. This required an extra docking port because the Soviets did not want to risk stranding a crew on a station without a return spacecraft, if a resupply mission failed to dock.

Docking failures were the most common problem for the Soviets in the 1960s and 1970s. Therefore, the Salyut was redesigned to have a docking port on both the rear and front ends. They did not put more docking ports on the Salyut because, as the civilian Salyut station designer Konstantin Feoktistov later stated, there was too much equipment on the outside of the station. The stations' size was limited by the power of its Proton booster, which was then the largest Soviet launch vehicle. More importantly, adding docking ports to the side of the station would throw off the center of gravity of the station and require the use of gimballed rockets or some gyroscopic system to control the station efficiently. The Soviets were apparently unable or unwilling to develop these systems for the Salyut.

The Salyut 6 and Salyut 7 stations were the only space stations to be equipped with two docking ports (Skylab also had two ports, but only one was used for various reasons, including lack of funding). The new Salyut's design was probably related to both the Salyut 4 civilian-type, and the Salyut 3 military-type stations. The military design featured docking at the aft end of the station and a propulsion module surrounding the space station. These features were apparently adopted and modified for the second-generation stations. Salyut 4 featured a forward docking unit and three large solar panels, which were used in the second-generation Salyut. The military and civilian space stations before Salyut 6 had tested atmosphere regeneration and water recovery equipment, which became standard on the second-generation stations.

The three other spacecraft developed for the new Salyut station included a new automatic supply ship, a new version of the Soyuz manned transport, and a large add-on module that almost doubled the size of the Salyut. The development of a robot cargo transport spacecraft, the Progress, began by at least 1974. The Soyuz T was a modified version of the Soyuz Ferry, and again enabled three crewmen to fly to space stations. The Star module was developed from the military Salyut space station to solve shortcomings in the Salyut design. The Soyuz was very limited in its return capability and the Salyut had very limited electrical power, especially when refueling. Refueling compressors required at least 25 % of the power generated by the station. The Star module helped solve these problems by providing extra experiment space, electrical power, and a return capsule for the return of bulk materials and experiment results to Earth. The new Soyuz T was also equipped with solar panels to provide extra power to the Salyut.

The new series of spacecraft was to put Soviets in space for long periods. During long-duration missions, the Soviets would determine the manufacturing capabilities of space stations; further explore the military applications of manned space stations; exploit the propaganda value of spaceflight by flying cosmonauts from Soviet-bloc countries, France, and India; and further space research to enable decisions to be made about future flights to the planets.

The Intercosmonaut program was the Soviet reaction to the United States planned flights of European astronauts on shuttle Spacelab missions in the 1980s. The Soviets' guest countries usually sent six to ten candidates to be trained and tested at Star City. After evaluation, a prime and backup cosmonaut were chosen from the group. During their flights, they were often told not to touch anything in the Soyuz or the space station. On rest days for the long-duration crew, some international crews were not even allowed in the station. Some countries did do research with their own experiments like the French medical experiments (Soyuz T-6). The price for flying the experiments was to leave the equipment on the station for the Soviets to use later. This was necessary because the weight that could be returned to Earth in the Soyuz series transports was very limited, and carefully allocated for the most important information and samples to be returned to Earth. During the first eight international flights, 60 astrophysical experiments were carried out, thousands of photographs were taken, and 200 materials processing experiments were carried out.[3]

Salyut 6 was exceptionally successful in fulfilling its mission. It operated for well beyond it designed lifetime and suffered no major failures. This enabled the Soviets to repeatedly break records for the time spent in space. Salyut 6 also hosted several international crews and proved the abilities of the Progress robot cargo ship.

Salyut 7 was not so lucky. While it was nearly a duplicate of Salyut 6, Salyut 7 suffered major systems failures. Yet, the station hosted additional international missions and two propaganda flights of Svetlana Savitskaya. Her flights drew some attention to the Soviet space program while the Western media were concentrating on upcoming flights of women mission specialists on the U.S. shuttle. Even with the frequent failures on Salyut 7, the long-duration crews were able to perform two record-duration flights of 211 and 237 days. A third potentially record flight, in 1985, was ended by medical problems. They also gained valuable but unwanted experience in repairing the space station in two missions.

During this time, it was easy to see the Soviets gain confidence in the equipment and the cosmonauts at performing difficult missions, which would not have been considered in the middle 1970s. Konstantin Feoktistov said about the repair missions, "If they had told us about five years ago that this was possible, we wouldn't have believed it ourselves."[4] But for most of the daring repair missions, the Soviets could look back on previous NASA flights of Skylab and the Shuttle to see the possibilities of in flight repair by cosmonauts. Without these NASA demonstrations, like the shuttle repair of the Solar Maximum satellite, the Soviets would probably have been much less inclined to propose new and potentially dangerous missions.

The Progress turned out to be the most successful spacecraft of the Soviet manned space program to date. As of October 1989, a Progress has never failed to dock with a space station and has never been known to have suffered a launch failure. Meanwhile, the Soyuz series in the same time period (1975 to 1989) suffered two launch failures and five docking failures. In the end, the Soyuz T proved to be more capable then its predecessor, but it still pressed the limits of Soviet technology to perform its mission. Still, the problems encountered in docking the Soyuz T were mainly the result of inadequate crew training and an inadequately equipped spacecraft. This was the result of the Soviet philosophy of a totally automatic spacecraft. The Star module never seemed to reach its full potential with the Salyut station. Only two were used for manned missions to Salyut 7, and both were not used to their full abilities. This doesn't necessarily indicate a problem with the spacecraft but was just a consequence of the many other failures during Salyut 7's mission. The fact that more Star modules were not launched is easy to visualize when frequent failures wreaked havoc with planned schedules of missions. Also, the Star module was nearly as large and complex as a Salyut station and undoubtedly nearly as expensive.

One of the main goals of this period, beyond the mere scientific tests and political missions, was the extension of mission duration to the point at which a decision on the feasibility of manned interplanetary flights could be made. During Salyut 6's lifetime, long-duration missions proceeded relatively quickly—first 96 days, then 140 days, 175 days, and 185 days. A short 74-day mission was flown to fulfill the international mission agreements before the delay until Salyut 7's launch. Salyut 7, as mentioned earlier, was not very successful in being maintenance free. The first long-duration mission went well, extending duration to 211 days. The next failed to dock. The next ended after 149 days when a relief crew's booster exploded. This was probably the first attempt at an eight-month mission. The next mission lasted 237 days completing the planned eight-month mission. The next crew was to attempt a ten-month mission, but it had to be postponed when the Salyut 7 lost all power and drifted for months. The repair mission was successful and turned into the planned ten-month mission, but the crew was forced to land short of eight months due to medical problems. Salyut 7 was not used for long-duration missions again. Thus, the ten-month mission was set back about two years until the third-generation space station, Mir, was fully equipped, in 1987.

Space Shuttle Program

During the period covered by this chapter, the long awaited U.S. shuttle was also beginning flights. The Soviets efforts in developing a small space plane from early 1970s continued with two more test flights launched by the Proton booster. The small shuttle program had been begun by 1975. In 1978, the Soviets reported that work on a shuttle was under way with flights planned in the

1980s.[5] A long shuttle landing runway was completed by 1978, northwest of the Baykonur launch complexes and was connected to large assembly buildings by wide roadways. Drop tests of the shuttle or an aerodynamic model from a Tu-95 Bear bomber took place in 1978.[6] General Beregovoi, cosmonauts Popovich, Filipchenko, Shonin, Khrunov, and others all made statements confirming development of a reusable shuttle was underway, and that they were not going to be left behind by the U.S. Also in 1978, a group of cosmonauts was formed to fly the space shuttle. They included Igor Volk, Anatoliy Levchenko, and Anatoliy Shchukin.[7]

The landing tests of the small space plane reportedly were not successful and the program was apparently canceled in 1979. This is evident from a statement made by Shatalov in June 1980 that reusable spacecraft had been studied, but were not economically justified yet, considering more proven methods. Numerous others in the Soviet space program made statements confirming that the well tested Salyut-Soyuz systems would continue to be used until the mid-1990s.[8] In light of an impending U.S. shuttle launch, the Soviets ended all public talk of a reusable space plane.

Some analysts report that the Soviets did not anticipate the launch of the U.S. shuttle until later than 1981, due to the technical complexities of the program.[9] No components of the U.S. shuttle were ever tested in space, and the propulsion systems and structures were never tested in flight before the first manned launch. The Soviets would never have considered this approach to manned spaceflight, and this undoubtedly contributed to the impression that the United States had completely abandoned manned spaceflight after 1975. The Soviets would have expected flight tests before making such a spacecraft and booster man-rated, as the U.S. had done in all previous manned space programs. When the first flight of the shuttle Columbia approached, the Soviets were reportedly surprised and worried about their misjudgment of the U.S. effort. The Soviets were caught years behind in the development of a shuttle. After years of progress in extending space flight duration, the world's attention was on the amazing U.S. shuttle.

The Soviets small shuttle was intended to help supply Salyut stations in the 1980s. With apparently no hope of salvaging any of the small shuttle plan, the Soviets apparently returned to the super booster program, which had been a low priority since being canceled in the middle 1970s. The super booster became the launch vehicle for a large new space shuttle that would be a copy of the U.S. space shuttle orbiter.[10] By 1982, Soviet officials again started to talk of reusable spacecraft, stating that in a few years, Soviet shuttle activity would be more apparent and flight tests would begin around 1986.[11,12]

The efforts of the Soviets coincidentally parallel early U.S. efforts to develop a shuttle. In the U.S., in the 1960s and 1970s, many different efforts would eventually play a part in the development of the NASA space shuttle. Drop tests of lifting bodies like the HL-10, M2F2, M2F3, X-24A, and X-24B tested landing characteristics of wingless vehicles. Tests of miniature spacecraft like ASSET and the X-23 tested hypersonic flight of lifting bodies. The X-15 tested manned control of hypersonic and sub-orbital spaceflight and reentry. The X-20 Dynasoar program initiated most of the tests, but was canceled in 1963 when the first spacecraft was being built. The work done to support the X-20 contributed greatly to development of the NASA space shuttle in the 1970s. The Soviets duplicated many of these efforts in the 1960s and 1970s.

In the early 1960s, the Soviets developed a shuttle-related test vehicle that was launched by a large aircraft at 10,000 meters altitude. Current shuttle cosmonaut Igor Volk participated in the 1960s program.[13] The vehicle's designer, Gleb Lozino-Lozinskiy, is now chief designer at the Buran assembly center near Moscow. The Soviet efforts of the 1970s also included development of a small space plane. The space plane was flight tested on the double Kosmos flights of 881/881, 997/998, and 1100/1101. Drop tests of the shuttle testing low-altitude flying abilities were apparently unsatisfactory and lead to the cancellation of the program by 1979.

All Soviet shuttle efforts then turned to developing the new redesigned heavy booster and a large shuttle not unlike the NASA shuttle orbiter. The large Soviet shuttle was developed in parallel with the redesigned super booster, later named Energia. The old G-1 super booster facilities were slowly rebuilt to support the new Energia booster and large shuttle. The runway built at Baykonur to support the small shuttle would now be used for transporting large shuttle orbiters and booster segments, and serve as an orbiter test landing site and primary recovery runway.

Beyond the purely propaganda usage of building a shuttle to match U.S. technical abilities, the Soviet military probably wanted to develop a large shuttle to counter any possible U.S. threat. The Soviets always closely tied the U.S. shuttle to military activities, claiming that it would be used for everything from an anti-satellite weapon, a space bomber, to a pirate spacecraft that would capture innocent foreign satellites.[14] In the late 1960s, the Soviets also claimed that the proposed U.S. manned orbiting laboratory was an anti-satellite weapon.[15] In the 1970s the Soviets also erroneously predicted that the U.S. lunar module would be used as a anti-satellite weapon.[16] Of course, the U.S. Department of Defense also points out that the Soviet shuttle can be used as an anti-satellite weapon.[17] In 1987, the Soviets admitted that the Energia had been developed by a military design bureau. Similarly, NASA also designed the U.S. shuttle to meet military specifications, in return for some financial support from the military.

During the 1980s, the Soviets conducted new reentry tests of a small space plane. At least 12 orbital or sub-orbital flights were made up to 1989, six of which are publicly known and listed here (see Figure 5-1). The spacecraft

was of a different design and much smaller than the earlier dual launch
Kosmos missions. The small space plane was used to test reentry techniques
and heat-shield tiles for the large shuttle orbiter being developed. Many

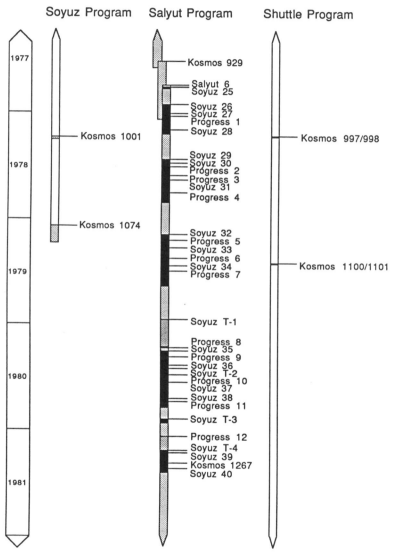

Figure 5-1. The Salyut and Shuttle programs were continued from 1977 to 1985, and
the Soyuz test program ended with the last development flights of the Soyuz T version.
All known flights related to these programs are shown here and are described in this
chapter. The gray portions show space station spacecraft lifetimes and black portions
show manned mission durations.

analysts, including the U.S. Department of Defense, used the flights as evidence that the small shuttle program of the 1970s was still underway, but this is not supported by any Soviet statements. The shuttle orbiter was under construction by the middle 1980s and the launch facilities and Energia booster were under final development to support not the Salyut series, but the third-generation Mir space stations, and the fourth-generation Mir 2 station of the 1990s.

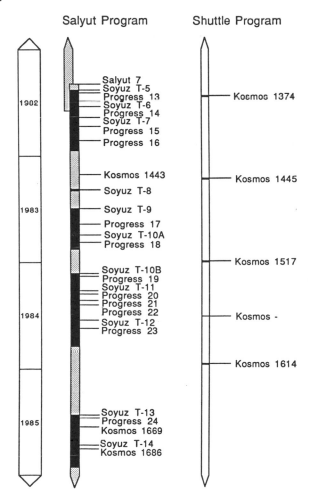

Figure 5-1. Continued.

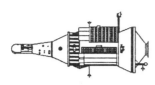

Kosmos 929

First Flight of Star Module

Launched: July 17, 1977, 12:00 p.m.
Capsule Landed: August 18, 1977
Module Reentry: February 2, 1978
Altitude: 214 × 278 km @ 51.6°

This first test flight of the Star module was a step toward expanding the Salyut 6 class stations, which had two docking ports. The module would enable a crew on the station to use the module's additional equipment and its reentry capsule (see Figure 5-2).

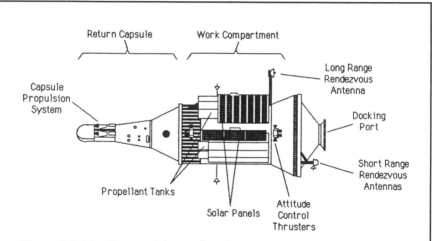

Figure 5-2. The Star module was based on the military Salyut design and provided second-generation Salyut stations added electrical power, experiments, and a return capsule. The Soyuz was the only other way to land experiment results, and it was severely limited in weight (50 kg) and space available. The Star module return capsule could return ten times more weight than a manned Soyuz.

Star Module

As with the military Salyuts, the Soviets are very secretive about the Star module. It was made to dock to the forward port of a Salyut 6 or 7 type space station and provide extra experiment space, electrical power, and maneuvering ability. The total length was about 14 meters, and the maximum diameter for launch was 4.15 meters.

The module normally carried a reentry module that was about the size of a Gemini capsule, and had a hatch through its heatshield, similar to the MOL-Gemini B, so that it could be loaded by the space station crew in the module. The capsule could return 500 kg of cargo compared to 50 kg for a manned Soyuz T. The module could carry 2,800 kg of cargo to a station, and add 50 cubic meters of habitable volume, not including the landing capsule. The Star module also had two Salyut-type solar arrays spanning 16 meters and totaling 40 square meters to produce 3 kilowatts of electricity and had a smaller solar array covering the propellant tanks. The module carried 3,000 kg of propellant to make orbital corrections to the station complex using its engines. The propellant could not be transferred to the Salyut. The empty module section weighed about 13,500 kg and the capsule weighed about 2,500 kg.

Star modules have flown under the Kosmos name on flights 929, 1267, 1443, and 1686. The later versions of the module weighed more than 20,000 kg fully loaded. Kosmos 1686 tested elements of systems to be used on the highly customized Star modules that would be attached to the Mir space station.

The Star module's design was based on the military Salyut space stations. It provided additional services for the civilian stations, which needed additional electricity, backup engine, and orientation systems. The Star module was first mentioned by the Soviets in July 1975 and was then expected to be in operation by 1978 or 1979, along with the Soyuz T and Progress spacecraft. Since this was the first flight of the spacecraft, it is unlikely that it was meant to be docked to a Salyut station, however, the maneuvers it made suggest a simulated rendezvous and docking were carried out.[18]

After launch, the Proton booster's third stage carrying the Star module entered an orbit at 211 × 260 km. The module then separated leaving the third stage, which reentered on July 29. Kosmos 929 made four small orbital changes during the first month in orbit to maintain an 89-minute orbit. On August 18, the module was in a 306 × 330-km orbit when the reentry capsule was jettisoned for landing in the U.S.S.R. The next day, the module changed its orbit to 312 × 318 km, and after four small maneuvers over the next two weeks, the orbit was raised to 329 km and 90.95 minutes. Between August 16 and August 20, telemetry, which had been monitored on 166 MHz, ended. This was probably associated with the reentry module separation. The telemetry was initiated by Soviet ground stations as the module passed overhead. The module then sent real time data followed by recorded data from the modules systems.

Four small maneuvers occurred from November 28 to December 19, and on December 19, the orbit was raised to 438 × 447 km. On February 1, on orbit 3,146, the module made a retroburn to lower the orbit to 336 × 437 km. On February 2, the module deorbited and made a destructive reentry over the Pacific.[19]

Kosmos 997/998A

Failed

Launched: August 5, 1977, 2:18? a.m.
Landed: —
Altitude: ?

It was reported that a Proton booster launch failed on August 5, carrying a dual payload similar to the Kosmos 881/882 flight. The launch time is estimated from the other dual Kosmos flights. These series of flights were probably tests related to the small shuttle program that was later canceled.[20–22]

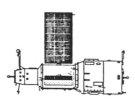

Salyut 6

Second-generation Station—
Manned 676 Days
Launched: September 29, 1977, 9:50 a.m.
Reentry: July 29, 1982
Altitude: 214 × 256 km @ 51.6°

Salyut 6 was similar to previous Salyuts, except that it had two docking ports, one forward as in earlier civilian Salyuts and one aft, which was equipped with propellant line connections for station refueling by Progress transports. Salyut 6's weight at launch was about 18,900 kg (see Figure 5-3).

The station was equipped with many experiments and many more would be added over the years. Normal operational altitude for the Salyut 6 type stations was 300 to 400 km. The operational lifetime was planned for 22 months, and it far exceeded that. The extended mission plan, using Progress transports, was expected to be five years.[23] The station was occupied by long-duration crews that were visited by short-duration crews and Progress transports.

After separating from the Proton booster third stage in orbit at 209 × 225 km, Salyut 6 used a transfer orbit of 348 × 229 km. The Protons third stage entered on October 5. By October 7, Salyut 6 was in a 336 × 352-km orbit, ready to receive the first manned mission.

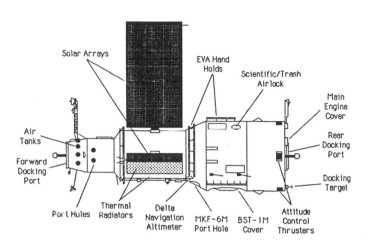

Figure 5-3. The Salyut 6 was the first second-generation space station. The new features included a second docking port on the rear of the station and a new refuelable unified propellant propulsion system. The second docking port allowed refueling by Progress cargo ships allowing missions of indefinite length.

Salyut 6 Space Station

Salyut 6 incorporated the features of both the military and civilian Salyut stations that preceded it. Taken from the military design was a rear docking port, and taken from the civilian stations were their larger solar panels and the forward transfer compartment. The length of Salyut 6 was 13.5 meters not including antennas (14.4 meters inclusive). The large end of the work compartment was 4.15 meters in diameter and 2.7 meters long; the small section was 2.9 meters in diameter and 3.5 meters long and a cone connecting the two was 1.2 meters long. Internal volume was 90 cubic meters. The solar arrays deployed to 17 meters maximum diameter. The transfer compartment was 3 meters long and 2 meters in diameter.

The station's work section was very similar to Salyut 4. In the forward section were the main control panels for station systems, communication equipment, and attitude control stick, and the optical sight. Forward of the consoles was the astro-navigation post from which the station could be oriented using instruments in 2 portholes. Forward of the control posts was the galley with a table with food heating units and a water dispenser. To the left of the galley was the station's computer. Aft of the main controls were the

(continued on next page)

(continued from page 169)

medical instruments, hand-held cameras, treadmill, ergometer, shower, and the Chibis suit. The treadmill was a belt type (90 by 40 cm), and elastic belts provided 50 kg down force, although at the high force levels the belts could hurt the runners. The crew carried personal radiation dosimeters during any flight. The spacecraft also transmitted information on radiation levels from other sensors. On the floor was the MKF-6M multispectral camera fitted to a porthole. Also in the aft work section were controls for experiments and communications, solar array controls, telemetry systems, and the telescope housing containing the Yelana telescope. On the aft end, at the ceiling, were two scientific and trash airlocks, food storage, water storage, a vacuum cleaner, dust filters, and linens. The toilet was in a recessed area just forward of the hatch to the intermediate section.

The intermediate compartment was smaller than the transfer compartment, but one person could fit into it if necessary and it could function as an airlock. It was mainly used as a scientific airlock for experiments like the Kristall furnace, which was installed in the compartment to keep radiated heat out of the living section of the station.

The main propulsion and attitude control systems were both supplied with propellant from three UDMH and three nitrogen tetraoxide tanks by a pressure feed system. The station was equipped with 14-kg thrust attitude control thrusters, located in clusters of eight at four locations 90° apart around the equipment section at the rear. The main engines, both a primary and a reserve, were located on both sides of the docking port. They both provided 300 kg of thrust and were normally covered by movable caps.

Soyuz 25

Aborted

Launched: October 9, 1977, 5:40 a.m.
Landed: October 11, 1977, 6:25 a.m.
Altitude: 194 × 240 km @ 51.6°
Crew: Vladimir Kovalyonok and
             Valeriy Ryumin
Backup: Yuri Romanenko and
             Aleksandr Ivanchenko
Call sign: Foton

Soyuz 25 was launched from the same launch pad as Sputnik into low clouds at 5:40 a.m. Television coverage in the U.S.S.R. showed the lift-off and the crew riding in the reentry module during launch. The transfer orbit to Salyut 6 was 265 × 309 km, and the Soyuz was initially 27 minutes, and 13,500 km behind Salyut 6. On the 17th orbit, the Soyuz was ready to dock at 339 × 352 km.

Kovalyonok took manual control at 120 meters distance and attempted to dock with the Salyut.[24] The Soyuz contacted the station's forward port but failed to hard dock at 7:09 a.m., October 10.[25] At 8:50 a.m., the cosmonauts reported to mission control that they had tried docking four times; but that the "force was not sufficient and the contact light did not come on."[26] The cosmonauts were then told to wait while ground controllers considered the problem. At the time, the Soyuz was soft docked to the Salyut with only the docking system probe latched onto the Salyut docking drogue. The docking probe normally retracts, pulling the spacecraft together. In this case, the force of the contraction was not enough to make the docking collar latches engage, or the sensors malfunctioned. The Soyuz remained soft docked while the controllers attempted to resolve the problem. On orbit 20, Kovalenko undocked the Soyuz, and kept with the Salyut until orbit 23 when he tried to dock again. The attempt also failed and the cosmonauts began preparations returning to Earth the next day.

Mission rules dictated immediate return to Earth because the Soyuz Ferry had only two days' battery power and no solar arrays. They also had no propellant left to try to dock at the aft port, and they probably suspected the problem was in the Soyuz probe. The capsule landed 185 km northwest of Tselinograd, at 6:25 a.m., October 11.[27,28]

Soyuz 26

Long-Duration Crew—96 Days
Launched: December 10, 1977, 4:19 a.m.
Landed: January 16, 1978, 2:25 p.m.
Altitude: 195 × 235 km @ 51.6°
Crew: Yuri Romanenko and Georgiy Grechko
Backup: Vladimir Kovalyonok and Aleksandr Ivanchenko
Call sign: Tamyr

On November 28, Salyut 6 boosted its orbit to 345 × 360 km in preparation for the Soyuz 26 launch. Fearing the forward docking port might be unusable or might have been damaged after the Soyuz 25 docking attempt, the Soyuz 26 spacecraft was to dock at the aft port. The launch of Soyuz 26 on December 10 was nearly delayed due to bad weather in the Atlantic Ocean, where the tracking ship *Cosmonaut Yuri Gagarin* was stationed. The ship's large dish antennas could only be stabilized in winds under than 24 knots. Winds of 40 knots forced the captain to lock the antenna into position, and use the communication equipment's power to overcome the loss of pointing ability. The ship was scheduled to be used for relay of the Soyuz docking operations, which were especially critical after the recent Soyuz 25 docking failure. The launch proceeded and the ship was able to perform the relay operation.[29]

The transfer orbit from initial orbit to Salyut 6 was 251 × 321 km. Docking was accomplished under manual control from a few meters distance, at 6:02 a.m., December 11, at 337 × 354 km (see Figure 5-4). The cosmonauts spent three hours verifying seals with the station before opening the hatches. For the mission, the tracking and communication ships *Komarov* and *Gagarin* were used along with a Molniya satellite to communicate with mission control.

By December 13, the station was demothballed and ready for operation, and the Soyuz was deactivated for in orbit storage. On the 14th, the crew gave a tour of the station using the Salyut's portable television camera. The cosmonauts worked a normal daily schedule based on Moscow time that

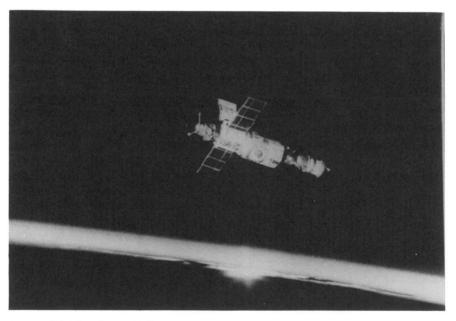

Figure 5-4. This photo of Salyut 6 was taken by a departing Soyuz. The forward sections of the Salyut are covered by dark (green) insulation blankets while the rear is painted orange to aid visibility. The middle section of the station is covered by white thermal radiators. (Source: Tass from Sovfoto.)

allowed for 2.5 to 3 hours exercise, 8 hours work and 2.5 hours for meals per day. They got a day off from official work every 5 or 6 days. The crew later reported that during the first days of the mission they got little sleep, probably because of the heavy work load of activating the station.[30]

On December 20, at 12:36 a.m., the crew performed an EVA to inspect the forward docking port to access any possible damage that might have prevented the Soyuz 25 docking. The station was in a 335 × 365-km orbit at the time and in the day-time portion of the orbit. Only Grechko left the Salyut and Romanenko stayed in the airlock supporting him (see box). During his work, Grechko used the portable color television camera to show mission control the stations exterior, and reported that there was no damage to the Salyut's docking drogue. During the EVA, Grechko also placed the Medusa materials exposure experiment on the transfer compartment of the station to be retrieved by a later crew. The EVA had been practiced in the Star City hydrobasin water tank and in weightlessness training flights on aircraft.[31,32]

Salyut Spacesuits

Salyut 6's transfer compartment/airlock contained two new-type EVA spacesuits described as semi-rigid. Each suit provided 3.5 hours independent life support separate from the station. An inner cooling garment was worn under the spacesuit along with a sensor belt that allowed mission control to monitor the cosmonaut's pulse rate, temperature, respiration, and electrocardiogram in real time. Body temperature was monitored by a sensor placed behind the ear. The suit was water cooled with manual temperature control. To prevent decompression sickness there was a period of prebreathing in the airlock. The suit's atmosphere was almost pure oxygen at 280 to 300 mm of pressure.

The suit had an entrance hatch covering the back of it that also served as the container of the life support systems. Once in the suit the cosmonaut closed the rear door with a lever on the chest of the suit. The Soviets claimed that a cosmonaut could touch his hands behind his back wearing the suit, which is a great achievement in spacesuit design, although this probably required lowering the suit pressure much below normal. The suits could be more or less adjusted to fit any potential crew member by using straps to take up or let down the sleeves and legs. A tether was almost always used to tie the cosmonaut to the space station during EVA's. An electrical connection was also maintained by the tether for communications or power, but only used as a backup.

(continued on next page)

Salyut 6 Experiments and Equipment

Medusa: Exposed materials to space on hull of station.

BST-1M: 1.5 meter infrared submillimeter telescope.

Kristall: Two versions of the furnace were used on Salyut 6 for semiconductor production.

MKF-6M: East German camera for Earth photography. Imaged in 6 bands with 10–30 meter resolution.

KATE-140: Mapping camera, could photograph 450×450-km areas on 600-frame film cassettes.

SPLAV: "Alloy" furnace, could heat samples from $650°$ to $990°C$ with hot, cold, and gradient heating.

Isparitel: An electron gun that could cause 1-mm thick deposits on metal, glass, or plastics of vaporized sample materials.

Yelena-f: Portable gamma ray detector to measure background gamma ray flux.

Chibis: Lower body low-pressure suit for use during work and exercise.

Penguin: Suit with elastic bands to simulate the force normally needed to remain upright on Earth.

Biogravistat: Plant growth unit using one disc as a control and another rotating to produce artificial gravity.

Svetobloc: Plant growth unit.

Oasis: Plant growth unit.

Malakhit: Plant growth unit.

Polynom-2: Electrocardiograph

Beta: Combination instrument to record electrocardiograms, lung capacity, and seismographic action of the heart.

Rheograph: Measured blood flow in the head, arms, legs, and torso.

VPA-1: 10 MeV radiation detector.

VTL-3: Treadmill.

Spektr-15: Bulgarian Earth resources spectrograph recorded in 15 bands

After Grechko was back in the airlock, shortly before local sunset, Romanenko said he wanted a look outside, so Grechko moved over in the airlock to let him through the small hatch. Romanenko pushed out hard against the airlock and lost his grip on the station, and started to thrash about. Grechko then saw that Romanenko's safety line was unfastened. As Romanenko drifted out of reach of the station, Grechko caught the line and pulled him back to the station. The crew did not reveal this event to mission control until the flight debriefing after landing.[33,34] Grechko spent only 20 minutes outside the station, but total EVA time was 88 minutes. The crew spent 30 of those minutes in the airlock when the air-pressure indicator read there was no air refilling the airlock. They waited until they were in communication with mission control to consult with them on the problem. After determining that the valves had to be working, they assumed the pressure gauge failed, which was correct.[35]

The crew then settled down to routine work including medical tests, exercise, biological experiments, technical experiments, and observing ocean currents and noctilucent clouds. The crew switched on a backup voice transmitter when one of the four normally used broke down. This was one of the few anomalies of the flight.

On December 21, the crew made Earth observations including viewing glaciers, snow cover of the U.S.S.R., the oceans, and forest fires in Africa. The next day, they started biological experiments on specimens brought from Earth including tadpoles, algae, and plants. The next days were devoted to medical checks and exercise. On December 25, they tested the Delta automatic navigation system, which kept track of the station's orbital position with only periodic updates from the crew. The crew used the Soyuz to boost the orbit of the station to 344 × 371 km, on December 29.

Through January, the crew continued astrophysical, biological, and Earth resources experiments. On January 3, the crew requested more work to be scheduled since work activating and setting up the space station was ending. The cosmonauts had also finished adapting to weightlessness and were working more easily.[36]

By the middle of January, the station would need refueling by Progress transport, which needed to dock at the aft port to use the propellant line connections, but Soyuz 26 was docked to the aft port. A mission to dock a Soyuz to the forward port would free the aft port, because the short-duration crew would depart in the Soyuz 26. This was Soyuz 27's mission.

Soyuz 27

First Triple Docking

Launched: January 10, 1978,
3:26 p.m.
Landed: March 16, 1978,
2:19 p.m.
Altitude: 190 × 237 km @ 51.6°
Crew: Vladimir Dzhanibekov and Oleg Makarov
Backup: Vladimir Kovalyonok and Aleksandr Ivanchenko
Call sign: Pamir

Soyuz 27 was a resupply mission and a final test for the forward docking port, after the Soyuz 25 docking failure. It also provided a fresh spacecraft for the long-duration crew, and by docking to the forward port and leaving in the ship at the aft port, the aft port would be open to receive a Progress transport.

Soyuz 27 was launched 17 minutes after Salyut 6 passed overhead of the Baykonur Cosmodrome. Soyuz 27's transfer orbit to the station was 241 × 304 km. Dzhanibekov first saw the station at 1.5 km distance, while the Soyuz continued its automatic approach. At 300 meters, the Salyut was again sighted with its running lights easily visible. The Soyuz docked at 5:06 p.m., January 11, at 330 × 350 km, as the station passed into Earth's shadow.[37] At mission control, there was a sigh of relief when the docking was successful. Feoktistov said that there was some concern that the force of the docking might jolt lose the docking latches of the Soyuz 26 on the other end of the station. As a precaution, during the docking, Grechko and Romanenko had moved into Soyuz 26 and closed the hatches in case of a docking accident.[38,39]

After the usual three-hour wait, checking the docking system, and shutting down Soyuz systems, the Soyuz 27 crew entered the station. They delivered to the station supplies including food, film, books, letters, newspapers, equipment, and the French Cytos biological experiment. Cytos was a biological experiment in which micro-organisms were delivered to the station cooled to inhibit growth and heated on the station to allow growth in weightlessness. The cells were again cooled before landing to preserve their state.

While on the station, Dzhanibekov, an electronics specialist, inspected the Salyut's electrical systems. The next day, the station was in an orbit at 334 × 367 km and the crews performed Earth resources work and made television reports. On January 13, the crews exchanged seat liners in preparation for the Soyuz 27 crew's return in the Soyuz 26 spacecraft. Trading spacecraft was necessary since the Soyuz spacecraft had a limited stay time in orbit. After thousands of freeze-thaw cycles caused by passing through intense sunlight and shadow on every orbit, the spacecraft systems degrade, especially the engine systems and propellant tanks that can leak small amounts of highly toxic propellant through propellant line seals. For safety reasons, the Soyuz manned transports were replaced regularly. The only mechanical changes to trade spacecraft were to exchange seat liners, which are formed for individual cosmonauts, and to exchange centering weights. The weights were essential to ensure that the capsule's center of gravity was near expected. Changes of the center of gravity could have a significant effect resulting in a undershoot or overshoot. The cosmonauts' weight and the weight of the experiment results must be considered when positioning the centering weights, which are used to adjust the center of mass of the capsule.[40]

The crews also perform Resonance tests of the docked vehicles to monitor possible dangerous resonances of the combined structure. The Resonance experiments involved having the crew jump up and down on the treadmill to the beat of a prerecorded clock rate. While testing at different rates, the resulting vibrations were recorded at various points in the station. Unexpected resonances could at least disturb experiments and at worst cause a complete structural failure of the space station. The Pamir crew undocked in Soyuz 26 at 11:05 a.m., January 16, and landed 310 km west of Tselinograd at 2:24 p.m. The capsule also returned film and data to Earth from the long-duration crew. In addition to trading Soyuz, the visiting flight also was beneficial for the long-duration crew. After spending hundreds of days in close quarters with the same person, the cosmonauts like the added variety a visiting crew provides and say it is a positive psychological factor. But, the visiting crews also throw the regular schedule off pace.[41-43]

☆ ☆ ☆ ☆ ☆

On January 18 and 19, the long-duration crew continued Earth resources work using the Salyut's MKF-6M. The MKF-6M camera was a modified version of the multispectral camera carried on Soyuz 22. It weighed 175 kg and used a film cartridge with 1,200 frames of film recording a different part of the spectrum. Only two of the large, 13-kg film cartridges could be carried in the Soyuz capsule during return to Earth. The altitude and time of the photos were also recorded on the film as it was exposed. To analyze the photos on the ground, a projector called the MSP-4 was used to view four of the pieces of the spectrum together.[44] Meanwhile, as the January launch window to Salyut 6 was ending, the first Progress flight was being prepared at Baykonur to deliver cargo to the station that would allow the record-breaking mission attempt to continue.

Progress 1

First Automatic Transport

Launched: January 20, 1978,
                11:25 a.m.
Reentry: February 8, 1978,
                5:54 a.m.
Altitude: 173 × 256 km @ 51.6°

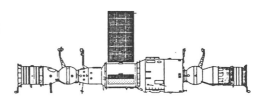

The Progress spacecraft was an unmanned cargo ship designed to refuel Salyut stations and resupply cosmonauts with air, water, food, and other dry goods and experiments. The Progress was based on the same design as the Soyuz. It used a lengthened Soyuz T-type service module, with tankage instead of a reentry module, and an orbital module that was equipped with racks to hold bulk cargo (see Figure 5-5).

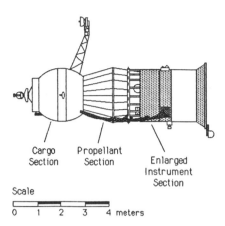

Figure 5-5. The Progress was a modified version of the Soyuz spacecraft. The Soyuz reentry module was replaced on the Progress by an unpressurized tankage section, and refueling connections were made through the docking collar.

Progress Spacecraft

The Progress was a highly modified version of the Soyuz, used to automatically deliver cargo to the space stations. The Soyuz design was altered by replacing the capsule with a tankage section. The Progress usually weighed about 7,020 kg at launch depending on the amount of cargo. Payload capacity was 2,300 kg, of which about 998 kg was the maximum weight of propellants, and 1,287 kg was the maximum weights of orbital module cargo. The Progress was 7.94 meters long and 2.7 meters in diameter at the base. Internal volume of the cargo module was 6.6 cubic meters.

The Progress docked only at the aft space station port because it was the only port with propellant line connections. Propellant was transferred from the Progress to the station by closing off the station's tanks and using a pump to lower the pressure of the tanks from an operational pressure of 220 atm to about 3 atm by extracting nitrogen from the tanks flexible bellows. This took several hours because the pumping consumed about 1 kW of power and the Salyut's batteries had to be recharged periodically during the process. The Progress tanks were pressurized to 8 atm using its own nitrogen supply. When the valves were opened, most of the propellant flowed to the station's tanks,

(continued on next page)

which were then closed off refilled. The lines in the docking collar were then purged with nitrogen after refueling, to prevent the volatile fuel and oxidizer residue from spilling out onto the station after undocking.

The Progress service module used a highly modified version of the Salyut OUD pressure-feed engine system. The Progress was also used to test the new engine system before its use on Soyuz T flights, which used the same system. The instrument section of the service module was lengthened and equipment for automatic docking was apparently installed in this additional space. The Progress carried twice as much instrumentation for rendezvous and docking as the Soyuz Ferry.

A feature of the Progress (and probably the Soyuz and Kosmos spacecraft) was the ability to remove the entire docking hatch and probe from its mounting to facilitate movement of cargo. Many items like air regeneration tanks, clothes, and human wastes were disposed of using Progress ships that burn up on reentry. When a Progress was not available, the cosmonauts put the small trash in bags and put it out through the scientific airlocks. These bags fell into the atmosphere after about a month, and burned up upon reentry.

After undocking, the Progress drifted away from the station and then lowered its orbit to about 200 km. Only when the spacecraft passed over the U.S.S.R. was the command given for retrofire. During reentry over the Pacific, the entire vehicle was destroyed, or at least most of it.

The transfer orbit to Salyut 6 was 250 × 334 km. The cosmonauts prepared for the docking on January 21 and 22. Since the Progress was unmanned, the crew did not retreat to the Soyuz during the docking as when the Soyuz 27 docked, they instead manned the station's controls ready to maneuver away from the approaching Progress in case of a malfunction. The Soviets allowed this change from the Soyuz 27 docking because the Progress was unmanned and expendable if damaged by the rocket exhaust of the Salyut in an emergency separation maneuver.

The crew spotted Progress 1 at 10 km distance as it performed a fully automatic approach. Progress 1 docked automatically at 1:12 p.m., January 22, at 329 × 348 km. It carried 1,000 kg of propellant and 1,300 kg of supplies (see Figure 5-6). Among the supplies delivered were some replacement parts for vital systems, clothes, air-regeneration units, linens, the Splav materials processing furnace, new orientation and movement control sensors, heavily stressed Penguin suits, replacement air for that lost during the EVA, and food, including apples, garlic, and onions[45] (see box).

Figure 5-6. A Progress mockup is shown here docked to a Salyut mockup. The Progress has no solar panels and has a lengthened service module to contain the automatic flight controls. (Source: Tass from Sovfoto.)

Soviet Space Food

Food on Soviet spacecraft has remained about the same for many years. Most food is freeze dried and prepared using hot water. Five different types of bread, fresh vegetables, honey and freeze-dried strawberries are routinely delivered by cargo spacecraft. The food reportedly tastes good on Earth, but loses some of its taste in space, so for supply flights the crews usually request onions, garlic, and other tasty foods to spice up their bland meals.

The menu rotated every 6 days and provided each cosmonaut with 3,100 calories per day. A supplement consisting of 10 vitamins, methionine, and glutamic acid was taken every 15 days. On Salyut 7 and later on Mir, cosmonauts had 70 food items to choose from and they could choose which prepared meals they wanted as long as the required caloric intake of up to 3,100 calories was met. The menu included 20 meats, fish, poultry, potatoes, peas, cabbage, porridge, and cheese in 100-mg cans, 10 pureed second courses, and 14 dehydrated first and second courses including nuts, yogurt, and acidophilus paste. Drinks included powdered milk, tea, coffee, and fruit juices. The crew used around 1.7 liters per day per person of hot water for coffee, tea, and rehydrating food averages. Use of cold water was only about 0.1 liters per day per person. The cosmonauts also drank a preparation of eleuthera bark that was believed to have medicinal value. At least on missions to the later Salyut stations, the cosmonauts were allowed small quantities of vodka. The Soviets usually ate four meals a day. A typical day's menu included:

Breakfast: Pork with sweet pepper, Russian cheese, honey cake, prunes, and coffee.
Lunch: Jellied beef tongue, praline candies, and cherry juice.
Dinner: Ham, borsche with smoked foods, tallin beef, potatoes, cookies with cheese, and apple juice.
Supper: Cottage cheese with nuts, assorted meats, wheat bread, plum and cherry dessert, and tea.

The crew started unloading the Progress January 23, and on January 29, they released air from the Progress to replenish the Salyut's air. In preparation for refueling, they carefully checked the propellant lines for days. The Salyut had three sets of oxidizer and fuel tanks, of which one set was still full from launch. The Progress could refill any two sets of tanks. Preparation for oxidizer transfer was carried out on January 26, and they installed new air-regeneration units in the Salyut environmental control system.[46] On February 2 and 3, the refueling operation was completed by transferring fuel into the Salyut tanks. Later in the day, oxidizer transfer was completed by 12:00 p.m.

On February 5, the refueling lines were purged with the Progress nitrogen supply to prevent spillage of the toxic propellants when undocking. The Progress was then filled with used cloths, water and food containers, air-regeneration units, and other used equipment to be destroyed. The cosmonauts were careful not to shift the center of gravity of the Progress much in loading the trash since the Progress engine was not designed to gimbal to account for a changing center of gravity. The Progress was used to make orbit adjustments to the space station on February 5 and 6.

The ship was undocked by cosmonaut command from the Salyut at 8:53 a.m., February 6, and was allowed to drift at least 13 km away. From this distance, Progress 1 made a second approach using back up docking systems. The spacecraft then drifted away from the station and performed retrofire at 5:39 a.m., making a destructive reentry at about 5:45 a.m., February 8, over the southern Pacific.[47]

After the Progress departed, the Tamyr crew started work with the MKF-6M camera, and they installed the 23-kg Splav-01 in the intermediate compartment airlock. The temperature of the air in the station was limited to 40°C, and the Splav was put in the airlock so the heat it generated could be dumped overboard. The Splav weighed 23 kg, consumed 300 watts of power, and could heat samples from 650° to 990°C with linear heating by computer control to an accuracy of 10°C. Samples measured 170 mm long and 20.6 mm in diameter.[48,49] During use of the Splav, the space station was allowed to drift without attitude control adjustments, improving the micro-gravity environment on the station. Normal attitude control thruster firings would create small disturbances that would lower the quality of the materials undergoing processing. Grechko noticed that some of the samples were degraded by slight movements of the space station despite the precautions. By the middle of February, the first samples were being processed in the Splav. The alloy samples produced included copper/indium, indium/antimonide, aluminum/tungsten, molybdenum/gallium and aluminum/antimony.

In late February, the crew experimented extensively with the BST-1M telescope and its cooling unit. The BST-1M's sensors were cooled to −273°C by a helium cooling system.[50] By February 25, the crew was making periodic observations with the BST-1M of Earth's atmosphere, the Orion nebula, and other celestial objects. The BST-1M observations of Earth's atmosphere studied the amount of water vapor and ozone in the atmosphere. The Soviets reported that the BST-1M required three cosmonauts' attention for best operation. Obviously, the Soviets hoped to introduce the improved Soyuz T spacecraft to carry three person crews to Salyut 6 sooner than actually occurred.[51] The crew also continued normal Earth observations, exercise, and routine medical checks. The Salyut maneuvered to 332 × 347 km on February 20, and to 335 × 364 km on February 23, in preparation for a launch on February 27.

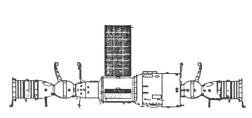

Soyuz 28

First International
Crew—Czechoslovakia
Launched: March 2, 1978,
6:28 p.m.
Landed: March 10, 1978,
4:44 p.m.
Altitude: 192 × 246 km @ 51.6°
Crew: Alesksei Gubarev and
Vladimir Remek
Backup: Nikolay Rukavishnikov
and Oldrich Pelczak
Call sign: Zenit

This mission was the first of the Intercosmonaut Program in which military pilots from Soviet bloc nations, and other countries later, were flown on short missions to Soviet space stations. The program was initiated as a reaction to U.S. plans to fly European researchers on space shuttle flights in the 1980s.

The team of Gubarev and Remek was chosen to make the flight one month before the launch after the two crews underwent final examinations and tests (see Figure 5-7). The launch was delayed for three days for unspecified reasons. Once launched, Soyuz 28's transfer orbit to Salyut 6 was 251 × 306 km. The Soyuz docked at 8:10 p.m., March 3, at 334 × 353 km. The crew opened the hatch to the Salyut after the usual three-hour checks of the docking system. The flight delivered mail, food, and other consumables. Soyuz 28 also delivered pictures of Earth taken by the long-duration Tamyr crew, which were returned to Earth by the Soyuz 27 crew and developed. These were to aid the crew in determining where additional photography was needed.[52]

A day after docking, the crews began performing Czechoslovakian experiments including Chorella, which monitored the growth of Chlorella seaweed

Figure 5-7. The Salyut 6 space station was about the same size as the previous stations, but it had improved systems and facilities. Here, the first international crew of Alesksei Gubarev and Czech cosmonaut Vladimir Remek are eating at Salyut 6's dining table. The station control panels are behind the cosmonauts and the hatch to the transfer compartment is open in the background. (Source: Tass from Sovfoto.)

in weightlessness; Morava, which used the Splav furnace to melt glass, lead, silver and copper chlorides; and Oxymeter, which measured oxygen in human tissue. The same day they participated in a press conference with Soviet and Soviet bloc journalists. On March 6, the crew continued with experiments including Extinctia, which measured high-altitude atmospheric dust by observing star light through the horizon.

By March 7, operations with the Splav were completed. The crew then began use of the Chibis lower-body, negative-pressure suit and the Polynom-2M cardiovascular monitoring unit. The Chibis suit lowered the pressure by 10 to 40 mm Hg on the lower body to put stress on the cardiovascular system, mimicking gravity's pull on the body's blood. An advantage over the similar U.S. Skylab Lower Body Negative Pressure device was that the Chibis allowed some mobility during its use. Polynom-2M was a electrocardiograph that could measure regular cardiograms, cardiograms of the major blood vessels, and the phase of the cardiographic cycle.[53] On March 8, the combined crews had a day of rest, entertainment, and direct television talks with their families. On March 8, the station was in orbit at 338×357 km.

During their stay, the joint crews celebrated the Tamyr crew's breaking of the U.S. Skylab record of 84 days in space, set in 1974. On March 10, the visiting crew prepared the Soyuz 28 for return to Earth, packing experiment results and testing spacecraft systems. On this mission, the crews did not exchange Soyuz Ferry spacecraft since the long-duration crew was scheduled to land soon. Soyuz 28 undocked around 1:25 p.m. the same day, over the western U.S.S.R. After reentry, the capsule was quickly spotted by helicopters during the descent. It landed at 4:44 p.m., shortly after local sunset, March 10, 310 km west of Tselinograd in light snow cover.[54]

☆ ☆ ☆ ☆ ☆

During the stay of the visiting crew, one of the Soyuz 28 crew relayed a message from mission controllers to Romanenko that Grechko's father had died. Due to the sensitive nature of the information, Romanenko was told to use his best judgment whether or not to tell Grechko. He decided to wait until the end of the mission.[55] During the Soyuz 28 mission the Tamyr crew had increased the amount of daily exercise in preparation for their upcoming landing.

On March 9, the long-duration crew's exercise schedule was increased from one to three hours in the Chibis suits to ten to 12 hours a day. On March 11 and 12, medical examinations and exercise with the Chibis suits was performed. On the 13th, they began deactivating Salyut systems while continuing some Earth resources work.

On March 14 and 15, the crew intensified exercise again. As in the first days of the mission, the crew got little sleep during the closing days of the flight, probably because of the intense preparation for returning to Earth, and setting the Salyut into automatic flight mode. On the 15th, the Soyuz's engines were tested and loading of experiment results was completed. They undocked from Salyut 6 at 11:00 a.m., March 16. After setting a new endurance record of 96 days, the Soyuz 26 crew returned to Earth in Soyuz 27, at 2:19 p.m., landing in the snow, 265 km west of Tselinograd. The landing was shown later on Soviet television.[56]

After landing, the crew was given immediate medical checks. The Soviets announced that medical examinations had indicated that Greckho's heart had changed position slightly during the flight, and after returning to Earth he experienced some chest pains as his body returned to normal. The crew was then put into isolation at Baykonur to guard against catching any illness in their weakened state of recovery, and the crew debriefing began.

During their initial recovery, the cosmonauts commented on the heaviness of their bodies and other objects. They had lost four kilograms of weight on average during the mission. Leg volume decreased by 16.8 and 9.2% due mostly to loss of muscle from disuse. It was later stated that the crew did not fully follow the recommended program of exercises. It was later noted by James

Oberg that Romanenko became argumentative and anxious during the flight. Romanenko also was said to have taken large doses of pain killers for a toothache without telling mission controllers for two weeks during the flight.[57,58]

After three days of recovery on Earth, the crew adapted to all aspects of life slowly. Both men still tried to swim out of their beds when they woke up, as they had for the 96 days of the mission. By the fourth day, both had regained half of the lost weight and walked through a nearby park. After one week, they still felt weak and unsteady.[59] After ten days, the cosmonauts said they felt well, but they continued to wear inflatable leggings to restrict blood flow into their lower body. After two weeks, they were fully recovered from the immediate effects of the flight. Their recovery took longer than any previous mission. They returned to Star City on April 22, accompanied by the Soyuz 28 crew, which landed six days before them.[60] They were given the standard awards for a successful mission.

The crew's mission report was started during the recovery a month before leaving Baykonur. Over the course of the flight they had conducted more than 100 experiments and observed many objects with the Salyut's telescopes including Mars, Venus, various stars, and nebulae. Testing of the Salyut's water regeneration system unexpectedly showed the system losing water over the course of the mission. The cause was determined to be that the wood paneling in the Salyut was absorbing the water from the air, because it was installed during a hot, dry summer in 1977, at the Proton assembly building at Baykonur.[61]

Kosmos 997/998B

Launched: March 30, 1978, 3:06 a.m.
Landed: March 30, 1978, 4:37 a.m.?
Altitude: 210 × 238 km @ 51.6°

This flight was the same as the Kosmos 881/882 flight in late 1976. The twin payloads separated from the Proton booster third stage in orbit at 188 × 213 km. The test was of two spacecraft for one orbit, ending in the same area as the 882/881 flight. Again, the landing was at about the same time and lighting conditions as the previous test.[62] These flights were probably tests related to the small shuttle program.

Kosmos 1001

Soyuz T Test Flight

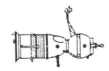

Launched: April 4, 1978, 6:07 p.m.
Landed: April 15, 1978, 3:00 p.m.
Altitude: 199 × 228 km @ 51.6°
Crew: none

This was the second Soyuz T test flight and was very similar to Kosmos 869. The initial orbit was 199 × 228 km, and after two days, it maneuvered to

196 × 228 km, and then before April 11, to 307 × 318 km. On April 15, the spacecraft jettisoned four objects, which probably included the orbital module. A new feature of the Soyuz T was jettisoning of the orbital module before retrofire, to lighten the spacecraft and conserve propellant needed for the de-orbit maneuver. The capsule made a normal reentry, landing after only 11 days.[63,64]

On April 21, Salyut 6 was in an orbit at 319 × 352 km. On May 16, the station raised its orbit, and by June 11, Salyut 6 was in orbit to 340 × 356 km in preparation for the Soyuz 29 launch.

Soyuz 29

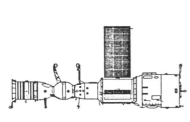

Second Long-Duration Crew— 140 Days

Launched: June 15, 1978, 11:16 p.m.
Landed: September 3, 1978, 2:40 p.m.
Altitude: 193 × 248 km @ 51.6°
Crew: Vladimir Kovalenok and
　　　　Aleksandr Ivanchenkov
Backup: Vladimir Lyakhov and
　　　　　Valeriy Ryumin
Call sign: Photon

This was the second long-duration crew to Salyut 6, which had been unmanned for three months after the record flight of Soyuz 26. After separating from the booster's upper stage, Soyuz 29's transfer orbit to the station was 258 × 309 km. At a distance of 22 km from the station, the Soyuz Igla automatic docking system was switched on, and the crew docked at 12:58 a.m., June 17. Kovalenok became the first man to fly to the same space station (see Soyuz 25). The cosmonauts waited until passing into a communications zone to televise entering the station to mission control. For the initial days of activating of the station, the crew work schedule was light to avoid overworking the crew as was done during the Soyuz 26 mission. The crew also wore Penguin elastic exercise suits during the first days on the Salyut for up to 16 hours at a time.

On June 19, the station was in a 338 × 368 km orbit. By this time, the cosmonauts had completed their initial adaptation to weightlessness. They also weighed themselves for the first time using the station's mass meter. Mass measurements were taken almost daily by the second week of the mission, but then decreased to only once every two weeks. For the remainder of the mission, the crew followed a normal Earth work-week schedule, with Saturday and Sunday reserved for housekeeping and administration.[65,66]

By the 21st, they were completing demothballing the station, shutting down the Soyuz, and activating Salyut's experiments. The crew had completed checking the station's portholes, propulsion system, and control systems. They also repaired the airlock pressure indicator and replaced a ventilator on the Splav furnace. The Salyut was also maneuvered, boosting its orbit, under ground control for the first time with the cosmonauts on board only observing. At the time, the station was in an orbit exposing it to sunlight for 24 hours a day. This occurs twice a year when the plane of the station's orbit faces the sun, allowing sunlight to illuminate the normally shaded portion of the orbit by passing over either of Earth's poles.

The cosmonauts also began normal medical checks and prepared to use the MKF-6M. Earth observation during the all sunlight orbits enabled greater use of low angle of illumination of Earth's surface. During the observations, the cosmonauts reported seeing underwater ocean ridges punctuated by visible ocean currents. The crew, as a part of standard training, had been instructed on Tu-134 aircraft by geologists on recognizing important features to be photographed by the station's cameras.

On June 23, they tested the Kaskad station orientation system that was used to keep the station in gravity gradient mode for three days of materials processing experiments using the Splav to produce mercury telluride and cadmium telluride alloys. Medical checks continued during this work, along with documentation work and routine rest days. On June 26, the crew checked the BST-1M telescope cooling system and checked on plant growth experiments.[67]

Soyuz 30

Second International Crew—Poland
Launched: June 27, 1978, 6:27 p.m.
Landed: July 5, 1978, 4:30 p.m.
Altitude: 294 × 244 km @ 51.6°
Crew: Pyotr Klimuk and Miroslaw Hermaszewski
Backup: Valeri Kubasov and Zenon Jankowski
Call sign: Kavkas

This was the second international mission to Salyut 6. Before the flight, the crew was trained, as all international crews, in the use of the MKF-6M camera on a Tu-134 flying at 10,000 meters, in preparation for using the Salyut station's MKF-6M in photographing Poland and other countries.

The Soyuz 30's transfer orbit to Salyut 6 was 264 × 310 km. They docked at the aft docking port, at 8:08 p.m., June 29th, and opened the hatch to the Salyut three hours later, joining the long-duration crew. While on the station, the international crew was severely limited in their activities on the station. The long-duration crew's schedule could not be interrupted and on their rest days, the international crew had to stay in their Soyuz and perform experiments.[68]

During the flight, Hermaszewski conducted many experiments despite the restrictions, including 46- and 14-hour-long Sirena crystallization experiments, each producing 47 grams of cadmium tellurium mercury semiconductor for use in infra-red detectors in the Salyut's Splav furnace.[69] A yield of 50% was attained in the space experiment compared to 15% in comparable Soviet ground experiments.[70] He also used the MKF-6M to take pictures of Poland in coordination with aircraft taking close-up pictures as usual, but bad weather over Poland limited the planned photography sessions.

Hermaszewski also participated in routine medical experiments with the Polynom-2M, Beta, and Rheograph units. Beta was a combination instrument that recorded electrocardiograms, lung capacity, and seismographic action of the heart during exercise or use of the Chibis low-pressure suit. Rheograph measured blood flow in the head, arms, legs, and torso. It could also measure the volume changes in the areas measured.[71] During the joint mission, both crews tried the Smak taste experiment, which stimulated the taste buds with electric currents in an attempt to understand the reduced palatability of some foods in weightlessness.

The international crew packed their experiment results in the Soyuz 30 capsule and undocked from the station at 1:15 p.m., July 5. The capsule landed at 4:31 p.m. July 5, 300 km west of Tselinograd in a Rostov state farm field where they were greeted by farmers. They were flown back to Baykonur immediately for a press conference before returning the Star City.[72]

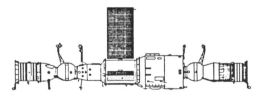

Progress 2

Launched: July 7, 1978,
2:26 p.m.
Reentry: August 4, 1978
Altitude: 182 × 234 km @ 51.6°

After a normal two-day rendezvous using a transfer orbit of 245 × 308 km, Progress 2 docked at 3:59 p.m., July 9. The flight delivered 50 days of supplies consisting of 200 liters of water, 250 kg of food, the Kristall materials processing furnace, 600 kg of propellant, Salyut air regenerators, Delta computer sub-systems, replacement parts, a new Globus panel, mail and raw materials for processing, 100 kg of film and a cardio-monitor device that could be worn by a cosmonaut to record cardiograms for 24 hours at a time. The Progress also provided air to replenish Salyut 6, after the July 29 EVA.

The cosmonauts began unloading on the 10th, and continued until the 17th. On the 12th, they replaced the Salyut air-regeneration canisters and preparation for refueling was underway. The refueling was done by ground control starting July 19. The Progress was loaded with used equipment, air regenerators and other trash. After undocking at 7:57 a.m., August 2, Progress 2 tested its rendezvous system again before performing retrofire at 4:32 a.m. It then made a destructive reentry over the Pacific, as was normal for all Progress flights.[73]

Following the evaluation of previous Resonance test data, the crew was told not to operate the Salyut treadmill at certain speeds, due to dangerous vibrations it could produce. On July 17, the cosmonauts tested the portable cardio-monitor continuously for 24 hours. At the same time, the crew experimented with making glass and gallium arsenide semiconductor material in the Kristall furnace. The Kristall furnace processed ampules of material 10 mm in diameter by moving them through a heating zone. This differed from the Splav furnace, which varied temperatures across the entire ampule of material. More samples of mercury telluride and cadmium telluride were processed on July 18. The crew also continued Earth resources photography and normal rest activities watching taped movies and talking with family and friends. On July 24, the Splav was used to process aluminum, tin, and molybdenum alloys. The next day was dedicated to medical exams.[74]

The crew complained of headaches for a few days until they discovered the carbon dioxide detectors had failed to alert them to change the air purifiers. This happened three times during the mission. Normal carbon dioxide level was 8.8 mm Hg. To cause headaches, the level probably reached about 62 mm Hg.[75,76]

On July 27, the cosmonauts rehearsed the upcoming EVA to the point of hatch opening. They tested the spacesuit's functions, adjusted them, and cleared some equipment out of the airlock to give them more space. Salyut 6's orbit was then 328×346 km.

The crew performed the EVA at 6:57 a.m., July 29, to retrieve materials from the Medusa experiment package on the exterior of the station, which included micro-meteoroid sensors, aluminum, titanium, steel, rubber, ceramics, glass, protective coatings, and radiation sensors. The crew's normal work schedule was changed to allow mission control to monitor preparation activities of the crew in the airlock. Grechko acted as the capcom during the EVA, being the only cosmonaut with recent EVA experience, from the Soyuz 26 EVA. Some of the Medusa samples to be retrieved had been placed on the outside of Salyut 6 by Grechko during that EVA, and the rest were in place at the launch of Salyut 6. A spotlight on the outside of the station allowed work in darkness, and mission control watched using the Salyut's portable color television camera held by Kovalenok.[77]

The EVA lasted two hours, during which they saw a meteor pass quickly below them that blinded them for a few seconds.[78] Ivanchenkov secured himself to the stations hull by "anchors," which were similar to ski bindings. Ivanchenkov used special tools to remove and replace the sample blocks and reported no large meteor hits on the 6 cm² micro-meteoroid sensor. Feoktistov said that later examination revealed two hundred small craters created by small orbital debris, which was much more than anticipated. Similar results were found by NASA on retrieved equipment from the Solar Maximum satellite in 1984. A large amount of the small debris was determined to be paint chips and aluminum particles (a propellant residue) from the solid rocket motors of some satellite upper stages. Ivanchenkov reported that working in space was difficult but interesting. They returned to the airlock and closed the hatch at 9:02 p.m. Five similar EVAs were planned at this time for Salyut 6 crews.[79-81]

The crew replaced air lost for the EVA, and periodic use of the scientific/trash airlocks, with air from Progress 2's tanks. On July 30, the crew rested and stowed equipment used for the EVA. The cosmonauts then continued their experiment program, using the Kristall to make germanium and indium antimonide crystals on August 4. They made medical checks the next day, and raised the station's orbit to 328 × 359 km.

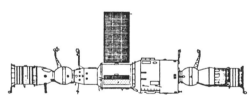

Progress 3

Launched: August 8, 1978,
1:31 a.m.
Reentry: August 23, 1978
Altitude: 190 × 232 km @ 51.6°

Progress 3 separated from its booster's upper stage, entered a transfer orbit of 243 × 249 km, and docked at 3:00 a.m., August 10. The docking was fully automatic with the cosmonauts observing. After docking, the Progress was used to boost the station's orbit to 244 × 262 km.

The flight delivered 280 kg of food including strawberries, onions, garlic, and milk; 450 kg of air, 190 liters of water, fur boots, alloys and semiconductor material for processing, film, letters, newspapers, biological and medical equipment, and Ivanchenkov's guitar, which he had requested. No propellant was carried by Progress 3 and unloading began shortly after docking. The Progress raised the station's orbit on August 17, to 343 × 359 km. Unloading was completed by August 19 and the Progress undocked August 21. Retrofire was at 8:30 p.m., August 23, sending the spacecraft into the atmosphere over the Pacific.[82,83]

On August 11, the crew put the station into gravity gradient stabilized flight for materials processing experiments using both the Kristall and Splav. Processed materials included gallium arsenide, tin, tellurium, aluminum-bismuth alloy, cadmium and zinc sulfide and selenides. On August 14 the crew continued unloading the Progress and continued experiments with the Kristall and Splav. On the 16th, the crew underwent medical checks using the Beta, Rheograph, and Polynom-2M. The next day, the continued material processing with silicon and sulfide compounds.[84]

Soyuz 31

Third International
Crew—East Germany
Launched: August 26, 1978,
        5:51 p.m.
Landed: November 2, 1978, 2:05 p.m.
Altitude: 193×243 km @ $51.6°$
Crew: Valeriy Bykovsky and Sigmund Jahn
Backup: Viktor Gorbatko and Eberhard Kollner
Call sign: Yastreb

Soyuz 31's transfer orbit to Salyut 6 was 256×322 km. Docking to the aft docking port occurred at 7:37 p.m., August 7, with the station in orbit at 339×354 km. Soyuz 31 delivered 4 kg of food including fresh onions, garlic, lemons, apples, milk, soup, honey, pork, Bulgarian peppers, and ginger bread.

The usual international flight medical and biology experiments, using the Polynom-2M, Beta, and Rheograph, were carried out by the crew. The audio experiment tested sound and noise perception limits. Berolina was a program to use the Splav materials processing furnace. In Berolina experiments, an ampule of bismuth and antimonide was processed with the material between two plates in the ampule. The resulting tree structure crystal was four to six times larger than obtained in similar experiments on the ground.[85] Other samples were of lead and tellurium, which were vaporized by the heat in one end of the ampule with a seed crystal in the cool end. Other experiments included Reporter, which tested using different films for photography in the station's interior; Vremya, which tested cosmonaut reaction times; and Syomka, which was a program for using the MKF-6M Earth resources camera.

The Soyuz 31 crew traded Soyuz spacecraft to provide a fresh return craft for the long-duration crew. The visiting crew tested Soyuz 29's main engine for five seconds on September 2. The crew also transferred 25 containers of experiment samples, totaling 100 various experiment results, to the capsule and packed a MKF-6M film cartridge containing photos taken by the crews. Finishing the exchange of the Soyuz spacecraft, the crews exchanged centering weights and the seat liners on September 3.

The Soyuz 31 crew returned in Soyuz 29, undocking at 11:23 a.m., September 3. Retrofire was at 1:52 p.m., with landing at 2:40 p.m., 140 km southeast of Dzhezkagan. The crew started climbing out of the capsule as the recovery forces reached the capsule two minutes after the landing. On September 11 the crew returned to Star City.[86]

☆ ☆ ☆ ☆ ☆

On September 7, after the return of the Soyuz 31 crew, the long-duration crew undocked the new Soyuz 31 at 1:53 p.m., and pulled away from the Salyut one to two hundred meters. The Salyut, then in gravity gradient mode, was commanded to activate its forward docking port. When the Soyuz activated its docking radar, the Salyut responded by turning around to point the forward port at the Soyuz. About an hour after undocking the Soyuz was now ready for redocking, but due to orbital mechanics it was now pointed 90° away from the station. After correcting this, the crew redocked at the forward port, clearing the way for a new Progress to refuel the station. The redocking operation was completed in one orbit.[87]

On September 8, the cosmonauts underwent routine medical exams and continued Earth observation and photography, materials processing, and astronomical observations. On September 15, they took their second shower on the station. The next day, they observed a lunar eclipse using the BST-1M telescope. They also reported seeing high-altitude clouds over the equator, which were first sighted by Grechko and Romanenko during their record-breaking mission. On September 22, they started operations using the Kristall, after putting the station gravity gradient stabilized mode.[88]

By October, the Photons had taken about 3,000 pictures and operated 50 experiments. On October 2, they performed tests of the station's guidance and control systems in manual and automatic modes. The next day they performed routine medical checks.

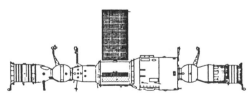

Progress 4

Launched: October 4, 1978, 2:09 a.m.
Reentry: October 26, 1978
Altitude: 185 × 247 km @ 51.6°

The final transfer orbit of Progress 4 after maneuvering from its initial orbit was 325 × 347 km. The Progress docked at 4:00 a.m., October 6. It delivered

1,300 kg of equipment including air-regeneration canisters, magazines, newspapers, replacement Penguin exercise suits, clothes, a tape recorder, music tapes, film, partitions for privacy, special luxury foods and fur-lined boots, which the crew requested. Among the 124 kg of food were various meats, strawberries, fresh milk, onions, garlic, curds and nuts. Also delivered were 176 kg of water, 1,000 kg of propellant, and 46 kg of air.[89] Ivanchenkov's wife also sent brandy-filled chocolates in an ordinary chocolate box so the officials at Baykonur would not notice. Normally, only small amounts of alcohol are allowed on the station. When opening the box, the small candies flew out and into the station. The crew spent two hours collecting them, when they discovered the brandy fillings.[90] Despite the special chocolates, the crew later reported that the luxury food items was not as good as normal fresh foods.

By October 11, the crew had unloaded most of the cargo and installed the new air-regenerator cartridges. Refueling was completed by October 13, and Progress 4 was used to change the Salyut's orbit with two burns on October 20 and 21, to 359 × 362 km. The orbital adjustment was made to adjust the return time of the long-duration crew. Progress 4 undocked at 4:07 p.m., October 24. Retrofire was at 7:28 p.m., August 26.[91]

☆ ☆ ☆ ☆ ☆

After the departure of the Progress, the crew resumed their normal work routine. They also observed another lunar eclipse on October 7. On October 12 the crew started using the Chibis low-pressure suits every four days to prepare for return to Earth. On October 17 and 18, the crew underwent extensive medical checks at the same time as conducting a Kristall processing experiment. On October 22, the Kristall experiments included testing processing lead telluride, cadmium sulfide, and cadmium mercury telluride samples in the Splav furnace.

They also continued use of the MKF-6M camera for agricultural data in conjunction with an AN-30 aircraft flying at over 20,000 ft. Plant growth experiments of the mission were coming to an end with plant samples of wheat, onions, tomatoes, garlic, carrots, and cucumbers.[92]

During their last month in orbit, the crew exercised about three hours a day to prepare for return to Earth. During the last week, they used the Chibis suit every other day to exercise their cardiovascular systems. On October 30, they transferred experiment results and samples of the station's air to the Soyuz 31 capsule, test fired the Soyuz 31's engines, and cleaned the station's interior. The next day they continued to set the Salyut up for unmanned flight and drank a saline solution to elevate their blood pressure for return to Earth. Their blood pressure was still lower than normal after landing.[93,94]

Undocking was at 10:46 a.m., November 2, with retrofire at 1:15 p.m. They landed 180 km southeast of Dzhezkagan at 2:05 p.m. The landing was covered live on Soviet television, showing the cosmonauts climbing out of the capsule and walking unaided for a short time. Despite this show of strength, they could not function normally for five days, but were mostly recovered within ten days, and completely recovered in 25. They were also the first people to have difficulty talking immediately after landing. They spent 139 days in orbit and had lost 3.1 kg of weight on average. Their hemoglobin and red blood cell count was also lower than pre-flight levels.[95–97]

To ease their recovery they tested wearing pants that restricted blood from pooling in their legs. Despite their slow recovery, their condition was still slightly better than that of the last long-duration crew. By November 14, they returned to Star City to write their mission report. Five tracking ships that had been providing communications for the flight returned to ports after the landing to prepare for the next mission. The ship *Cosmonaut Pavel Belyayev* had been at sea for more than 200 days supporting the manned mission.[98]

After the return of the second long-duration mission to Salyut 6, Soviet officials stated that Salyut 6 had completed its primary mission. The designer of the Salyut, Feoktistov, said on November 16, that Salyut 6 would be checked for its technical and operational condition over the next few months. The results would determine the feasibility of the station's future use, although the results were a mere formality since new missions were already planned to use the station. This evaluation continued until at least January 15. On December 29, the station's orbit was 337×357 km.[99]

Kosmos 1074

Soyuz T Test

Launched: January 31, 1979, 12:07 p.m.
Landed: April 1, 1979, 3:00 p.m.
Altitude: 195×238 km @ $51.7°$
Crew: none

This flight was the third test of the Soyuz T spacecraft. Its initial orbit was 195×238 km. On February 3, the spacecraft boosted its orbit to 308×322 km. Five days later it maneuvered to 364×383 km. It remained in space for 60 days in storage mode to verify its ability to remain safe for use by long-duration space station crews. Before the retrofire maneuver, the orbital module was jettisoned as normal Soyuz T procedure. The spacecraft then made a normal reentry and landing.[100]

Soyuz 32

Third Long-Duration Crew—
175 Days

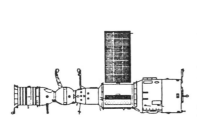

Launched: February 25, 1979, 2:54 p.m.
Landed: June 13, 1979, 7:18 p.m.
Altitude: 193 × 256 km @ 51.6°
Crew: Vladimir Lyakhov and
         Valeriy Ryumin
Backup: Leonid Popov and
         Valentin Lebedev
Call sign: Proton

In the first weeks of February, the Soviet tracking ships left their ports to take up positions to support another manned mission. As the crew waited for launch, music was played over the radio for their enjoyment. This was a standard practice for the Soviets and it denotes the routine and automatic nature of the launch operations.

After launch and separation from the booster's upper stage, the Soyuz entered an initial orbit of 193 × 256 km. The Soyuz was soon boosted into its 240 × 281 km transfer orbit to Salyut 6. Soyuz 32 docked at 4:30 p.m., February 26, with the station in orbit at 296 × 309 km. Salyut 6 was pushed to a higher orbit at 308 × 338 km by Soyuz 32 after docking. When opening the hatch to the Salyut, the cosmonauts smelled burnt steel coming from the hatch, which had been exposed to the sun before docking. Ryumin called this the odor of space.

After four months of unmanned flight, the Salyut was demothballed. The crew's basic mission was to perform an overhaul of the station's systems and prepare it for further long-duration missions. At the same time, they would attempt a new record duration flight. They brought a special tool kit for the repairs, which included electrical meters, screw drivers, pliers, vises, clamps, nuts, bolts, and a soldering iron.[101,102]

Along with the reactivation and repairs, the crew began routine activities and medicals. Two types of medical exams were routinely carried out by the crew. These were a short daily checks of the crew's mood, their daily accomplishments, and food intake. The second type checked their psychological condition by observing regular communications sessions. Every 8 to 10 days the crew performed thorough medical checks of their cardiovascular systems using the Polynom-2M, Beta, PCM cardio-monitor and Rheograph instruments, and their body mass was measured. Lyakhov expressed no problems after being forced to stop smoking because of the flight and adapted to weightlessness after two days.

On the second week, they increased the exercise period to 2.5 hours a day. This included the equivalent of 4.5 km walking and 4 km jogging on the treadmill. The crew also used the Penguin elastic exercise suits for up to 10 hours a day for additional exercise. Their average daily food intake was 3,100 calories and 2.5 liters of water. For the initial months of the mission the crew was allowed to sleep up to 10 hours a day.

The Soyuz also delivered a biological experiment to the Salyut similar to one built for a routine biosatellite mission. Lyakhov and Ryumin reportedly requested that a similar experiment be flown on their mission after becoming interested in the biosatellite experiment. The experiment was to place 8 quail eggs in an incubator and observe any changes in development of the chicks in weightlessness. The control samples on Earth developed normally, but the chicks in space developed much slower than normal, and had no heads.[103] On March 1, the Soyuz was used to boost the station's orbit to 308 × 338 km.

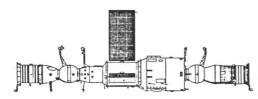

Progress 5

Launched: March 12, 1979,
8:47 a.m.
Reentry: April 5, 1979
Altitude: 183 × 256 km @ 51.6°

As the Progress approached the Salyut, the crew strained to see the spacecraft because the sun appeared behind the Progress. The crew had to control the station in case of a malfunction of the Progress, to maneuver away preventing a collision. The docking proceeded normally as Progress 5 docked at 10:20 a.m., March 14, at 293 × 325 km The crew did not open the hatch to the Progress until early the next day and spent the next 4 days unloading the cargo.

The flight delivered parts requested by the crew for the repairs to the station, 1,000 kg of propellant, 1,300 kg of cargo including an extra storage battery to complement a deteriorating electrical system, a television monitor for the crew to see Earth communications, a new signal command unit, a new Kristall furnace (the old one broke), the Yelena gamma ray telescope, a new teletype, a linen dryer, shampoo, the Biogravistat plant growth centrifuge, a second specially filtered Fiton plant growth unit, a new science instrument control panel, bread, vegetables, apples, onions, honey, and strawberries, letters, newspapers, air-regeneration units, clothes, 23 kg of water, a carbon dioxide and smoke detector for each work station and the Koltso radio communication system for the crew to use within the station. The Progress delivered a total of 300 items in 27 containers.

On March 30, the Progress raised the station's orbit to 284 × 357 km. Progress 5 undocked April 3, at 7:10 p.m., and de-orbited at 4:04 a.m., April 5.[104]

☆ ☆ ☆ ☆ ☆

Shortly after the Progress docked, mission control discovered a problem with a Salyut fuel tank. Fuel was leaking into the nitrogen bellows that pressurized the fuel. The problem did not interfere with the Salyut's engines operation, but it could have effected valves and regulators in the pressurization system. The crew cut off the bad tank and used the reserve tank. On the ground, a procedure was developed for fixing the tank or at least making it safe. Salyut 6 was equipped with three fuel and oxidizer tanks, of which one set acted as a reserve.

The crew tried to fix the problem in the fuel tank by emptying the tank to space. First, they started the station rotating to create a little artificial gravity to pull the usable fuel to an end of the tank. They transferred the good fuel to one of the other Salyut tanks and then transferred the mixture of fuel and nitrogen in the bellows to an empty Progress tank. The Progress was used to stop the station's rotation. They then purged the tank and fuel lines with nitrogen and closed it off. This procedure was completed by March 23. Also on that day, the crew used the Salyut's shower for the first time. They had been supplied with scuba masks to keep water out of their eyes.

The crew continued spending most of their time on repairs and listing defective parts to be delivered by the next Progress. They replaced the read/write head on the VCR, the shower curtain, the mass meter, carbon dioxide detectors, the linen dryer, the clocks, electrical cables, the Salyut command signalling unit, the instrument control panel, and replaced light bulbs for the entire station. They also added new batteries and tested the new Koltso cordless intercom system.[105]

They installed the Yelena gamma-ray telescope, replaced the Kristall material processing furnace, and replaced the old EVA spacesuits. While making repairs, they found a solution for removing tight screws in weightlessness, one person pushed on the screwdriver, while the other turned it, otherwise the cosmonauts turned themselves and not the screws. Much of the repairs were said to be preventive maintenance, such as replacing the light bulbs and electrical cords. The crew was able to watch a video of their launch, inaugurating the new Earth-to-orbit television link. For entertainment, the crew had 64 audio cassettes for their tape recorder and 50 video cassettes of films, concerts, and Soviet cartoons.

On March 30, Progress 5 boosted the station's orbit to 284 × 357 km. On April 6, Soyuz 32 was used to raise the station's orbit to adjust for the next launch of a visiting crew on Soyuz 33.[106]

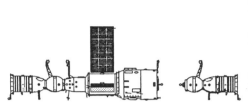

Soyuz 33

Fourth International
Crew—Bulgaria—Aborted
Launched: April 10, 1979,
8:34 p.m.
Landed: April 12, 1979,
7:35 p.m.
Altitude: 194 × 261 km @ 51.6°
Crew: Nikolay Rukavishnikov
and Georgi Ivanov
Backup: Yuri Romanenko and
Alexander Alexandrov
Call sign: Saturn

This launch was delayed for two days because of a wind storm at Baykonur. The flight was eventually launched into 40 km/hr wind gusts. A Bulgarian delegation was at Baykonur to observe the launch, as usual with an international mission. This delay turned out to be the least of the flights difficulties. Soyuz 33 was said to be the most perfect ship to be launched, and that it was received from the assembly factory with only four minor flaws, but it would provide the most excitement and apprehension since the aborted Soyuz 18A flight. Another oddity of the flight was that Rukavishnikov became the first civilian cosmonaut to command a Soviet spacecraft, although Lyakhov, an officer in the Soviet Air Force, was in command of Salyut 6, which was the destination of Soyuz 33.[107,108]

The flight proceeded normally after launch entering a transfer orbit at 245 × 312 km and then 273 × 330 km. The automatic rendezvous system was used as normal, bringing the Soyuz into an orbit of 292 × 353 km near the space station. At 9 km distance from the station, the Igla automatic docking system was activated. Docking was planned to occur at 10:15 p.m. When closing to 1,000 meters from Salyut 6, the Soyuz main engine started automatically for a burn of six seconds. But, the engine failed and automatically shut down after only three seconds. During the firing, the ship shook so strongly Rukavishnikov held the instrument panel to steady it. At first, Rukavishnikov thought the engine control system had malfunctioned.[109]

After consulting with mission control, the cosmonauts were cleared to try activating the Igla docking system again. The docking system tried to fire the engine again but shut down the engine immediately. Ryumin, observing from the Salyut, reported to mission control that a lateral glow emanated from behind the Soyuz during the engine burn, which was not normal (see Figure 5-8). Mission control then ordered the crew to end attempts to dock, and prepare for landing at the first normal opportunity.

Figure 5-8. The Soyuz main engine nozzle is in the center and the twin backup engine nozzles are next to the main engine. Four attitude control rockets surround the main and backup engines.

The malfunction was determined to be a failure of the Soyuz main rocket engine. A pressure sensor in the rocket's combustion chamber was shutting the engine off when it sensed normal combustion pressure was not being attained. This safeguard prevented propellants from being pumped into a damaged engine risking more damage and/or an explosion.[110]

Normally, the flight director told the crew the duration of a burn needed for rendezvous. The crew then timed the spacecraft's automatic burn on a stopwatch and reported the result, to double-check the automatic system. The crew asked to try another engine burn, but mission control denied the request and they were told to sleep.

Rukavishnikov did not sleep, he thought about the book *Marooned* in which an American spacecraft is stranded in orbit after an engine failure.[111] The Soyuz had a backup engine, but it was situated next to the main engine and could easily be damaged if the main engine failed. If the backup engine was damaged, this would mean the crew could be marooned with five days of supplies in an orbit that would not decay for ten days. If this occurred, the Soviets said that there were other options to return the crew. Among these possibilities would be the use of attitude control thrusters to slow the Soyuz below orbital velocity. This would have required many individual firings to prevent engine overheating, and there might not have been enough propellant left for this option (the thrusters were less efficient for this task than the main engines). This would also have made the landing point more difficult to predict.

The Salyut could also have been maneuvered within one km of the disabled Soyuz, which could then dock normally using the small thrusters.[112] But Soyuz 33 and Salyut 6 were drifting apart at 28 meters/sec., and time would be needed to compute the needed maneuvers. Also, if Soyuz 33 docked with the station, there would then be four cosmonauts on board with only room for two to return in the now questionable Soyuz 32 (its engine was the same type as Soyuz 33's, and carried the same risk of failure). In any event, the first option was to try and return the crew using the Soyuz backup engine. To pass the time the cosmonauts ate rations and opened a bottle of cognac they intended to give the Salyut crew. Small amounts of alcoholic drinks were sometimes carried by the visiting crews to long-duration crews.

The next day, the Soyuz was in orbit at 298×346 km, and as it passed through the communication zone of the Borovichiy tracking ship at 6:47 p.m., the backup engine was fired as planned. However, it fired for 213 seconds, 25 longer than necessary, before Rukavishnikov manually shut it off. Mission rules stated that the cosmonauts should allow an engine to burn for several seconds before attempting a manual shutdown. Using the backup system meant making a ballistic reentry.

This excessive retrofire put the Soyuz in a steep trajectory, which resulted in loads of up to 10 gravities from the 298×346-km high orbit.[113] The Soviets said this was the second ballistic reentry of a Soyuz. The first was Soyuz 1, although Soyuz 18A (not officially recognized by the Soviets) also performed a ballistic reentry, and Soyuz 24 reportedly used a ballistic reentry.[114]

They successfully landed at 7:35 p.m., April 12, 320 km southeast of Dzhezkazgan, only 15 km from the estimated point, and the capsule rolled out its side as usual. This was Rukavishnikov's second failed attempt to visit a Salyut (see Soyuz 10). The crew was also supposed to have traded spacecraft with the long-duration crew.

An investigation of the failure would take a month. The part that failed was later said to have been tested 8,000 times without failure. The Soyuz engine had performed roughly 2,000 firings in the course of the Soyuz flights in space since 1967 without failure.[115] A new version was made and tested for future flights. The crew returned to Star City on April 23, and later received the standard awards.

☆ ☆ ☆ ☆ ☆

After the Soyuz 33 mission, the long-duration crew on Salyut 6 was left with a doubtful spacecraft, considering the Soyuz 33 failure. The Soyuz main engine must be used for the retrofire to return to Earth. The Soviets did not want the two men to have to use a Soyuz with only one reliable engine system (the backup) and no new spacecraft could be launched until the problem with the design was corrected. But, for the present, the crew was safe on the Salyut, and could use Soyuz 32 to return in an emergency. The fifth international flight, scheduled for June 5, was postponed because of the failure as well.

The Proton's continued their work using the MKF-6M, BST-1M and Yelena, which was placed in the Soyuz 32 orbital module. The crew was given nearly 5 days off official work for the May Day holiday, although there was no rest from 2 hours of exercise a day. Mission control fed television coverage of the May Day celebrations to the crew.[116] The cosmonauts also carried out some experiments intended for the Bulgarian mission, including Pirin. This materials processing experiment investigated the formation of metal whiskers on zinc crystals to strengthen the material, and it produced foam aluminum.[117] Other Bulgarian experiments carried out included using the Spektr-15K to make multi-spectral measurements of the daylight atmosphere. Duga was a similar instrument made for viewing the night atmosphere.[118]

Progress 6

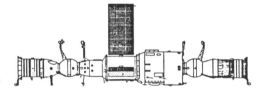

Launched: May 13, 1979,
          7:17 a.m.
Reentry: June 9, 1979
Altitude: 190×247 km @ $51.6°$

Progress 6 carried supplies to the long-duration crew on Salyut 6, while they continued their try at breaking the previous long-duration record and waited for a replacement Soyuz to be launched. The Progress was not limited by the engine failure experienced by Soyuz 33 because the Progress was equipped with the new Soyuz T version of the service module, which differed greatly in internal design from the Soyuz Ferry.

The transfer orbit to Salyut 6 was 313×334 km. The Progress docked at 9:19 a.m., May 15, as the cosmonauts manually oriented the space station toward the approaching Progress. The flight delivered 100 items including parts, food, air, propellant, water, new experiments, new television lights, light bulbs, a new Stroka teleprinter, a new navigation system control panel, a third specially filtered Fiton plant unit, biological samples (plants and seeds), letters, and newspapers. While unloading the Progress, the crew said the training in the hydrobasin weightlessness trainer for the unloading was not helpful for the actual operation. The Progress was unloaded after only 2 days. Shortly after the docking, the crew installed the new navigation unit.[119]

The Progress raised the station's orbit May 22, to 333 × 352 km and on May 27, air was released from the Progress to raise Salyut air pressure from 756 mm to 800 mm Hg. Refueling of the Salyut was completed by May 28. Progress 6 made orbit adjustments on June 4 and 5, adjusting the station's orbit to 358 × 371 km before undocking at 11:00 a.m., June 8, carrying 150 kg of waste and equipment to clear the way for Soyuz 34, which docked 12 hours later. Retrofire was at 9:51 p.m., June 9.[120]

Kosmos 1100/1101

Launched: May 23, 1979, 1:56 a.m.
Landed: May 23, 1979, 4:58 a.m.?
Altitude: 193 × 222 km @ 51.6°

This flight was the same as the Kosmos 881/882 and 997/998B flights, except that one of the spacecraft made two orbits before reentry and landing. The upper stage of the booster entered an orbit of 182 × 229 km after separating from the payloads. The stage reentered on May 26.[121,122]

The small shuttle program was apparently canceled shortly after this mission. The long periods between launches of this series cannot be taken as an indicator of failure. The Soyuz T, Progress, and Star module were also tested over a period of years before being used in the manned program. It is more likely that these unusual tests were just part of the research and development program and were not flight tests of actual equipment of the small shuttle.

☆ ☆ ☆ ☆ ☆

On May 30, the Salyut crew replaced the Delta control panel with the one delivered by Progress 6. By the third month in orbit, the crew was sleeping only 7 to 8 hours, compared to the beginning of the mission when 10 hours were normal.

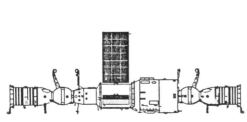

Soyuz 34

Launched Unmanned to Replace Soyuz 32

Launched: June 6, 1979, 9:16 p.m.
Reentry: August 19, 1979, 3:30 p.m.
Altitude: 192 × 254 km @ 51.6°
Crew: none

After investigating the failure of Soyuz 33, the Soviets launched Soyuz 34 to test the engine systems, and if they worked, to supply the long-duration crew on Salyut 6 with a new reliable and tested spacecraft. The transfer orbit to Salyut 6 was 289 × 364 km. The main engine performed normally and Soyuz 34 docked at the aft port at 11:02 p.m., June 9. The crew opened the hatch within one hour and inspected the ship. The opportunity of the unmanned Soyuz 34 flight was also used to deliver some biological samples to the crew for experiments.

Soyuz 32 undocked, unmanned, at 12:51 p.m., June 13. It carried 50 kg of experiment results, and because the craft was returning unmanned, the crew loaded 130 kg of the Salyut's replaced instruments for inspection on Earth. The cargo included 29 ampules of materials processed by the Kristall and Splav, 50 rolls of film from the MKF-6M, KATE-140, and hand-held cameras; cucumber, lettuce, radish plant specimens; burned-out light bulbs, used filters, vacuum cleaner bags, and a mold of a micrometeoroid impact crater found on one of the station's docking drogues. The total was equivalent to the cosmonauts' weight. After 3 orbits, the Soyuz landed 295 km northwest of Dzhezkagan, at 7:18 p.m.[123]

The Salyut crew undocked in Soyuz 34 at 7:18 p.m., June 14, and moved 100 meters away from Salyut 6, while the station was turned 180°. They redocked shortly afterward to the forward port clearing the aft port for another Progress.

Progress 7

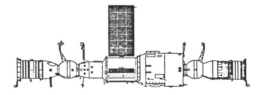

Launched: June 28, 1979,
    12:25 p.m.
Reentry: July 20, 1979
Altitude: 186 × 251 km @ 51.6°

Progress 7's transfer orbit to Salyut 6 was 275 × 345 km. The Progress docked to Salyut 6 at 2:18 p.m., June 30. The flight delivered 1,230 kg of cargo including 50 kg of food, 500 kg of propellant, film, clothes, air-regeneration units, plants, mail, a picture book of Moscow, Resistance experiments, the 350-kg 10-meter-diameter KRT-10 radio telescope, new Penguin suits, and Isparital, which was a 24-kg electron gun that vaporized aluminum deposited on a titanium disk. The crew unloaded the Progress in 4 days. Progress 7 also carried a special detector to measure the amount of upper atmospheric drag on the space station.[124]

The Progress raised the orbit of the station on July 3 and 4, from 353 × 360 km to 399 × 411 km, at the time the highest of a Soviet manned spacecraft ever. On July 12, the Progress began transfer of propellant to the Salyut. This continued until July 17. The cosmonauts assembled the KRT-10 wire mesh parabolic antenna in the aft docking port. It was attached by 3 claws to the fixed edge of the Salyut drogue, which also served as the station's aft hatch.

Progress 7 undocked at 6:50 a.m., July 18, with its forward hatch open (or most probably removed) to accommodate the half-meter-diameter folded telescope antenna, which protruded into the orbital module section.[125] Mission control observed the antenna deployment using the Progress docking television camera while the Progress station kept with the Salyut. The Progress then drifted away from the station and performed retrofire on July 20, at 4:57 a.m.

While preparing for KRT-10 experiment, the crew continued with other experiments. When Progress 7 was still docked, the station was turned so that one side was always in sunlight during the day portion of the orbit for a week. The sun heated the station's structure on one side while the other side cooled in the shadow causing the station's metal hull to expand and contract. The cosmonauts used one of the station's telescopes to measure the warping of the structure, which could cause optical, navigational, and scientific equipment to be misaligned. The experiment was later conducted during the Soyuz 36 mission. The crew also continued routine Earth observations and sighted a group of plankton 1,000 km long in the north Pacific.[126]

They also viewed instructions on the operation of the KRT-10 telescope, which were on a video tape sent with the telescope. The telescope consisted of five radiometers, a timing unit, a data recorder, and a control panel. During some observations, the telescope was used in conjunction with a 70-meter telescope in the Crimea. Several days were spent calibrating the antenna and observing Earth, the sun, and other objects.[127] The telescope's performance was not very good, but this may be due to poor design, or the possibility that it did not deploy completely. After the telescope's experiment program was completed August 9, the cosmonauts jettisoned it to clear the aft port. However, the telescope didn't depart the station with enough velocity and got entangled with the station's aft docking target, blocking the docking port.

Mission control suggested that the crew fire the Salyut engines to try and dislodge the antenna. When this failed, they tried pitching the Salyut, but this also failed. The only alternative, for continued operation of the station, was an EVA to clear the hatch. With the aft port blocked, Progress craft could not dock and refuel the station, shortening the station's lifetime to a matter of months. The crew favored the EVA idea, and the procedures were worked out at Star City, in the hydrobasin water tank.[128]

The crew performed the EVA on August 15, at 5:16 p.m., to free the antenna and retrieve materials left outside the station earlier. Ryumin at first could not open the airlock hatch, but when using more force it opened. The cosmonauts then had to wait for 36 minutes to come into communication range before exiting the airlock. Due to the failure of the spacesuit radio relay in the airlock, the communication was only one way, space to ground. Ryumin was tied to the station by a 20-meter tether and using hand rails, he worked his way to the rear of the station. He cut away the part of the antenna that was snagged to the docking target and pushed it away, while Lyakhov watched his tether from the airlock. They reportedly only used their helmet's sun visors for 10 minutes, for fear they might freeze over leaving them blind after sunset. Lyakhov joined Ryumin outside the station, and they retrieved the Medusa materials experiment from the station's exterior. Lyakhov had some difficulty getting back through the hatch. The EVA lasted 83 minutes—20 less than expected. James Oberg also reported that the crew was apprehensive about the EVA so late in the mission, when they were not in prime condition. They wrote letters to their families and packed them with experiment results, in the Soyuz, so it could return automatically in case of an accident.[129-131]

On August 16, the Soyuz boosted the station to 386×411 km. On August 18, the crew tested the Soyuz engines and prepared to return. They had been exercising for 2.5 hours a day in preparation for the return. Among the experiment results packed in the capsule was the second Kristall furnace to be used on the station to investigate the failures of the first and second units. They also returned a used air-regeneration canister that had begun to corrode to determine the cause.[132]

They undocked in Soyuz 34 at 12:08 p.m., August 19. Shatalov and Yeliseyev briefed the crew on weather conditions at the landing site, where clouds were at 1,000 ft, temperature was 20°C, and visibility was 20 km. The capsule landed at 3:30 p.m., 170 km southeast of Dzhezkazgan. They became the first crew to use the ground recovery system of platforms and slides to get down from the capsule, although Ryumin could walk. The cosmonauts got their first medical check 30 minutes after landing. Medicals would occupy most of their time for the next week. It was reported that they had some difficulty in speaking for a time after landing.[133] Lyakhov lost 5.5 kg of weight during the mission, while Ryumin's weight remained the same, which was quite an achievement since nearly all people lose weight, mostly muscle mass, during spaceflight. Lower leg volume decreased by about 20% for both cosmonauts.[134]

The acute period of re-adaptation lasted for 3 days. The doctors let them swim in the cosmonaut hotel pool with nets under them, to try easing readaptation. Their recovery took only 7 days, 2 to 3 days less than expected, and Ryumin was in much better condition than Lyakhov. Ryumin walked a half a mile a day after landing, and was jogging after three days. After a week, the post flight medical tests were completed and they returned to Star City September 9.

Soviet medical results from the mission showed that the bone decalcification leveled off after the first few months in weightlessness. But, these results were later proved, by NASA specialists, to be too insensitive to show any change if it ever occurred.[135] During the mission, the crew made many interesting Earth observations and directed fire fighters to forest fires. African brush-fire smoke was seen to drift to South America, and African dust storms carried dust to North America in the upper atmosphere. Two volcanic eruptions was also witnessed. They also directed Soviet fishing fleets to schools of fish by seeing thin films of oil on the surface that was produced by the fish. Other oceanic observations included waves meters high and hundreds of kilometers long.

Plant growth experiments conducted during the mission yielded a variety of results. Using a series of Fiton filtered plant growth units, the Oasis and Biogravitsat units, the crew grew a tree, flowers, wheat, peas, cucumber, garlic, parsley, lettuce, and borage. The Biogravitsat had two identical disks for plant growth, one of which rotated to produce artificial gravity. Seeds planted on the rotating disk usually produced roots radiating away from the center of rotation. Once the plants had developed sufficiently, they were transferred to the Oasis unit where they grew in weightlessness. Some of the samples were eventually eaten by the crew, but most were returned to Earth for study. A total of 54 experiments using the Kristall and Splav materials processing units were conducted during the mission. Crystal samples obtained were generally at least a five to six times better than produced on Earth. Superconducting niobium/aluminum/germanium was also produced late in the mission. Other materials processed included gallium arsenide, indium antimonide, germanium vanadium oxide, magnesium alloy, cobalt alloy, aluminum-tin aluminum-copper, aluminum-tin aluminum-lead, zinc alloy, selenium alloy, and telluride alloys.[136]

The Vaporizer experiment using the Isparitel electron gun was set up late in the mission after initial problems. Using its low-voltage electron beam, silver coatings were deposited on titanium and carbon disks in exposures from 1 to 200 seconds. The experiment mounted in the Splav airlock in the intermediate section was left set up for the next long-duration crew to use.[137]

By August 22, Salyut 6 was in orbit at 384 × 409 km. In the middle of September 1979, Feoktistov said that specialists were checking the Salyut 6 space station to see if further manned flight missions could be made.[138] The Soviets announced the results of the investigation of space adaptation sickness by the first three Salyut 6 long-duration crews. All of the crews reported the same symptoms: the sensation of tumbling backwards, sensation of blood rushing to the head, stuffed nose, puffiness of the face, decreased appetite, and discomfort when moving the head. In three cosmonauts, the effect was enough to cause vomiting after eating. Most symptoms disappeared by the fourth day, but the sensation of blood rushing to the head persisted up to 10 days, and when doing heavy physical work for the rest of the mission.[139] On November 29, Salyut 6 was in orbit at 352 × 372 km.

Soyuz T-1

Unmanned
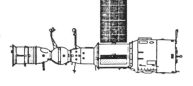
Launched: December 16, 1979, 3:30 p.m.
Landed: March 26, 1980, 12:47 a.m.
Altitude: 194 × 205 km @ 51.6°
Crew: none

Soyuz T-1 was the fourth test flight of the modified Soyuz, but the first test to be given the Soyuz designation (see Figure 5-9). The Soyuz was launched 73 minutes after Salyut 6 passed over the Baykonur Cosmodrome. The transfer orbit to the station was 201 × 232 km, and 231 × 259 km. On December 18, the T-1 approached Salyut 6, but overshot the station. The spacecraft was maneuvered back to its original position, ready for another try at docking by the next day.

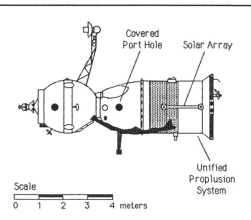

Figure 5-9. The Soyuz T was developed to replace the Soyuz Ferry. Improvements included new computers, solar arrays, new pressure suits, a new engine system similar to the Progress, and slightly increased payload capability.

Soyuz T Spacecraft

The Soyuz T was an improved version of the Soyuz Ferry spacecraft. Total spacecraft weight was about 6,850 kg and dimensions were unchanged except for the solar arrays. The solar arrays were similar to the ASTP designed arrays, and deployed to make the total spacecraft width 10.6 meters.

(continued on next page)

(continued from page 207)

A new development of the T version was the ability to jettison the orbital module before the retrofire maneuver. This enabled the crew to leave the orbital module attached to the station for use as an extra compartment (although it was never used for this purpose). It also decreased the amount of fuel needed for the reentry burn by 10%. With these savings, 250 kg of propellant were required for retrofire to return to Earth. These developments increased the ability of the Soyuz to carry three crewmen, or two and 100 kg of extra cargo.

The improvements included an engine system with a unified propellant system for both main engine and four attitude thruster units using nitrogen tetroxide and UDMH. The capsule's landing rockets were also enlarged to provide softer landings, and the escape tower was enlarged for added safety and to account for the enlarged cargo capacity of the spacecraft. The fully automatic flight systems included a new computer called Argon with 16 Kbytes of random access memory and improved telemetry systems. New windows with jettisonable covers allowed cosmonauts to look outside the capsule after reentry when the old windows would be burned black. New 8-kg pressure suits were made using more plastic for lightness and flexibility, and a new helmet had increased visibility, but the suit was not designed for EVA.

The T-1 finally docked at the forward port of Salyut 6 at 5:05 p.m., December 19, at 342 × 362 km. On December 25, the Soyuz raised the station's orbit to 370 × 382 km. The spacecraft was docked to Salyut 6 for 95 days. Salyut 6 was not manned during this period. On February 29, the orbit was 343 × 369 km. At 12:04 a.m., March 24, the Soyuz undocked at 344 × 363 km and after 2 days of tests, landed at 12:47 a.m., March 26 in the Kazakhstan.[140,141]

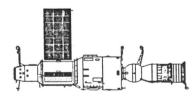

Progress 8

Launched: March 27, 1980, 9:53 p.m.
Reentry: April 26, 1980
Altitude: 198 × 275 km @ 51.6°

Another long-duration mission was about to be launched to Salyut 6, and Progress 8 was launched to provide supplies and materials to the cosmonauts upon their arrival. The Progress carried propellant, food, air, clothes, air regenerators, water, replacement control panels, batteries, temperature control system components, and new equipment for the Salyut.

Twenty hours before docking, the Progress' orbit was 230 × 242 km. The spacecraft docked (2 days after the Soyuz T-1 undocked) at 11:01 p.m., March 29. The next day, the Progress raised the station's orbit to 348 × 360 km for the next long-duration mission. Refueling was performed automatically by mission control. After being unloaded by the Soyuz 35 crew, on April 24, the Progress raised the station's orbit to 349 × 353 km. The Progress undocked at 11:04 a.m., April 25, and retrofire was at 9:54 a.m., the next day.[142]

Soyuz 35

Long-Duration Crew—
185 Days
Launched: April 9, 1980,
        4:38 p.m.
Landed: June 3, 1980, 6:07 p.m.
Altitude: 197 × 247 km @ 51.6°
Crew: Leonid Popov and Valeriy Ryumin
Backup: Vyacheslav Zudov and Boris Andreyev
Call sign: Dneiper

Ryumin was originally instructing the crew of Popov and Lebedev when Lebedev injured his knee in a trampoline accident during training, and had to undergo surgery.[143] When this happened the backup crew was unprepared and Ryumin, being the most experienced with Salyut 6's systems after his 175-day mission, was given the choice to fill in for Lebedev or delay the flight. Ryumin had been cleared for flight 8 months before, after fully recovering from his recent 6-month flight.[144] Long-term studies of Ryumin's recovery must also have been over or close to ending to allow for a new mission. Ryumin's family was quite upset when they learned that he would be gone again so soon for another long flight. In training for the flight, Popov had been preparing for weightlessness by sleeping with his head below his feet, and training on a rotating, moving chair to resist space sickness.

During the launch, Rymuin's pulse was 178 and Popov's was 133. This was the reverse of the U.S. norm, which was the commander having the highest rate. But, Ryumin certainly had more spaceflight experience than his commander on this mission. The Soyuz docked under Igla automatic control from 10 km distance, at 6:16 p.m., April 10.

Ryumin found the letter he had left on the station for the next crew, not expecting he would be on the next crew. On April 11, they began unloading the Progress 8 supply ship that had docked 13 days before them, and they started to activate the station. They replaced one attitude control unit, the air regenerators and added another 80-kg storage battery to counter the degrading electrical system, and they replaced another battery in the steering unit. They also replaced units of the communication system, ventilation system, and installed new command and warning systems, units of the water regenerators and installed the Yelena-F, which had been used on Soyuz 32, in the Salyut. Fifty items in total were replaced by the crew.

Air was released from the Progress into the station and clocks were synchronized with mission control. The crew later surprised mission controllers by repairing the Yelena-F by replacing a broken pin in its mechanism, without consulting mission control before disassembling the telescope for the repairs. The crew also set up the Malakhit plant growth unit to begin investigating the possible use of two types of orchids to supplement environmental control systems on long-duration space flights. The Malakhit used a synthetic soil that consisted of time release capsules of fertilizer. The plants grown in the unit were returned to Earth for analysis after 60, 110, and 170 days.[145–147]

After one week, Ryumin reported that his ability to see fine detail on Earth's surface was back to the level that took two months to develop during his first long-duration flight. On April 24, the Progress raised the station to 349 × 353 km then at 11:04 a.m., the next day it undocked. Retrofire was at 9:54 a.m., April 26.[148]

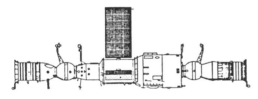

Progress 9

Launched: April 27, 1980,
9:24 a.m.
Reentry: May 22, 1980
Altitude: 185 × 255 km @ 51.6°

Progress 9's transfer orbit to Salyut 6 was 244 × 350 km. The Progress docked at 11:09 a.m., April 29. It carried mail, food, air, clothes, redesigned air-regeneration canisters, a new motor for the Biogravistat plant experiment, gas analyzer filters, VCR parts and a plastic molding device. With this flight, the resupply of Salyut 6 now was complete for the long-duration crew's mission.[149]

After Progress 9 docked, the crew continued repairs and began Earth resources work. They were given 2 days off from official work for the May Day holiday. After the holiday, they began unloading the Progress. A new system for water transfer was used on this Progress. The Rodnik system used a pipe from the Progress to the Salyut's water tanks through which 180 kg was transferred. The old system was to carry 5-kg water canisters to the Salyut tanks.

On May 16, the Progress boosted the station to 349 × 369 km, and undocked at 9:51 p.m., May 20. After 2 more days, the Progress deorbited at 3:44 a.m.[150]

Soyuz 36
Fifth International Crew
—Hungary
Launched: May 26, 1980,
    9:21 p.m.
Landed: July 31, 1980, 6:15 p.m.
Altitude: 191 × 265 km @ 51.6°
Crew: Valeri Kubasov and Bertalan Farkas
Backup: Vladimir Dzhanibekov and Bela Magyari
Call sign: Orion

The Soyuz 36 mission was rolled out to the launch pad on May 22, while in orbit above, Soyuz 35's main engine was used to adjust the station's orbit for rendezvous with the new mission. The flight was originally scheduled to begin June 5, 1979, but was canceled after the Soyuz 33 failure.

The crew boarded the Soyuz about 2.5 hours before launch as usual. Soyuz 36 entered an initial orbit of 191 × 265 km after separating from the booster's upper stage and then maneuvered into a transfer orbit of 251 × 314 km. The next day at 6:00 p.m., the crew donned their pressure suits again in preparation for docking. Twenty minutes later the Soyuz maneuvered to approach the Salyut. At 10 km distance, the Igla automatic rendezvous system activated. Soyuz 36 then docked to the aft port at 10:56 p.m., May 27, in orbit at 340 × 360 km. After the normal 3 hours of checks of the docking systems and attending the Soyuz systems, the crew opened the hatch to join the Dnieper crew.

So many experiments were conducted by the visiting crew that they sometimes only got about 3 hours of sleep a day. The crew performed the usual international crew experiments, including taking samples of the station's air, cardiovascular exams, psychological tests, and Earth resources work using the MKF-6M and KATE-140, in coordination with an AN-30 photography aircraft flying at 6,000 to 7,000 meters and an AN-2 aircraft taking 4 spectral band photos flying at around 2,000 meters. In addition to the planes, a helicopter took infrared photography, and crews on the ground took direct measurements of the surface conditions.

They also conducted 3 biological and pharmaceutical experiments that investigated the production of interferon. This included blood tests determining the effect of weightlessness on the body's normal interferon production. An experiment also injected human white corpuscles into interferon producing a medium to test production in weightlessness. Another experiment investigated the effects of spaceflight on already produced interferon pharmaceuticals.

Materials processing experiments using the Splav produced copper aluminum mixtures. The Kristall was used to produce gallium arsenide/chromium, indium antimonide, and gallium antimonide. Other experiments included measuring the deformation of the Salyut's structure due to 300°C temperature variation on the outer hull in sunlight and its effects on optical navigation instruments. The results showed changes of more than 0.1° over a few hours. The crews also took pictures of the station's portholes to access damage from micrometeroids, and routine studies of the upper atmosphere were performed.

On May 29, Soyuz 36 was used to boost the station's orbit to 344 × 360 km. The international crew traded spacecraft with the long-duration crew, exchanging seat liners, pressure suits, and personal items. Used equipment and trash was packed in the Soyuz 35 orbital module for disposal when the module is jettisoned during reentry.

The visiting crew undocked at 2:47 p.m., June 3. Retrofire occurred at 5:16 p.m. over the South Atlantic and 11 minutes later the capsule separated from the orbital and service modules about 170 km over the Sudan. The capsule landed 140 km southeast of Dzhezkazgan, at 6:06 p.m.[151]

☆ ☆ ☆ ☆ ☆

The long-duration crew moved the new Soyuz 36 to clear the aft docking port for future Progress flights. They undocked the spacecraft from the aft port at 7:38 p.m., June 4, moving away about 180 meters. The Salyut was then commanded to orient its forward docking port, turning the station to face the Soyuz for redocking. At the time, the crew exercised about 1.5 hours a day. The crew also directed efforts against forest fires sometimes in direct communication with those on the ground fighting the fire. They also made observations to find better fishing grounds on coastal areas.

Soyuz T-2

First Manned Soyuz T Flight

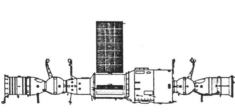

Launched: June 5, 1980,
          5:19 p.m.
Landed: June 9, 1980, 3:40 p.m.
Altitude: 195 × 231 km @ 51.6°
Crew: Yuri Malyshev and
          Vladimir Aksenov
Backup: Leonid Kizim and
          Oleg Makarov
Call sign: Jupiter

This flight was the first manned test flight of the modified version of the Soyuz Ferry. The Soyuz T had new engine systems, could launch up to 3 cosmonauts, and had new control systems. In preparation for the mission, the crew had to study computer programming to better understand the operation of the new Soyuz T Argon computer, which controlled docking and reentry.

Launch sequence for the Soyuz T was identical to the old Soyuz Ferry. After separating from the booster's upper stage, the Soyuz T made a small maneuver to enter an initial orbit of 195 × 231 km. The Soyuz T then maneuvered into a transfer orbit of 262 × 318 km to rendezvous with Salyut 6. During the final docking operation, the new Argon docking computer failed 180 meters from the station, leaving the T-2 turned perpendicular to the Salyut. Malyshev then took control and docked using the manual system at 6:58 p.m., June 6.[152] Yeliseyev explained that after launch, the Argon computer chooses which of several possible approaches to fly to the station, and that the crew and controllers did not practice the particular approach the computer ultimately selected on the mission. The crew decided to switch to manual systems to ensure they could cope with an emergency if they had to. They said that the computer would have docked successfully, if they allowed it to continue. The computer also selected which of several descent paths to use during landing, depending on flight conditions.

The crew entered the station a few hours later to join the Dnieper crew. They stayed on the station only 2 days before returning, during which they unloaded supplies for the station from the Soyuz T, participated in some medical tests, and used the Salyut's MKF-6M camera.

Soyuz T-2 undocked at 12:20 p.m., June 9. The Soyuz then took pictures and inspected the Salyut as it turned around for them at 267 × 316 km. The ships then separated and about 3 hours later, the cosmonauts separated from the orbital module before retrofire, as was normal procedure for the Soyuz T (see box). The capsule landed around 3:40 p.m. in the normal landing area.[153]

Typical Landing Sequence

+ 0 min — Soyuz separates from the station.

+ 60 min — Soyuz orbital module is jettisoned followed by a 200-sec retrofire.

+ 71 min — Soyuz capsule separates from the service module.

+ 82 min — Atmospheric breaking begins.

+ 96 min — Capsule's main parachutes open at 8 km altitude; light beacon is turned on and recovery antennas deployed.

+ 102 min — Radio contact is established with recovery forces.

+ 110 min — Capsule lands at 3 meters per second velocity within 5 hours before local sunset.

+ 113 min — Crew egress or wait for recovery. They sign the charred capsule with chalk.

— The crew is flown to the cosmonauts' hotel at Baykonur for recovery in isolation and debriefing before returning to Star City within days or weeks.

After the departure of Soyuz T-2, the long-duration crew carried out time consuming repairs of a redundant Kaskad attitude control system. During the repairs, the crew used an unexpectedly large amount of propellant. The crew also continued materials processing experiments. Other studies included using the BST-1M telescope and the Yelena-F gamma ray detector.[154,155]

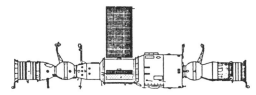

Progress 10

Launched: June 29, 1980,
7:41 a.m.
Reentry: July 19, 1980
Altitude: 200 × 260 km @ 51.6°

Progress 10 docked at 8:53 a.m., July 1. Many items of supplies for the crew were carried including a Polaroid camera, a 9-inch color television monitor, a zoom lens telescope, tapes of Soviet pop music, seeds for plant growth experiments, letters, newspapers, an image intensifier for the BST-1M, and food including onions, dill, peas, parsley, cucumber, radish, and canned fish.

Refueling was completed by July 8, and the Progress boosted the stations orbit to 328 × 355 km. The Progress undocked at 1:21 a.m., July 18. Retrofire was at 4:47 a.m., July 19.[156]

☆ ☆ ☆ ☆ ☆

In the short time between the departure of the Progress and the arrival of the next international mission, the long-duration crew reported that the windows of Salyut 6 were beginning to cloud over. This can be a indication that the environmental control system was malfunctioning. They also broadcast a prepared statement for the opening of the Olympic games in Moscow.

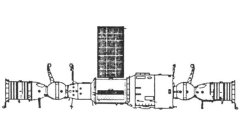

Soyuz 37

Sixth International
Crew—Vietnam

Launched: July 23, 1980,
9:33 p.m.
Landed: October 11, 1980,
6:15 p.m.
Altitude: 190 × 272 km @ 51.6°
Crew: Viktor Gorbatko and
Pham Tuan
Backup: Vaeriy Bykovsky and
Bui Thanh Liem

Call sign: Terek

The crew boarded the Soyuz on schedule at about 7:00 p.m., 2.5 hours before launch. The transfer orbit to Salyut 6 was 258 × 294 km. While approaching the station in darkness, the Salyut cosmonauts supervising the docking asked the Soyuz to turn on its docking lights at 2 km instead of the planned 100 meters to make the Soyuz more easily visible. The Soyuz docked to the aft Salyut port at 11:02 p.m., July 24.

Tuan's military career was highlighted by the Soviet press. He was said to have shot down the first B-52 over North Vietnam, after dodging F-4 fighters, and on returning to base had to crash land on a bombed runway. The fact was disputed by the U.S. Air Force, which maintained that all B-52's downed were hit by surface-to-air missiles.[157]

Tuan reportedly experienced some space sickness on the first days of the flight. The cosmonauts operated 30 routine international mission experiments including the usual medical and Earth observation work. Photos were taken of Vietnam using the KATE-140 and MKF-6M cameras. Medical experiments were conducted using the Chibis low-pressure suit, and cardiovascular monitoring instruments. Experiments using the Kristall to produce gallium phosphide, and bismuth antimony telluride materials. They also conducted a plant growth experiment with a Vietnamese azolla water fern for possible use in closed-loop life support systems.

On July 28, the crews transferred the seats, pressure suits, and personal items from the Soyuz 37 to the 36 in preparation to depart in the old Soyuz 36, leaving the new Soyuz 37 for the long-duration crew to use. The Soyuz capsule was packed with experiment results from the short mission and long-duration crews, including samples of orchids grown in the Malakhit unit.

The international crew undocked in the Soyuz 36, at 2:55 p.m., July 31. At 5:25 p.m., the command for retrofire was relayed from mission control by the tracking ship *Borovichi* in the South Atlantic. The capsule landed at 6:15 p.m., 180 km southeast of Dzhezkazgan. Visibility on landing was 10 km.[158,159]

☆ ☆ ☆ ☆ ☆

On August 1, at 7:43 p.m., the long-duration crew moved the new Soyuz 37 from the rear docking port to the forward port to allow the next Progress to dock. The Soyuz undocked and moved away from the station. The station then enabled its forward docking port, which automatically turned the station around, to face the Soyuz for redocking.[160]

The long-duration crew participated in a joint experiment with scientists on the ground, using their Yelena-F gamma ray telescope on the Salyut. The scientists on the ground released a high-altitude balloon equipped with another Yelena-F to take measurements to compare with the Salyut's. After the balloon observations were completed, the balloon malfunctioned and did not release its instrument package by parachute as planned. When the ground forces found where the balloon had landed, they found that the instrument package had been dismantled and its parts stolen by local thieves, along with the unused parachute silk. The thieves did, however, leave some data tapes.

On September 2, the crew reported that the arabdopsis plants growing in the Svetobloc plant growth unit had flowered, only a few days after their controls on Earth. They also observed that the best plant growth was when the units were placed near portholes. On September 4, the station's orbit was raised to 343 × 355 km. On September 12, the crew replaced ventilation fans, which were consistently wearing out, and making noise. The orbit was changed again September 16, using the Salyut 6 reserve main engine to adjust the station's orbit for the next launch. The reserve engine was being tested for the first time in two years.[161,162]

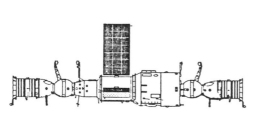

Soyuz 38

Seventh International
Crew—Cuba

Launched: September 18, 1980,
          10:11 p.m.
Landed: September 26, 1980,
          6:54 p.m.
Altitude: 195 × 257 km @ 51.6°
Crew: Yuri Romanenko and
          Arnaldo Tamayo-Mendez
Backup: Yevgeniy Khrunov and
          Jose Lopez-Falcon
Call sign: Tamyr

The Tamyr crew boarded the spacecraft two hours before launch, a bit behind schedule. The launch came unusually early for a normal Salyut rendezvous launch window, which might have been because the primary objective of international flights was to photograph the country with the Salyut's cameras during the mission.[163] Normally, one of Mendez's main duties would be taking pictures of Cuba with the MKF-6M camera, but Salyut 6 never passed over Cuba in daylight during the flight. So, there was no need to adhere to the usual schedule of international mission launches because the photography could not be accomplished. The long-duration crew promised to send Mendez some pictures after the mission.[164]

Eight minutes and 47 seconds after launch, the spacecraft separated from the booster upper stage and maneuvered into an initial orbit at 195 × 257 km. Soyuz 38 then maneuvered on the fourth and fifth orbits into a transfer orbit of 263 × 313 km. The Soyuz automatic rendezvous system brought the spacecraft to a distance 23 km from the space station. The Tamyr crew then docked using the automatic system at 11:49 p.m., September 20. About 3 hours later, after the usual checks of the docking system and shutting down Soyuz systems, the visiting crew entered the space station. This crew did not trade spacecraft with the long-duration crew because they would return soon.

The crew performed 27 material processing and medical experiments. Other experiments included Cortex, which measured brain activity with various stimulus; Support, which was a shoe that placed force on the arch muscles to test the effect on walking after return to Earth; and Anthropometry, which measured changes in muscle and bone structure. Various medical tests also investigated the circulatory system and blood. Sugar and Zone were experiments to test the growth of sugar crystals. Carribe was a series of material processing experiments using the Kristall and Splav units to produce gallium arsenide/aluminum alloy, tin telluride/germanium telluride alloy, germanium indium alloy, and zinc indium sulfide alloy. Routine psychological tests were also performed as on all international missions.[165]

On September 25, the Soyuz engine was tested as usual before landing, boosting the station's orbit a little. The Soyuz capsule was packed with experiment results from the short mission and long-duration crews, including samples of orchids grown in the Malakhit unit.[166] The next day, the Tamyr crew sealed off in the Soyuz 38, donned their pressure suits, and undocked at 3:34 p.m., September 26. Before landing, Shatalov reminded the crew to be patient because the landing was at night and it might take recovery forces longer to locate them. The combination of the early launch window of the Soyuz 38 produced the unusual night landing, but did allow the next Progress mission to be launched during the same launch window.

Retrofire was at 6:05 p.m. and separation of the capsule from the orbital and service modules was 21 minutes later at 144 km altitude. Five minutes later, the capsule entered the dense layers of the atmosphere with the deceleration eventually reaching 4 gravities. Nineteen minutes later, radio contact was made with the capsule and 10 minutes later the parachutes opened. The landing at 6:54 p.m. was 3 km from the target point and was said to be the most precise landing in the history of the Soviet space program to date, 175 km southeast of Dzhezkazgan. The crew was presented with the usual awards at the Kremlin on October 1, and the capsule was sent to a Cuban museum.[167,168]

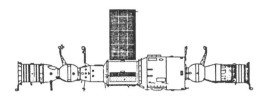

Progress 11
Launched: September 28, 1980,
6:10 p.m.
Reentry: December 11, 1980
Altitude: 189 × 247 km @ 51.6°

Progress 11's transfer orbit to Salyut 6 was 260 × 340 km. The spacecraft docked at 8:03 p.m., September 30. The Soviets said that the last Progress flight (Progress 10), had over-stocked Salyut 6, and thus produced the long period between Progress flights. This flight carried the necessary cargo to prepare Salyut 6 for a period of unmanned flight and other items to complete the long-duration crew's mission including water, oxygen, food, and Amplituda equipment, which monitored vibrations in the station (similar to Resonance). The ship was unloaded by October 4.[169]

On October 8 and 11, the Progress' engines boosted the orbit and adjusted it for the returning Soyuz 35 crew. It remained docked until the next manned mission and was used again then. On November 16, the Progress refueled the Salyut automatically under ground control. On November 18, the Progress boosted the station's orbit to 299 × 315 km. It eventually undocked on December 9, at 1:23 p.m., after boosting the station's orbit to 290 × 370 km. Retrofire was at 5:00 p.m., December 11.

☆ ☆ ☆ ☆ ☆

During the unloading of the Progress, the crew also started cleaning the station, preparing for the next long-duration crew's arrival. They also started taking increased amounts of salt and water to reverse the effects of dehydration in weightlessness and wearing the Chibis low-pressure suit along with the usual exercises in preparation for return to Earth.

On October 10, the crew tested the Soyuz engines and continued packing data and results in the Soyuz capsule, including samples of orchids grown in the Malakhit unit.[170] On October 11, the crew completed packing and undocked at 9:30 p.m. Retrofire was about 2 1/2 hours later.

After a record flight of 185 days, the long-duration crew returned to Earth and landed at 12:50 p.m., 180 km southeast of Dzhezkazgan. Ryumin was the first out of the capsule, walking unaided. After landing, Popov commented that they had experienced 4 gravities of acceleration during reentry, which was a little higher than the nominal 3 gravities for a Soyuz capsule. Popov said that the force had felt like 8 gravities after more than a half a year in weightlessness.[171]

For the first time the crew gained an average of 2 kg of weight during the flight. The crew explained this was the result of very strict adherence to the scheduled exercises and diet. They both grew about 2.75 cm during the flight, as usual, due to expansion of the spine. They gradually returned to their normal height after landing. Arm and hand strength increased for both cosmonauts due to the need to move large items when unloading the Progress cargo ships. Their cardiovascular condition and calf muscles were relatively unaffected by the long flight because of exercise on the treadmill, but muscles in the back, chest, and shoulders degraded. Extensive medical checks had been performed by the crew 24 times during the mission.[172]

The cosmonauts recovered soon and were able to walk for 30 minutes the next day. By October 15, they were able to play tennis. Among the experiments performed on the mission was material processing of germanium, indium antimonide, indium arsenide, gallium arsenide, gallium bismuth, gallium arsenide, cadmium sulfide, gadolinium cobalt, and bismuth oxyllaride. Using the Isparitel device, the cosmonauts collected 196 samples of thin coatings of gold, copper, silver, aluminum, nickel, and aluminum alloys several microns thick on titanium and carbon disks. In Earth resources work, the crew took nearly 3,500 photographs with the MKF-6M and almost 1,000 with the KATE-140. More than 1,000 photos were taken with hand-held cameras and more than 40,000 spectrograms were made with the Spektr-15, RSS-2M and other instruments. Earth observations of the crew aided in the fighting of forest fires and guided Soviet fishing fleets to schools of fish feeding on masses of plankton easily visible from orbit. One hundred hours of data were obtained with the Yelena-F background gamma ray detector, including comparison data for high-altitude balloon flights.

The state of the Salyut was also assessed by the crew before they left. Ryumin commented that the portholes were covered by a thin film of dust, which could hamper future photography through the windows. He also noted that the insulating blankets on the transfer compartment and mid section of the station were torn in spots due to micrometeroid impacts and that some of the plastic covering on the solar cells in the Salyut's 3 solar panels had broken off, allowing the solar cells to be damaged by heat and sunlight.[173]

Soyuz T-3

Maintenance Mission—13 Days

Launched: November 27, 1980,
          5:18 p.m.
Landed: December 10, 1980,
          12:26 p.m.
Altitude: 196 × 236 km @ 51.6°
Crew: Leonid Kizim, Oleg Makarov, and Gennadiy Strekalov
Backup: Vasili Lazarev, Viktor Savinykh, and Valeri Polyakov
Call sign: Mayak

Soyuz T-3 was the first flight of a three-person crew in nine years, since the death of the Soyuz 11 three-man crew. The normal three-person crew consisted of the commander, flight engineer, and a research cosmonaut or guest cosmonaut. The commander sat in the center with the engineer on his left.

The orbit was 255 × 260 km 16 hours before docking. They docked using the Argon automatic system, from 5 km distance, at the forward port at 6:54 p.m., November 28. The crew found the traditional bread and salt left by the last crew to greet travelers, and activated Salyut 6 for a short stay. Progress 11 was still docked the aft docking port. It had accomplished refueling under ground control on November 16, and had boosted the station orbit on November 18.

One of the cosmonauts main tasks was to replaced a four pump hydraulic unit of the temperature control system. The pumps were not meant to be replaced in flight, but Salyut designer and cosmonaut, Feoktistov, devised a procedure to replace them. In the process, Feoktistov became so interested that he scheduled himself as flight engineer on Soyuz T-3, but the doctors told him that at the age of 53, he was to old to fly, but more importantly, he was too valuable to the space program to risk.[174] To replace the hydraulic unit, the crew had to use a hacksaw to cut away metal supports that were in the way of the unit, while avoiding any hydraulic leaks and collecting the dangerous metal filings. The repair was accomplished with little comment.[175,176]

They also repaired a commutator unit in the Salyut electrical system, and units of the telemetry system. They replaced and reprogrammed the Salyut timing and control system and replaced a refueling system compressor transducer, which failed during the Progress 11 refueling. While making these repairs, the crew often talked to the previous long-duration crew, which was vacationing in the Caucasus, to make sure they were repairing the correct equipment.

Along with the repair work, the crew also conducted material processing experiments and made a hologram of a salt crystal dissolving, using a helium-neon laser which weighed 5 kg. This was to have been delivered by Soyuz 38 but was delayed. They also performed Resonance tests, used the Splav furnace for three days to produce cadmium telluride mercury, used the Kristall to produce gallium bismuth, and conducted the usual medical experiments. Plants were also grown in the Oasis and Svetblok units during the short flight and the crew took air samples from the station for analysis on Earth.[177]

On December 8, the crew started packing items to be returned in the Soyuz, and Progress 11 was used to boost the station's orbit to 290 × 370 km. By the next day, the packing was complete and the Soyuz engine was tested. Progress 11 undocked at 1:23 p.m., the same day and reentered on December 11. The

Soyuz undocked from its orbital module at 9:10 a.m., December 10, at 255 × 260 km and the orbital module was left attached to the station.[178]

They landed at 12:26 p.m., 130 km east of Dzhezkazgan. Kizim and Strekalov had unexpected difficulty after landing because they had neglected exercise, during the relatively short, and very busy mission. It was noted that the cosmonauts were so busy that a free day and two television sessions had to be canceled. The Soyuz T orbital module they left attached to Salyut 6 was jettisoned later the same day and may have reentered January 26, 1980. On December 29, the Salyut was in an orbit at 308 × 349 km awaiting its next and final long-duration manned mission.

Progress 12

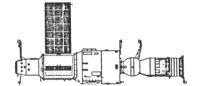

Launched: January 24, 1981, 5:18 p.m.
Reentry: March 21, 1981
Altitude: 181 × 282 km @ 51.6°

This resupply mission was flown before the next manned mission to ensure supplies would be awaiting the new crew when they arrived, and to maneuver the station into a higher orbit. Progress 12 used a transfer orbit at 247 × 310 km to reach the station, and docked at 6:56 p.m., January 26.

On January 28, at 8:00 p.m., the Progress raised Salyut 6's orbit to 307 × 359 km. The Progress then refueled the Salyut by command from mission control. The Progress delivered 1,200 kg of dry goods including air regenerators, filters, food, replacement parts, experiments, 200 kg of film and 200 kg of water for the next crew. After being unloaded by the T-4 crew, the Progress undocked at 9:14 p.m., March 19, leaving the station in orbit at 345 × 358 km. Retrofire was at 7:59 p.m., March 21.[179]

☆ ☆ ☆ ☆ ☆

Soviet space officials said before the Soyuz T-4 mission that Salyut 6 would not be the last Salyut station, implying that the launch of Salyut 7 would occur soon. While Salyut 6 was left unmanned, one of the solar arrays had failed to rotate and track the sun normally with the other two arrays. The total generated power available dropped. This caused the temperature dropped inside the station when the electrical power was reduced. Normally, with a crew on the station, electric heaters in the walls of the Salyut are not needed to heat the station, but without the added heat produced by a crew, the station's heaters are needed to maintain the temperature. This caused some condensation and high humidity in the station. Flight controllers spent two weeks orienting the station to the sun to maintain interior temperature for the Soyuz T-4 crews arrival when the problem would be repaired.[180]

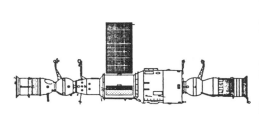

Soyuz T-4
Long Duration—74 Days
Launched: March 12, 1981,
10:00 p.m.
Landed: May 26, 1981,
4:38 p.m.
Altitude: 185 × 222 km @ 51.6°
Crew: Vladimir Kovalenok and
Viktor Savinykh
Backup: Vyacheslav Zudov and
Boris Andreyev
Call sign: Photon

This mission was not intended to be a record breaking flight. Savinykh later said the only purpose of the mission was to complete the Soviets' obligations for international (Soviet bloc) missions.[181] The crew also performed some work preparing Salyut 6 for later testing with the Star module spacecraft.

The Soyuz T-4 maneuvered into a transfer orbit of 245 × 315 km to approach the station. Kovalenok first sighted Salyut 6 in the periscope as a point of light at a distance of 5 km. At 500 meters, the station's structure could be seen through the periscope. With the Argon computer controlling the approach, the Soyuz halted 200 meters from the station and held that position until coming within communication range of mission control, which was being supplied by the ship *Cosmonaut Dobrovolski* in the South Atlantic and the *Gagarin* in the western Mediterranean. The Soyuz then approached the station in darkness with the spotlight illuminating the Salyut.[182] Docking occurred at 11:33 p.m., March 13.

In their first hours in the Salyut, the crew was busy with repairs to make the station habitable. After the last crew left Salyut 6, one of its solar arrays quit tracking the sun, reducing power to the station. The interior heaters in the station, used during unmanned periods to keep the station warm, could not be used as usual because of this failure. Temperature in the station had dropped to 10°C and moisture condensed on the inside walls. The crew replaced the orientation mechanism for the failed array and it began operating properly. The cosmonauts spent their first few days adjusting the environmental control system to reduce humidity levels and increase the temperature.

The other repairs included repairing the treadmill, a pump in the thermal regulation system, and a television camera. The crew also carried out Resonance experiments. On March 16, Progress 12 was used to raise the station's orbit to 339 × 353 km. The cosmonauts started some biological experiments after completing the repairs, planting arabidopsis in the Fiton unit, and planting other specimens in the Biogravistat and Malakhit.[183] Progress 12 was unloaded by March 18, and undocked March 19, at 9:14 p.m., at 343 × 358 km. The main objective of the mission was to host two international flights. The Mongolian flight was the first.

Soyuz 39

Eighth International
Crew—Mongolia

Launched: March 22, 1981,
       5:59 p.m.
Landed: March 30, 1981, 2:42 p.m.
Altitude: 195 × 261 km @ 51.6°
Crew: Vladimir Dzhanibekov and Judgerdemidiyin Gurragcha
Backup: Vladimir Lyakhov and Ganzoring
Call sign: Pamir

Soyuz 39's transfer orbit to Salyut 6 was 271 × 320 km. The international crew docked at 7:28 p.m., March 23, and entered the station 3 hours later. The crew's experiment schedule included experiments of the type usually conducted on international flights. They searched for possible ore and petroleum deposits in Mongolia using the MKF-6M, KATE-140 and hand-held cameras (with simultaneous aircraft and ground-based surveys as usual). They also conducted atmospheric studies taking spectrograms of the terminator and measuring atmospheric distortion. The VPA-1M unit was used to measure the polarization of light reflected by Earth's atmosphere. Results from the instrument were relayed by radio to a recorder in the Salyut telemetry system. Spectrograms were also taken of Earth in regions of high snow, sand, and water cover to better analyze MKF-6M multispectral photos.

They tended the Vazon and Oasis plant growth experiments, made a hologram of a porthole with a meteor hit, and assessed the amount of particles on the window's exterior. They also took samples of the station's air for analysis on the ground, and performed Resonance experiments. The Splav material processing furnace was used to study mass transfer and diffusion of liquid tin and lead in a temperature gradient. The Splav was also used to grow a vanadium pentoxide semiconductor. Copper sulfite and water were observed for diffusion qualities while being heated in the Salyut food heater. When recrystallized, the material was hoped to produce a good cleaning material. After finishing the Splav experiment, a radiation detector was placed in the Splav airlock to measure cosmic rays under 10 MeV. A similar detector was also placed in the station's interior was also used determine the amount of the radiation passing through the station's hull.

The crew also participated in the usual medical experiments using the Beta, Polynom 2M and Rheograph instruments. The crew also tested a device to restrict head movements during the first 3 days of adaptation to weightlessness to help curb space sickness.

The Soyuz capsule was loaded with experiment results, including plant specimens and tadpoles that were used to study vestibular function. They undocked at 11:22 a.m., March 30 and performed retrofire at 1:51 p.m., which was followed by normal capsule separation from the orbital and service modules at 2:17 p.m. The capsule landed in fog and rain at 2:42 p.m., 170 km east of Dzhezkazgan. They were located and recovered quickly and received the standard awards.[184]

☆ ☆ ☆ ☆ ☆

The Photon long-duration crew began a quiet period between visits; they cleaned the Salyut, replaced parts of the water recovery system, cleaned circulating fans, overhauled equipment including the Yelena-F, MKF-6M, and replaced a compressor switching block of the cryogenic cooling system on BST-1M. They continued exercising for 2 hours a day with medicals every 10 days. On May 3, Savinykh spent the day recording electrocardiograms with the portable Cardio-monitor device.

In mid-April, they made Earth observations of the U.S.S.R. and Bahama's weather. They commented on being able at times to see through the ocean and see underwater island formations. They also made measurements of noctilucent clouds with the Yelena-F and BST-1M. They also continued experiments with the Splav, producing gallium arsenide, gallium antimonide, gallium bismuth, germanium silicon, lead zinc, bismuth antimonide, and glasses. On April 23, they removed Splav from the aft airlock and installed the Isparital, which they tested for 5 days coating silver and copper onto titanium disks. They tested the holographic laser camera twice, and continued plant growth experiments with peas, onions, orchids, and arabdopsis using Oasis, Vazon, Svetoblok, Magnetogravistat, and Malakhit units.[185]

Kosmos 1267

Star Module

Launched: April 25, 1981, 5:01 a.m.
Capsule Landed: May 24, 1981
Module Reentry: July 29, 1982
Altitude: 192×259 km @ $51.6°$

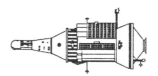

This was the second test flight of the Star module, the first was Kosmos 929. The Star module was almost as large as the Salyut, and was equipped with a cargo return capsule. The module was launched into an orbital path 10° to the east of Salyut 6. This offset made the eventual docking slower and more controlled than the usual Progress or Soyuz docking. Kosmos 1267 did not dock with Salyut 6 until the last manned mission had already returned to Earth. During that period, the module flew in low orbit to measure its atmospheric drag characteristics.[186] *Aviation Week* reported that U.S. intelligence once believed that the 1267 might have been equipped with infra-red anti-satellite missiles, but they were just externally mounted cylindrical propellant tanks.

☆ ☆ ☆ ☆ ☆

On May 7, the Salyut's orbit was raised to 339 × 367 km using the Soyuz T-4 engine. In mid-May, the crew experimented with exchanging the Soyuz probe with the Salyut's drogue, possibly indicating a plan to dock one Salyut to a another, or docking the Soyuz to the recently launched Kosmos 1267. During the mission, the crew found that mechanical watches did not keep time well on the station, so they used electronic watches instead. This worried the Salyut's designers because the station's systems were highly mechanical. To investigate the effects of a long spaceflight on equipment, the crew packed one of the station's control panels for return to Earth and examination. They also conducted an emergency station evacuation drill.

Soyuz 40

Ninth International
Crew—Romania
Launched: May 14, 1981,
       8:17 p.m.
Landed: May 22, 1981, 5:58 p.m.
Altitude: 191 × 269 km @ 51.6°
Crew: Leonid Popov and Dumitru Prunariu
Backup: Yuri Romanenko and Dumitru Dediu
Call sign: Dnieper

It has been reported that the crew for the Soyuz 40 mission underwent many changes before the flight. It was suggested that the initial crew might have originally been Bykovsky or Khrunov and Dediu. Then, the commander may have been replaced by Romanenko. Then, the crew of Romanenko and Dediu may have been replaced shortly before launch for unknown reasons, by Popov and Prunariu, which had been the backup crew. Romanenko and Dediu then were listed as the backup crew.

Soyuz 40's transfer orbit to Salyut 6 was 260 × 307 km. It docked to the rear Salyut docking port at 9:50 p.m., May 15. After the usual 3-hour check of the docking system and shutting down of the Soyuz, they were greeted with the traditional bread and salt by the Salyut crew as they entered the station.

The international crew conducted research of Earth's magnetic field and upper atmosphere in the Minidoza experiment. Nanovesi was an experiment that exposed structural materials and optical coatings to the space environment to assess any degradation. The Mongolian mission's Emission experiment was repeated, the crew measured microgravity disturbances in the station, and performed Resonance tests. They also performed the Kapillyar experiment, which used capillary action to form a germanium crystal in a molybdenum matrix. They also did the usual psychological and medical tests. The normal activity of photographing the visiting cosmonauts' country had to be delayed until the last day of the flight when the station began passing over Romania during daylight. On May 19, the combined crews held a press conference.[187,188]

Another unique experiment was carried out by the combined crews on the station to test the station's orientation systems. A simple telescopic device was mounted on one of the portholes in the floor of the work section of the station. An image of the sun over a half a meter in diameter was projected by the telescope on the ceiling of the station. By measuring the image's position, the cosmonauts determined that the station's automatic orientation system was accurate to 10° of arc. Over long periods, the cosmonauts could maintain orientation of the station to within 20° of arc.[189]

On May 20, the crew corrected the orbit of the station by firing the Soyuz 40 engine. Before undocking at 1:37 p.m., May 22, the Soyuz boosted the station's orbit, and they photographed the Salyut before landing at 4:58 p.m., 225 km east of Dzhezkazgan. After landing Popov praised the Soyuz's designers for their work as the Soyuz Ferry series was ended.

☆ ☆ ☆ ☆ ☆

With the departure of the international mission, the primary objective of the Salyut crew's flight was accomplished. The long-duration crew continued experiments with Pion and wore the Chibis suits, and concentrated on medical tests in the final days of the flight. They also operated the Pion heat exchange and mass transfer experiment that was delivered by the last Soyuz. They packed a station control panel in the Soyuz capsule and photographed the station's interior. Before leaving, the crew attached a device onto the forward docking port drogue, in preparation for the future Kosmos 1267 docking.[190]

The crew was the second to leave their orbital module docked to the station while departing with the reentry module and service modules of the Soyuz T. The T-4 crew undocked from their orbital module May 26, and landed at 4:38 p.m., 125 km east of Dzhezkazgan.[191] The orbital module attached to Salyut 6 was jettisoned on May 31, 1981.

The crew, which lost an average of 2.5 kg of weight during the flight, had performed around 150 types of experiments, made 60 periods of astrophysical observations, taken 2,000 photos with hand-held cameras, and about 2,000 with the MKF-6M and KATE-140 cameras. About 200 materials processing experiments were conducted and around 900 medical exams and biological experiments were performed.[192]

After the mission, the Soviets said that there would be no manned spaceflights for a few months while the condition of Salyut 6 was evaluated and its value for future use determined. At this time, the Soviets had 16 cosmonauts in training for Salyut missions out of a total 50 cosmonauts.[193]

While the Soyuz T-4 cosmonauts were the last crew Salyut 6 would host, the Salyut's mission was still not finished. The Kosmos 1267 Star module was still in a close orbit with Salyut 6 and the two spacecraft were to be docked to test their joined systems (see Figure 5-10). The Star module capsule landed on May 24, 1981 after the module had flown for a month in a low orbit.[194] The Soviets later said that the time in low orbit was used to test the aerodynamic drag characteristics of the module. After extensive maneuvers, the module docked with Salyut 6 on June 19, at 10:52 a.m., in orbit at 335×377 km.

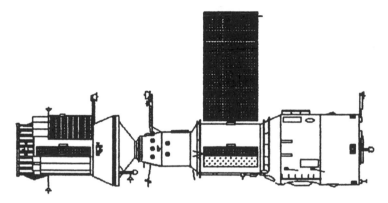

Figure 5-10. After the last crew left the Salyut 6 station, the Kosmos 1267 module (without its reentry module) docked to the station to test the combined spacecraft.

The module was used to boost the orbit of Salyut 6 twice to a final orbit of 362×382 km. The ground controllers tested the combination of the station and module for their thermal characteristics and experimented with the thermal control of the complex.

The Salyut 6 and Star module were finally pushed into a destructive reentry over the Pacific by the module on July 29, 1982, 3 months after Salyut 7 was launched. Salyut 6 was designed to last 22 months, but lasted 58 months with the constant repair and refurbishment by cosmonaut crews. The water recovery system recycled more than 600 liters of water and supplied the manned missions with hot water.[195]

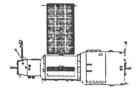

Salyut 7

Launched: April 19, 1982, 10:45 p.m.
Altitude: 212 × 260 km @ 51.6°

Salyut 7 was launched to continue the research begun by Salyut 6 and to further perfect space station operations, and study the docking of large modules to the Salyut. The delay between the last mission to Salyut 6 and the launch of Salyut 7 was to give time for personnel of the ground control teams and technicians to take vacations and time to update equipment that should have been replaced when the expected lifetime of Salyut 6 was reached. Extending operations of Salyut 6 made it necessary to keep old outdated equipment in operation. A third control room at Kaliningrad was also built to support the multiple spacecraft used for station operations.

Soviet spacecraft designer Konstantin Feoktistov said that there was little difference between the Salyut 6 and 7 stations (see Figure 5-11). Salyut 7 had a few minor changes from the Salyut 6 design and was probably the original backup to Salyut 6. Salyut 7 was intended to have a lifetime of 4 to 5 years instead of Salyut 6's 18 months (see Figure 5-12).

Salyut 7 was launched into orbit 70° displaced from that of the still-orbiting Salyut 6. By late April, Salyut 7's orbit was 308 × 352 km. Initial control of Salyut 7 was accomplished through the Soviet ground stations and the communications ship *Korolev*, which was stationed in the Atlantic Ocean.[196]

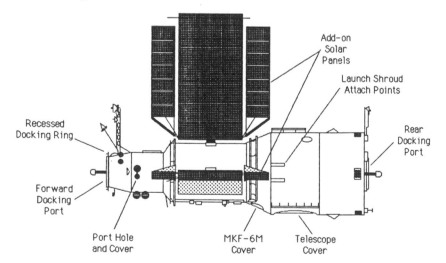

Figure 5-11. The Salyut 7 station was a modified version of Salyut 6. The front docking port was strengthened for docking large modules, the living accommodations were improved, and connections for additional solar panels were added.

Figure 5-12. Salyut 7 is seen here as technicians install systems and perform checks. The solar arrays have not been installed and the rear of the station had not been painted yet. The launch shroud is on the left. (Source: Tass from Sovfoto.)

Salyut 7 Space Station

Salyut 7 was probably a modified backup spacecraft to Salyut 6, and many minor changes from Salyut 6 were made. On the outside of the station modifications included outside covers for some portholes that could be opened and closed by the crew to help protect the windows over the years of use. The solar arrays were modified to accommodate additional arrays during EVAs, and permitted use of higher efficiency solar cells that increased generated power by 10% over that of the Salyut 6 space station. The forward docking port was strengthened for use with the massive Star modules by adding a recessed ring with 8 pins around the docking collar. The docking ring was 1.7 meters in diameter maximum and positioned 10 cm behind the docking collar.

Inside the station, changes included the addition of a 50-liter refrigerator, the ability to have hot water 24 hours a day, and a quieter environmental control system. Air-regenerator cartridges and filters were made more compact for easier handling and shipment. Some components of the thermal regulation system and the radio communications systems were made more accessible to ease in-flight replacement. Some hydraulic lines were equipped with special connectors that prevented leakage if they had to be replaced in flight. Ports were also installed to add fluids to the systems in flight to counter for any losses. Two, 400-liter water storage tanks were installed in the unpressurized engine section to save space in the interior of the station.

(continued on next page)

(continued from page 229)

The Delta navigation system tested on Salyut 6 was also put into standard operation on Salyut 7. Delta could automatically keep track of the station's position in orbit over Earth and during the cosmonauts' sleep would print out all communications opportunities, equator crossings, and sunrise and sunset times for the coming day. The Delta also could turn off and on the radio equipment before each pass into range of a ground station. The cosmonauts could also use the system to make note of the time of an event and the station's coordinates, which were printed on the teletype for future reference.

The spacesuits carried on Salyut 7 were also modified to allow connection to support units in the transfer compartment before an EVA to conserve the suits' air during prebreathing. Before an EVA, the air pressure in the Salyut is lowered to shorten the time needed to prebreath oxygen in the spacesuits. Normally, up to 1.5 hours could be spent prebreathing and preparing for an EVA before actually exiting the station. This improvement extended the capability of the spacesuits from 3.5 hours to 5 hours. Modified versions of the suits flown late in Salyut 7's operations were capable of 6.5 hours operation using an improved carbon dioxide filter system. Pressure in the spacesuits could be lowered from 760 to 250 mm Hg if extra mobility were required.

Salyut 7 Experiments and Equipment

(This is an incomplete list)

MKF-6M: East German camera for Earth photography. Imaged in 6 bands with 10–30 meter resolution.

Kristall: Two versions of the furnace were used on Salyut 6 for semiconductor production.

SPLAV: "Alloy" furnace, could heat samples from 650° to 990°C with hot, cold, and gradient heating.

KATE-140: Mapping camera, could photograph 450×450-km areas on 600-frame film cassettes.

Magma-F: Material processing furnace.

KGA-2: Holographic camera.

Tavria: Electrophoresis molecule separator.

Aelita: Replaces Polinom apparatus from Salyut 6.

SKR-02M: X-ray spectrometer (2–25 KeV).

Piramig: French visible near-infrared camera.

PCN: French low-intensity light camera.

(continued on next page)

Cytos-2: Antibiotics effects in weightlessness.
Biobloc: Cosmos ray effects on biological material.
Synthetic Aperture Radar: Submarine tracking experiment.
Bioluminescence: Submarine wake tracking experiment.
EFU-Robot: Electrophoresis pilot scale experiment.
Svetabloc-T: Synthetic gel for electrophoresis.
Laser Targeting & Tracking: Tests with ground lasers.
RT-4M: 500-kg X-ray telescope.
Kometa: Math models of ocean colors.
Pion: Heat and mass transfer in weightlessness.
Comet: French comet dust collecting device.

Soyuz T-5

First Long-Duration Crew of
Salyut 7—211 Days
Launched: May 13, 1982, 1:58 p.m.
Landed: August 27, 1982, 7:04 p.m.
Altitude: 192×231 km @ $51.6°$
Crew: Anatoliy Berezovoi and Valentin Lebedev
Backup: Vladimir Titov and Gennadiy Strekalov
Call sign: Elbrus

Berezovoi and Lebedev started training for the Soyuz T-5 mission a year
before the launch. Both of the cosmonauts had already been through the training
program for a long-duration mission twice before as backups.[197] Soyuz T-5's
transfer orbit to Salyut 7 was 269×325 km. Docking to Salyut 7's forward port
was at 3:36 p.m. May 14, at 343×360 km.

The crew activated Salyut 7's systems and installed an air duct from the
Salyut to the Soyuz as part of normal activation of the Salyut environmental
control system. Lebedev reportedly adapted quickly to weightlessness, and the
crew was said to have adapted by the third or fourth day. The crew continued
setting up equipment on the station for the first week of the mission before
starting any experiments.[198,199]

Salyut 7 was maneuvered by the crew fully automatically using the Delta
system. Previously on Salyut 6, cosmonauts were required to monitor any
maneuvers by the automatic systems. On May 17, Iskra 2, an amateur radio
satellite, was launched by a spring, from a scientific airlock, into a 342×357-km
orbit. Before the deployment, the cosmonauts tested the satellite's power supply
and radio. With the antennas folded, the satellite was put into the left scientific

airlock. The spherical airlock consisted of two spheres, one inside the other. The inner sphere had a hole that matched up with the opening on the outer hull of the station. To launch the satellite, the outer sphere was closed and the inner one rotated until the opening matched the outer opening. A spring then pushed the satellite away from the station. After launch, the satellite deployed its radio antennas.[200] The satellite was hexagonal and weighed 28 kg. Iskra 2 decayed from orbit on July 9, 1982. With the launch of Iskra 2, the Soviets became the first to launch an active satellite from another Earth orbiting manned spacecraft (Apollo 15 and 16 launched sub-satellites into lunar orbit). This was an attempt to up-stage the U.S. STS-5 flight in November, which would launch two large communication satellites. On May 22, the station was put into gravity gradient mode flight using the Delta unit in its first fully automatic maneuver of the station.[201]

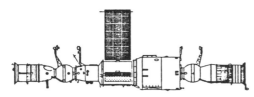

Progress 13

Launched: May 23, 1982,
             9:57 a.m.
Reentry: June 6, 1982
Altitude: 191 × 278 km @ 51.6°

Progress 13's transfer orbit to Salyut 7 was 290 × 347 km. The cargo craft docked at 11:57 a.m., May 25. The flight carried 290 liters of water, 660 kg of propellant, 900 kg of equipment including French experiments to be used during the next Soyuz international flight, the Kristall and Magma-F material processing furnaces, the Czechoslovakian EFO-1 electro-photometer, warm boots for the cosmonauts, letters, food, and the Oasis plant growth experiment.[202]

The Magma-F was an enlargement of the Kristall furnace. Magma-F could accept quartz ampules of samples up to 20 mm in diameter, and 200 mm long with the temperature monitored from 14 points in the furnace. A computer recorded the temperatures of up to 900°C and transmitted the data to mission control for analysis. The materials processing equipment delivered was not attached to the Progress by standard fittings because of its bulk, presenting the crew some extra difficulty in removing it.[203]

The Progress spacecraft now had been modified to carry water in tanks outside the orbital module (possibly in the tankage section like Salyut 7 itself), leaving more room for cargo inside the Progress. By May 26, the crew began unloading the cargo and finished the refueling process on June 1. They then used the Rodnik pumping system to transfer water to the Salyut tanks. On June 2, Progress 13 lowered Salyut 7's orbit from 335 × 341 km to 219 × 321 km in preparation to receive the next flight, which would carry a 3-person crew. A Soyuz with a crew of 3 could not reach a higher orbit. Progress 13 undocked at 10:31 a.m., June 4. Retrofire was at 4:05 a.m., June 6.[204]

Kosmos 1374

Shuttle Development Flight
Launched: June 3, 1982
Landed: June 3, 1982
Altitude: 167×222 km @ $50.6°$

This was the first of flights of space planes that are sometimes referred to as Kosmolyot's. They were launched from Kapustin Yar by the C-1 booster (see box). The flights were tests of heatshield materials and reentry techniques for the large shuttle that was in development at the time. Dimensions of the spacecraft were 2.28 meters wide, 2.86 meters long, with an orbital mass of 1,074 kg and dry mass of 795 kg.

C-Type Booster

The C-1 booster (vertical) consisted of an SS-5 first stage and a restartable second stage. The first stage had two RD-216 engines, each with two combustion chambers, that burned UDMH/IRFNA (inhibited red fuming nitric acid) producing 176,000 kg of thrust. The first stage was 21.5 meters long, and the second stage was 6.5 meters long. The standard launch shroud brought the total length to 31 meters, with a diameter of 2.5 meters. The C-type was used to launch the space plane tests of the 1980s, which were related to the development of the Soviet space shuttle. The C-1 was also used to launch numerous Soviet anti-satellite (ASAT) weapon test flights in the 1970s.

For many years, U.S. intelligence sources had reported that a theorized manned version of this space plane would weigh 15,000 kg, be four to five times larger, and be launched by the J type booster in the mid-1980s, but this is very unlikely considering the statements of the Soviets and their subsequent development of a large space shuttle.

To support communications for the test flight, the tracking ships *Patsayev*, *Dobrovolskiy*, and *Chumikan* were transferred from their normal Salyut support activities.[205] After one and a quarter orbits, the Kosmolyot reentered flying past the coast of India. The spacecraft then made a cross range maneuver heading about 600 km to the south toward the recovery fleet. The spacecraft landed by parasail, 560 km south of the Cocos Islands in the Indian Ocean, where the seven-ship recovery force retrieved the space plane.[206] A cone protruding from the top of the spacecraft was an inflatable float to help right the floating plane in the water. Along with the Soviet ships was an Australian

Orion patrol plane which photographed the recovery operation.[207] The most distinctive features revealed were small windows where a cockpit would be on a manned spacecraft, and a heatshield made of tiles, possibly like those used on the U.S. shuttle. The quick reaction of the Australians and photography of the spacecraft indicate that western intelligence agencies knew well before the launch that the short one-orbit flight was about to occur and where it was to terminate, giving the Australians time to get an aircraft in the recovery area.

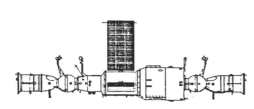

Soyuz T-6
Tenth International
Crew—France
Launched: June 24, 1982,
          8:30 p.m.
Reentry: July 2, 1982, 6:21 p.m.
Altitude: 189 × 233 km @ 51.6°
Crew: Vladimir Dzhanibekov,
          Aleksandr Ivanchenko, and
          Jean-Loup Chretien
Backup: Leonid Kizim,
          Vladimir Solovyev, and
          Patrick Baudry

Call sign: Pamir

The joint French-Soviet space mission was the result of a long cooperation between the two countries in the area of space science. In 1966, an agreement was reached that included launching French satellites on Soviet boosters and flying French instruments on Soviet space probes, high-altitude balloons, sounding rockets, and Earth satellites. The agreement to launch a French crewman on a Soviet Soyuz flight was reached when Soviet General Secretary Leonid Brezhnev and French President Giscard d'Estaing met in April 1979.[208]

Chretien and Baudry began training for the international mission in September 1980. A year later, Chretien was selected as the primary crewman for the flight. He had 7 years' experience as a test pilot and 5,000 hours' flight time. Some members of the French Academy of Sciences protested the joint spaceflight saying that the experiments were only afterthoughts to the political decision to fly with the Soviets.[209]

The original commander for this mission was Yuri Malyshev. During training for the flight, the Soviets said that he developed a heart irregularity, which grounded him.[210] Dzhanibekov was assigned as replacement commander of the flight.

The transfer orbit to Salyut 7 was 248 × 265 km. Dzhanibekov performed a manual rendezvous after the Argon computer failed 900 meters from the station.

The computer had just turned the Soyuz around and fired its main engine in the direction of flight to break its approach to the Salyut. It was then supposed to turn around again to begin a slow approach to the station. At this point, the Soyuz computer sensed its gyroscopes had reached their limits and stopped the maneuver. After disengaging the computer, Dzhanibekov could not see the Salyut and had to turn the Soyuz on all three axis to get the Soyuz pointed at the station again. Dzhanibekov then proceeded to dock at 9:46 p.m., June 25, 14 minutes earlier than planned, due to the failure in orbit at 288 × 311 km. Docking with a station at this distance under manual control was very unusual and a significant achievement for the Soviets.[211,212]

On the Soyuz T-6 and previous Progress missions, more than 500 kg of equipment had been delivered to the Salyut to be used for the international mission. Biological experiments were emphasized during the flight, including ultrasonic Doppler imaging of the heart and blood vessels with a standard French medical instrument called Echograph, which was left on the station for future Soviet use. Other experiments included testing the body's sensory systems and muscle activity required to maintain normal standing posture in weightlessness, the biological effects of cosmic radiation, and the effects of antibiotics on microorganisms. Experiments in materials processing were also conducted investigating capillary forces during crystallization of alloys at differing temperatures, producing aluminum indium alloy, and measuring the effects of very small accelerations on samples being processed and the thermal environment inside the Magma-F furnace. The cosmonauts also tested a device called Braslet, which restricted blood flow into their legs to easy adaptation to weightlessness. The elastic cuffs were worn around the thighs for only 30 to 60 minutes at a time, 5 times a day.[213-215]

Photography of the celestial objects was also carried out during night time using the Piramig image intensifier, while Berezovoi or Lebedev pointed the Salyut in the correct direction for the cameras. The station's interior lights and attitude control thrusters had to be turned off when using the Piramig camera. The Piramig operated between 200 and 1,000 nanometers wavelengths and used a fiber optic photomultiplier to enhance the image. Filters could also be used to take images at certain wavelengths from the near infrareds to the ultraviolet. Exposure, and the date and time on the image, could be changed by the operator. The cosmonauts took 350 pictures in 20 sessions with the Piramig. A camera called PCN was also used by the international crew to photograph zodiacal light, noctilucent clouds, and lightning in different wavelengths. The cosmonauts oriented the station for the astronomical observations using the Delta navigation unit. This was accurate to only one degree of arc, but using other instruments it was possible to achieve 5–10 arc seconds, and then pointing the photographic instruments themselves, achieve at least one arc second of arc accuracy. The station's attitude was controlled for observations for up to 6 hours a day.[216,217]

Berezovoi later commented that Chretien seemed to spend most of his time recording measurements of his heart and blood vessels using the Echograph ultrasonic imager. Chretien also complained that he was so busy with the experiments and reports that he didn't have time to view France when the station was in position. The cosmonauts then programmed the Delta navigation unit to warn them when passing over France so Chretien could take pictures. In order to complete the experiments later in the flight, Chretien got little sleep each night (see box).

Typical Crew Schedule

Radio Contact Times	Crew Schedule for June 30–July 1, 1982
0100–0230 a.m.	PCN camera experiments
0900–0908 a.m.	Morning toilet
0908–1044 a.m.	Breakfast
1140–1153 a.m.	Move PCN and Piramig to another porthole
	Electrotopograph experiment
	Space station maintenance
1313–1327 p.m.	Neptun and Tsitos-2 experiments
	Exercise
1445–1501 p.m.	Piramig and PCN experiments
	Orientation of station
	Lunch
1612–1635 p.m.	Orientation of station
	Piramig and PCN experiments
1741–1806 p.m.	Prepare television report
	Talk with reporter
	Orientation of station
	Piramig and PCN experiments
1910–1939 p.m.	Television report
	Film station activities
	Orientation of station
	Piramig, Tsitos-2, and PCN experiments
2033–2107 p.m.	Television report
	Dinner
	Likvatsiya experiment
2217–2238 p.m.	Preparation of next day's rations
	Familiarization with next day's schedule
2400–0900 a.m.	Sleep period

The Salyut cosmonauts also were instructed by Ivanchenkov, who had spent 140 days in space observing Earth. He showed them how to see atmospheric phenomenon, zodiacal light, and auroras. On June 28, the crew photographed a giant lightning flash south of Hawaii in the Pacific that lasted more than a minute and lit up an area roughly 30 × 50 km. On June 30, photos were taken of noctilucent clouds 80 km over Ireland.[218,219]

The T-6 and the international crew ended their mission after loading the Soyuz capsule with experiment results and undocking at 3:01 p.m., July 2. Orbital module separation occurred at 4:05 p.m., followed by retrofire at 5:35 p.m. Eleven minutes later, the capsule separated from the service module just above the dense layers of the atmosphere. At 6:10 p.m. the parachutes opened and the capsule landed at 6:21 p.m., about 4 km from the target point, 65 km northeast of Arkalyk.[220,221]

☆ ☆ ☆ ☆ ☆

On July 3, the Salyut maneuvered into an orbit at 309 × 344 km. The Salyut crew performed a 2¹/₂ hour EVA July 30. Leonov, the first person to perform an EVA on Voskhod 2 in 1965, was the capcom for the EVA and gave the cosmonauts advice. The cosmonauts had practiced the EVA many times during training, the last time being just before the launch. Lebedev positioned himself on a foot anchor on the transfer compartment and tested various tools including a power screw driver and cutting tools. He also removed some of the test samples from the Medusa fixture on the hull of the transfer compartment of Salyut 7 for return to Earth. Lebedev handed the Medusa fixture to Berezovoi who remained in the airlock. Berezovoi removed the samples already exposed to space conditions since Salyut 7's launch, and replaced them with new samples to be retrieved by a later crew. The tests were on components under evaluation for future use in spacecraft. These included pipe fittings conditioned to expand and contract significantly under varying temperatures to aid construction in space, various metals under constant stress, and types of threaded connectors.[222]

Progress 14

Launched: July 10, 1982,
1:58 p.m.
Reentry: August 14, 1982
Altitude: 192 × 258 km @ 51.6°

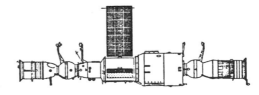

Progress 14 docked to the rear port of Salyut 7 at 3:41 p.m., which was in orbit at 301 × 325 km, July 12. The flight delivered propellant, water, mail, air-regenerator canisters, supplies, the Korund materials processing furnace, materials for processing, and food including tomatoes, cakes, apples, and lemons.

The Korund furnace was equipped with a revolving sample holder, which held up to 12 ampules that were processed one at a time. The unit could be controlled from mission control allowing processing to continue after the crew left the station. Each ampule measured 30 mm in diameter and 30 cm in length.[223] The samples could be heated to between 20°C and 1,270°C at a rate of between 0.1° and 10° per minute. The ampule could also be moved in the furnace at a rate of 100 mm per minute.[224]

The Progress was unloaded by July 16, and refueling was completed by July 20. The Progress undocked August 11, at 2:11 a.m., to clear the aft port for Soyuz T-7. Retrofire was at 5:29 a.m., August 13.

On July 10, Lebedev repaired a fault in the Oasis plant growth unit. For the next few days the cosmonauts unloaded the Progress. Letters to the cosmonauts were usually loaded near the hatch, but after opening the hatch, they found only one letter. Mission control told them that the rest of the letters were behind the other cargo, a tactic used to get the cosmonauts to unload the cargo more quickly. After completing the unloading that same day, the cosmonauts found that mission control was lying and that there were no more letters on the Progress. On July 23, the crew underwent a routine medical examination.[225]

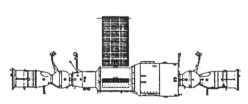

Soyuz T-7

Second Woman in Space

Launched: August 19, 1982,
　　　　　　9:12 p.m.
Landed: December 10, 1982,
　　　　　　10:03 p.m.
Altitude: 194 × 205 km @ 51.6°
　　　　　　(final booster stage)
Crew: Leonid Popov,
　　　　Aleksandr Serebrov, and
　　　　Svetlana Savitskaya
Backup: Vladimir Vasyutin,
　　　　Viktor Savinykh, and
　　　　Irina Pronina
Call sign: Dneip

Savitskaya became the second woman to fly in space on Soyuz T-7. She was a pilot and had 1,500 hours' flight time in 20 aircraft types as experience. Savitskaya was flown primarily as a propaganda stunt in response to plans of NASA to fly female astronauts on the shuttle beginning the next June. At this time, the commander of the Gagarin Training Center and former cosmonaut,

General Beregovoi, said that one other female was training as a pilot-flight engineer at Star City. The backup commander for the mission was probably either Vladimir Vasyutin or Yuri Romanenko.

Soyuz T-7's initial orbit was not released but would be similar to the orbit of the booster's upper stage, which was 194 × 205 km. After ten orbits, the Soyuz was in a transfer orbit of 228 × 280 km. Soyuz T-7 docked at 10:32 p.m., August 20, at 289 × 299 km. This flight emphasized medical experiments on female adaptation to weightlessness and used the French medical instruments from the T-6 flight. The visiting crew also tested the Tavriya electrophoresis device, and made observations of celestial objects with the French Piramig visible near-infrared camera and Czechoslovakian EFO-1 photometer (see Figure 5-13). Serebrov also tested the Braslet devices and French ultrasonic cardiovascular monitor. The crew also made the usual measurements of the station's atmosphere and took samples for analysis on Earth. During their stay on the station, Savitskaya was given full use of the Soyuz orbital module for privacy as required, although she still slept in the Salyut with the others.[226]

The T-7 crew traded spacecraft with the long-duration crew by switching seats and centering weights on August 22 and transferred experiment results to the T-5 capsule for return to Earth. They undocked on August 27, in the Soyuz T-5. The Soyuz capsule landed at 7:04 p.m., 70 km northeast of Arkalyk.[227]

Figure 5-13. During the Soyuz T-7 mission to Salyut 7, Savitskaya and Lebedev change the large film cartridges of the MKF-6M multispectral camera. The telescope housing behind the cosmonauts obscures a view of the aft intermediate compartment. The collapsible shower is stowed above Lebedev. (Source: Tass from Sovfoto.)

The Salyut crew transferred the new Soyuz T-7 to the forward port, undocking from the aft port at 6:47 p.m., August 29. After drifting away from the station, the docking system was activated again with the Salyut's forward port active. The station turned 180° to place the forward port facing the Soyuz and Berezovoi docked again normally.

On August 30, the crew resumed their research program with photographic sessions studying dust storms in the southern U.S.S.R. and natural gas fields near the Caspian Sea. Later that day, a maneuver was made using one of Salyut 7's main engines to change orbit. On September 3 the orbit of the station was 321 × 340 km.

On September 7, the crew completed their latest round of photographic duties and prepared to start a new series of astronomical observations using the X-ray telescope. By September 10, the cosmonauts were conducting routine medical tests using the Beta and Rheograph cardiovascular monitors. The next day they cleaned the station, took a shower, and made routine calls to their families over the two-way television link. These routine activities continued until the next supply mission was launched.[228]

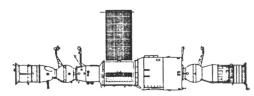

Progress 15

Launched: September 18, 1982, 8:59 a.m.
Reentry: October 17, 1982
Altitude: 195 × 258 km @ 51.6°

The Progress' transfer orbit to Salyut 7 was 214 × 372 km. The cargo ship docked at 10:12 a.m., September 20 at 302 × 372 km. The mission delivered food, supplies, and raw materials for processing. Refueling began on September 21, and was completed by September 24. The Progress was used to boost the station's orbit on September 28, to 312 × 379 km, and again on September 29, to 364 × 384 km. Undocking was at 4:46 p.m., October 14, and retrofire was at 8:08 p.m., October 17.[229]

☆ ☆ ☆ ☆ ☆

On September 24, the crew completed refueling the station and performed medical tests using the Chibis suit and the Echograph ultrasonic cardiovascular monitor. On September 28 and 29, the station was boosted by Progress 15 to an orbit of 364 × 384 km. On October 1 the crew started a new round of astronomical experiments after finishing another series of Earth resources photographic assignments. Objects they observed included the Crab Nebula and Earth's atmosphere using the EFO-1 photometer. Pravda noted that six groups of geologists were searching for mineral deposits in regions around the Caspian and Aral seas and Lake Balkhash previously photographed by Salyut 7.[230]

The cosmonauts also continued with a series of experiments using the Pion device. The Pion allowed viewing of the crystals growth while they were forming during heating. The particular liquid and gas being studied was contained inside a clear flattened disk. A light was shown in one side while the process was filmed from the other side. Temperature and time measurements were also recorded on the film of the experiment.[231]

Progress 16

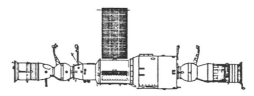

Launched: October 31, 1982,
    2:20 p.m.
Reentry: December 15, 1982
Altitude: 193 × 263 km @ 51.6°

The transfer orbit to Salyut 7 was 290 × 358 km. Progress 16 docked at 4:22 p.m., November 2, at 353 × 362 km. The flight delivered food, supplies, and propellant to continue the long-duration mission. The Progress was mostly unloaded by November 5. Progress 16 was used December 8 to make several maneuvers to raise and change the orbit of Salyut 7 to facilitate the return of the long-duration crew. The Progress undocked on December 13, and was destroyed on reentry as normal.

☆ ☆ ☆ ☆ ☆

On November 7 the crew had a communications session with their families. The next day was spent repairing the station's video recorder which had broken. On November 11, the cosmonauts probably recorded the infrared signature of space shuttle Columbia's launch (STS-5). The closest approach was 80 km at night, and the cosmonauts tried to look for the shuttle but could see nothing in the darkness. Crews on Salyuts often observed Soviet ICBM and ABM test launches. The same day, Berczovoi felt ill. The cosmonauts consulted with doctors who prescribed medication. Lebedev injected the medication and Berezovoi's condition improved. A decision about ending the mission was very near, but by the evening his condition improved more and the decision was postponed.[232]

On November 18, the crew launched a second sub-satellite, Iskra 3, into a 350 × 365-km orbit. This time the crew filmed the deployment. The satellite reentered less then a month later on December 16, 1986. On November 28, the crew used the Kristall furnace to produce alloys. On December 7, the crew cleaned and inspected the station. The mission was ending due to a failure of the Salyut's Delta navigation computer, which began writing data over its instructions. Without the Delta, the station's attitude had to be manually controlled by either the cosmonauts or mission control. Without any backup system other than the cosmonauts, the decision was to end the mission and repair the Delta on the next flight. They left the Korund materials processing furnace on automatic, to obtain samples, while there was no crew creating unwanted motions in the station. In previous experiments using the Kristall furnace, the cosmonauts detached the unit from the station's wall and let it hang unsupported while processing samples. This yielded better results, because it damped motions caused by the crew and equipment of the station that caused micro-gravitational disturbances effecting processing.[233] Among the samples obtained by the crew was a crystal of cadmium selenide weighing 800 grams, and measuring the full 30 cm by 30 mm.

The cosmonauts packed the Soyuz with log books, results from experiments, film, and video tapes. The crew returned to Earth on December 10, 1982, outside of the normal landing window, because of the Delta failure. This forced the crew to not follow mission rules and land at night. The predicted weather for landing was 21-km/hr winds, temperatures of 15°C, and 10-km visibility. They instead descended into low clouds and fog, at $-9°C$ temperatures, in a snow storm. They landed at 10:03 p.m., on a hill in deep snow and high winds, 150 km southeast of Dzhezkazgan. In the high winds, the parachute pulled the capsule on its side and it rolled several times down a hill. The recovery helicopters soon found the capsule by its radio beacon, since its light beacon was barely visible through the snow.[234-236]

Normally, the helicopters would have landed and taken the cosmonauts to a nearby airport for a trip back to Baykonur, but the helicopters were blowing up to much snow to see the ground to land. One helicopter pilot was given permission to land, at his own discretion, to give assistance to the cosmonauts. The pilot took the helicopter down and landed with enough forward speed to see the ground, but he broke a landing gear in the process. The 10-person recovery crew set out flares to guide another to land but the rest were sent away until the weather improved. Twenty minutes later, the cosmonauts were out of the capsule and a snow tractor reached the site. The cosmonauts spent the night in the cab of the tractor and were running out of fuel to keep warm when the helicopters returned in the morning.

The crew did not adjust to Earth as well as the last long-duration crew. They were not as prepared as they had planned, because of the Delta failure. But, they were walking in 3 days, went to the swimming pool often, and returned to Star City by New Year's Day.

During the mission, the cosmonauts had carried out the first fully successful experiments in plant growth with the plants reaching full maturity. Arabidopsis, a common weed, was grown in the Fiton plant growth unit, which used a special nutrient solution and specially filtered air to produced about 200 seeds. Other plants grown with less success were wheat, oats, borage, peas, radish, coriander, dill, and carrots. The Oasis plant growth unit was also tested using electric fields to stimulate plant growth.[237,238]

During the mission, the crew also took 2,500 photos with the MKF-6M camera and 200,000 separate spectrograms with various apparatus. Cosmonaut observations and photos from orbit were said to speed prediction of the spread of major crop disease by several days. The crew also made 40 reports to geologists about possible mineral deposits they had sighted. Fifty hours were spent performing astrophysical observations with the RT-4M telescope. More than 1,000 photographs were taken in astrophysical experiments. In all, 300 experiments were conducted by the crew.[239]

General Beregovoi said at this time that cosmonauts would be in orbit on a permanent basis in the near future. On January 1, 1983, Salyut 7 was in a 342 × 347-km orbit, awaiting its next cosmonaut crew.

Kosmos 1443

Star Module

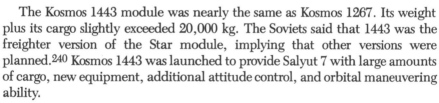

Launched: March 2, 1983,
      12:32 p.m.
Capsule Landed: August 23, 1983,
      3:02 p.m.
Module Reentry: September 19, 1983, 4:34 p.m.
Altitude: 191 × 259 km @ 51.6°

The Kosmos 1443 module was nearly the same as Kosmos 1267. Its weight plus its cargo slightly exceeded 20,000 kg. The Soviets said that 1443 was the freighter version of the Star module, implying that other versions were planned.[240] Kosmos 1443 was launched to provide Salyut 7 with large amounts of cargo, new equipment, additional attitude control, and orbital maneuvering ability.

The Star module was launched into an orbit that was shifted 1.25° longitudinally away from the Salyut's orbit and began a slow approach to the station. This slow approach saved propellant and provided for slower approach that was desirable for the Soviets, since they had little experience in docking the relatively large module. During the extra time before docking, the module's systems were tested and monitored by mission control. During the initial flight of the module, the tracking ships *Korolev* and *Dobrovolskiy* were stationed in the Atlantic Ocean.

As soon as the module docked to the station, it took over all attitude and orbital control of the space station complex. On March 2 the orbit was boosted to 199 × 269 km. The module maneuvered again on March 3 and 5 to 280 × 314 km and again on March 9 to prepare to dock at the forward port of Salyut 7. Docking was completed on March 10, at 12:20 p.m., at an orbit of 325 × 345 km. The total length of the station complex was 28 meters and massed 40,000 kg.[241,242]

The module delivered a 3,600-kg cargo of 600 items including new gallium arsenide solar arrays for Salyut 7, new Delta system memory units, water, air regeneration canisters, air filters, instruments, exercise and medical equipment, movies, film, a guitar, flash bulbs, cloths, spare parts, and foods including fruits, onion, garlic, and mustard. The module was equipped with a rail that could extend into the Salyut, on which bags of supplies were moved. This made the unloading much easier for the cosmonauts. On April 5 and 11, the module lowered Salyut 7's orbit to 293 × 305 km for the next manned mission.[243-245]

Kosmos 1445

Shuttle Development Flight

Launched: March 15, 1983, 1:34 a.m.
Landed: March 15, 1983, 3:27 a.m.?
Altitude: 176 × 217 km @ 50.7°

This flight was the same type as Kosmos 1374. The Academy of Sciences' tracking ships *Volkov* and *Belyayev* were used to support the test flight. The space plane splashed down 556 km south of the Cocos Island in the Indian Ocean and was successfully recovered by Soviet ships in the area. Detailed pictures of the spacecraft were again obtained by Australian aircraft.[246,247]

Soyuz T-8

Aborted

Launched: April 20, 1983,
                5:11 p.m.
Landed: April 22, 1983, 5:29 p.m.
Altitude: 196 × 213 km @ 51.6
Crew: Vladimir Titov,
             Gennadiy Strekalov, and
             Aleksandr Serebrov
Backup: Vladimir Lyakhov,
               Viktor Savinykh, and
               Aleksandr Aleksandrov
Call sign: Ocean

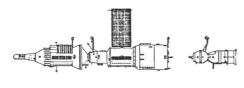

The first launch attempt of the Soyuz T-8 on April 11 was scrubbed because of technical problems, and the launch was again scrubbed on April 14. On April 20, the flight was finally launched, and all was well, until the launch shroud separated. *Flight International* magazine later reported that Titov said that the shroud accidentally tore off the rendezvous radar antenna. The crew, not knowing this at the time, tried to swing the apparently non-responsive arm holding the antenna into position by attitude thruster bursts. After this failed, the crew was told to sleep while mission control thought about the problem. The crew got little sleep. Titov later remarked he slept only a few hours.[248]

The next day, the crew went ahead with plans to try a manual docking with help from mission control. Mission rules previously used by the Soviet's stated that a mission must abort the docking attempt and return to Earth at this point. Flight directors decided to let Titov attempt a manual rendezvous. A normal rendezvous ended at a distance of 10 km from the station when Titov saw the station as a point of light ahead of the Soyuz. Titov later said that in order to dock manually, he had to determine the size of the Salyut in his periscope sight, and mission control would compute the closing velocity, so the crew could make the necessary burn to close on the station. An optical rendezvous from 10 km had never been attempted before by the Soviets and had a low chance of success.[249,250]

When within 330 meters of the station, the Soyuz passed out of communication range of the ground, and Titov lost his source of information on relative velocity. In darkness Titov was not sure of his depth perception to judge closing velocity. With the Salyut running lights on and illuminating the station with the Soyuz's floodlight, he closed the distance to 175 meters, but he feared a collision. He decided to fire the Soyuz thrusters and pass the station, sending the Soyuz into a slightly higher orbit and aborting the mission. Titov said later that he never trained for manual approach and docking before the T-8 flight.

At this point, the Soviet media moved the ranking of the story from the top of the news to just before discussing women's rights in Afghanistan, and dropped all mention of the Salyut 7 docking attempt. The propellant situation on the Soyuz was now critical. With most of the reserves used on the manual rendezvous, there was little extra left for retrofire if additional problems arose. To conserve propellant, the Soyuz attitude control system was shut off while awaiting the next pass over the landing area. The Soyuz was put in the spin-stabilized mode normally used on Soyuz flights in the early 1970s to save propellant. The crew landed safely although they had almost no reserve propellant for retrofire. Landing was at 5:29 p.m., April 22, 60 km northeast of Arkalyk. The crew would have been the first to occupy an expanded Salyut/Kosmos space station and would have performed EVAs to add solar panels to the Salyut.[251,252]

The Kosmos 1443 module raised Salyut 7's orbit back to a normal altitude of 291×347 km on April 28, after the T-8 failure to dock. On June 23, the module boosted the station's orbit from 315 × 328 km to 326 × 337 km to prepare for the next manned mission launch.

Soyuz T-9
Long-Duration Mission—149 Days

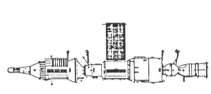

Launched: June 27, 1983, 1:12 p.m.
Landed: November 23, 1983, 10:58 p.m.
Altitude: 199 × 228 km @ 51.6°
Crew: Vladimir Lyakhov and Aleksandr Aleksandrov
Backup: Vladimir Titov and Gennadiy Strekalov
Call sign: Proton

This crew was the backup to the Soyuz T-8 crew. The T-8 backup crew originally consisted of Lyakhov, Aleksandrov, and A. Volkov, but Volkov had been dropped from the crew to allow the Soyuz to carry more propellant to dock with the Salyut at a higher orbit.[253] The Soviets had apparently been willing to use the Kosmos module's precious propellant to lower and raise the station's orbit once for the Soyuz T-8 docking, but not for the second attempt with the T-9 mission. Costly orbit changes to facilitate docking apparently could not be justified just to add a crewman to the mission. Soyuz T-9's transfer orbit to the Salyut 7 was 255 × 303 km. The Soyuz docked to the rear docking port of Salyut 7 on June 28, at 1:46 p.m., at 328 × 343 km, with the Kosmos module at the forward docking port holding the station in position. The combined weight of the Salyut, Star module, and Soyuz was 47,000 kg and the orbit was 325 × 337 km.[254,255]

The crew activated the Salyut and soon began Earth resources work. The Star module would enabled the crew to take more pictures in one week than on the entire 211-day mission. In actuality, the crew spread this work out over 8 weeks, but still took as many as 500 over only a 2-week period. The module's capsule was the only way to return the great amount of film this produced.

On June 30, they opened the hatch to the Kosmos module at 12:49 p.m. and began unloading the module. The crew reported discomfort adapting to weightlessness until the ninth day of the flight. The cosmonauts worked a slightly different schedule than previous crews by working half a day on Saturdays to reduce the amount of work during the rest of the week. The cosmonauts replaced the Delta navigation system memory unit and it was reprogrammed by telemetry from mission control. This procedure required a week to complete, but put the station back into routine operation.

On July 18, the cosmonauts began a series of experiments using the Yelena and Ryabine instruments to measure gamma radiation and changed particles. They also began use of the MKS-M spectrometer-camera. The MKS-M could be pointed with a higher degree of freedom than the other fixed cameras like the MKF-6M or KATE-140. The camera could also be fitted with two different instruments for photographing either the atmosphere or ocean. They also photographed lightning and auroras.[256-259]

The crew took part in a series of Earth resources experiments, focusing on the world's oceans and the Mediterranean Sea in particular. An AN-30 aircraft and a research ship took measurements, while the Meteor 30 satellite and the cosmonauts observed the release of colored dye in the Black Sea in an area 500 meters in diameter. The cosmonauts reported that the dyes could be seen easily with the naked eye. They also began use of the Tsvet-1 instrument. This unit replaced a printed catalog of 1,000 colors used for describing areas observed on Earth. With the Tsvet-1, the cosmonauts could tune a color with knobs to mix a nearly infinite range of colors. They also used the Astra-1 instrument, which measured the cloud of particles and debris surrounding the station created by rocket thruster firings and waste dumping.[260,261]

Another new piece of equipment on Salyut 7 was the Electrotopograph. This experiment was to expose various materials to the space environment in a scientific airlock. The materials would be sandwiched between two pieces of metal plates with a emulsion coated photographic plate next to the sample. After exposure to vacuum, a charge of several thousand volts would be applied to the metal plates. The electric field created would be contoured around any microscopic or invisible defects in the sample material caused during its manufacture, or exposure to vacuum and low temperatures.[262]

Material processing experiments also were conducted using the Kristall and Magma-F furnaces producing cadmium theioselinide for laser uses. The Tavriya electrophoresis unit was also used to continue tests separating biological materials. By July 23, the crew had unloaded about half of the cargo from the Star module. The crew had moved some equipment, including the Salyut's EVA spacesuits, into the module freeing up space in the Salyut transfer compartment.

On July 25, the crew started emergency procedures to evacuate the station when they heard a loud crack. After they realized there was no immediate danger, they investigated and found a impact crater 3.8 mm wide in a window.[263] Cosmonauts on Salyut 6 also found a 3-mm crater in a window before the station was destroyed in 1982.[264] An evacuation of the Salyut would take an estimated 15 minutes. The Soviets also estimated that a hole the size of a pin head in the hull of the Salyut would allow the atmosphere to escape in about one day. With a hole the size of a pencil it would take about 1.5 hours. The cosmonauts could also attempt to seal any leak with a special sealant.[265] Most space debris consists of very small particles, but the relative velocity makes even small particles very destructive.

On August 9, the crew started taking urine samples for 3 days, for later analysis on Earth. The samples were frozen and stored in the Plasma-1 apparatus.[266] In early August, the cosmonauts also began loading the Kosmos reentry capsule with a cargo of 317 kg of film, material samples from 45 experiments including results from the Electrotopograph experiment, used air regenerators, and the failed Delta memory unit. There reportedly was a major malfunction of the Star module that caused it to be undocked early with the reentry capsule only partly loaded. The module had performed more than 100 orbit corrections and attitude changes of the space station. The cosmonauts also commented that controlling the station with the Star module was very difficult and that a fully automatic system was not yet developed to control the space station complex.[267-270]

On August 14, at 6:04 p.m., Kosmos 1443 undocked and the Soyuz T-9 boosted the station's orbit from 314 × 330 km to 315 × 346 km. Kosmos 1443 boosted its orbit to 326 × 348 km and the Salyut lowered its to 313 × 326 km on August 18. The Star module reentry capsule undocked and landed with its cargo on August 23, at 3:02 p.m., 100 km southeast of Arkalyk. The Star module would continue to orbit for almost another month.

The long-duration crew then had to prepare for the next Progress mission by undocking the Soyuz T-9 from the aft port and moving it to the forward port where the Star module had been docked. At 6:25 p.m., August 16, the Soyuz was undocked and drifted back 250 meters from the station, which was then commanded to activate its forward docking system, causing the station to turn 180° to point the forward port at the Soyuz, which was then redocked 20 minutes after undocking.[271]

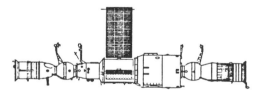

Progress 17

Launched: August 17, 1983,
4:08 p.m.
Reentry: September 18, 1983
Altitude: 186 × 228 km @ 51.6°

Progress 17 maneuvered into a transfer orbit of 257 × 321 km, which was raised to 313 × 326 km in the hours before docking with Salyut 7 at 5:47 p.m., August 19. The Progress was soon used to boost the station to 319 × 339 km. The flight delivered mail, food, and propellant. By August 26, the ship was unloaded and boosted the station's orbit again to 337 × 358 km. Refueling was begun by September 2. Before undocking, the Progress again boosted the station's orbit to 337 × 358 km. After undocking, on September 17, at 2:44 p.m., the Progress performed retrofire at 1:43 a.m. the next morning.[272]

On September 9, during refueling operations, Salyut 7 suffered a ruptured oxidizer line, which vented two of Salyut's three oxidizer tanks to space. The long-duration crew prepared to return to Earth immediately, but the ground controllers found the situation was stable and decided not to return the crew. This failure left the Salyut with only half of its 32 attitude control thrusters operational. It was not known if a malfunction of the Progress caused the failure, but the failure did occur during the refueling operation. The failure forced a change in the work done by the cosmonauts from Earth resources to material processing, since the attitude control system was nearly unusable and could not be used to point the Salyut's cameras.[273,274]

On September 16, the Star module, which had been undocked in August, lowered its orbit from 322 × 338 km, to 288 × 337 km in preparation for destructive reentry and disposal. On September 19, the Kosmos 1443 module, minus the reentry module, was deorbited over the Pacific with a retrofire at 4:34 a.m.

On September 20, the crew again started taking urine samples for three days in continuation of a previous experiment. The samples were frozen and stored in the Plasma-1 apparatus. On September 23, the crew replaced a pump in the water recovery system, set up the Pion to continue material processing experiments, and continued Earth resources work.[275]

Soyuz T-10A

Aborted

Launched: September 26, 1983, 11:36 p.m.
Landed: September 26, 1983, 11:40? p.m.
Altitude: 1 km
Crew: Vladimir Titov and Gennadiy Strekalov
Backup: Leonid Kizim and Vladimir Solovyev
Call sign: Ocean

The crew of Titov and Strekalov (from the aborted Soyuz T-8 flight) climbed into their new Soyuz capsule at T minus two hours in the countdown. Their crewmate Serebrov had been dropped from the mission, just as Volkov had been dropped from the Soyuz T-9 mission. This was again apparently to allow the Soyuz to carry more propellant, enabling it to dock with the Salyut, which was in a higher orbit than during the Soyuz T-8 docking attempt.[276] The Soyuz on its A-2 booster was on the primary A type launch pad, which was the same one that launched Sputnik and Vostok 1. The backup cosmonauts for the flight,

Kizim and Solovyev, were the capcom's (capsule communicators) for the launch, and were in the underground launch control bunker near the launch pad. Temperatures were 10°C with winds at 43 km/hr during the countdown. While waiting, the cosmonauts listened to music over the radio as most crews do before a launch.

At about 1:36 a.m. local Baykonur time, about 90 seconds before lift-off, a malfunction allowed fuel to spill out and flow around the base of the booster. Less than a minute before the scheduled launch, flames and smoke appeared around the booster as the spilled fuel caught fire and the ground controllers commanded an abort before the booster's engines started. The cosmonauts noticed the flames and smoke outside the porthole through the capsule and its launch shroud. A launch abort had been commanded, but the communication lines to the spacecraft were already burned through. Leaving the crew stranded on the booster, since there is no crew control of the abort system.

In the control rooms two launch safety officers operated a backup radio command system and simultaneously activated the abort sequence about 20 seconds after the first attempt. The command was sent by radio to the Soyuz, which started the escape tower solid rocket motors. By this time, flames and smoke had engulfed the entire booster, up to the top of the Soyuz's launch shroud.

Inside the Soyuz, Titov said that he felt the booster sway a little in the wind, and then felt two waves of vibration caused by the explosive bolts firing to separate the capsule from the Soyuz service module and the upper launch shroud surrounding the capsule from the lower shroud. This was immediately followed by a strong jerk caused by the solid propellant abort motor firing pulling the capsule, in its shroud, off the booster. Titov immediately knew that "we did not make it again (to Salyut 7)." The crew experienced a very short 17 gravities of acceleration during the tractor motor firing. Six seconds later, the booster exploded.[277]

The abort motor fired for only 5 seconds, making the high gravity forces easily survivable; then smaller sustainer motors fired, and the large square air brakes deployed to keep the launch shroud vertical. The cosmonauts then heard the pyrotechnics fire to separate the capsule from the orbital module, to allow the capsule to drop out of the launch shroud. This happened at about 650 meters altitude. Next, the heatshield separated and the emergency parachute opened, lowering the capsule to Earth within sight of the launch complex.[278] Titov said he tried to record observations on spacecraft performance during the abort procedure, but there was little time for this as they were contacted by Kizim in the control bunker over the landing radio link. Kizim told them that the abort system had been activated, but the cosmonauts were well aware of this as they looked out the left porthole to see the launch pad burning in the distance. The spacecraft landed hard, 4 km down range of the launch pad. During this abort, there was nothing for the crew to do since the sequence was automatically controlled after initiation by the launch crew.[279]

The cosmonauts were uninjured, but shaken. They had some vodka before getting their mandatory medical check-ups. The remains of the booster and launch pad burned for 20 hours. This flight had nearly the same crew as the Soyuz T-8 flight that also had to be aborted in orbit. This crew would have performed EVAs to place new solar arrays on the Salyut. The T-10A crew would have taken over as the new long-duration mission and the T-9 crew would have returned in November. This would have started the permanent manning of the Salyut 7 space station as General Beregovoi predicted a year earlier. It would be years before a crew exchange would be accomplished because of various problems.[280-283]

At about the same time as the Soyuz launch failure, the environmental control system malfunctioned on Salyut 7, probably because of the failure in the solar array power system, which caused an acrid odor that nearly overcame the crew. It was similar to a failure on Salyut 5 that brought its crew home early. In this case the crew remained on board with irritated eyes until the problem was corrected. The temperature in the station also dropped to 18°C because the solar array problem reduced the power to run the heaters. The cooling caused the humidity to rise to 100%.

By September 29, the crew had taken about 3,500 photographs of Earth. By the middle of October, the crew had gained about 0.5 kg each, and were working normal 10- to 12-hour days. On October 13, the Soviets announced that the mission was nearing the end.[284]

Progress 18

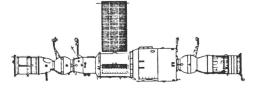

Launched: October 20, 1983, 12:59 p.m.
Reentry: November 16, 1983
Altitude: 193×269 km @ 51.6°

Because of the failure in the propulsion system of Salyut 7, a Progress was launched to provide additional propellant, mostly oxidizer, and the use of its engines to help control the station.

The cargo ship maneuvered away from its booster stage in an orbit of 185×237 km into an initial orbit of 193×269 km. Docking was at 2:34 p.m., October 22, in an orbit of 329×347 km. The flight delivered mail, propellant, materials, equipment, air-regenerator canisters, new components of Tavria experiment, and the second pair of add-on solar panels for Salyut 7. Unloading was completed by November 10. The Progress boosted the station's orbit November 5, to 326×342 km. It undocked on November 13, at 6:08 p.m., and retrofire was at 7:18 a.m., November 16, sending the Progress into the normal destructive reentry.285

After Progress 18 arrived, the crew was able to begin Earth resources work again because the Progress filled the one usable oxidizer tank of the Salyut to provide some attitude control ability. The Progress itself could also help the Salyut maneuver a little. The crew also began scrambling their radio transmissions for the first time in two months indicating possible military work by the crew.

On November 1 and 3, the cosmonauts performed EVAs, the first lasting 169 minutes and the second lasting 175 minutes, to add two small solar arrays to the center Salyut array. This EVA was originally to be conducted by the Soyuz T-10A crew, and changing the mission of the T-9 crew to perform the EVA showed a growing maturity of the Soviet space program. The EVA on November 1, started at 7:47 a.m., and on November 3, at 6:47 a.m. The gallium arsenide arrays measured 4.5 by 1.2 meters and a pair produced 800 watts of additional power.286,287

The same procedure was used for both EVAs. First, Alexandrov moved from the airlock to the center solar array where he secured himself to the station. Lyakhov then brought the tools and new solar array in its container from the airlock to Alexandrov. They then mounted the container onto pins on the center solar array and connected power lines to sockets built into the station. Then, the new array was uncovered and Lyakhov moved away while Alexandrov cranked a winch unfolding the new array. Pins on the top of the new array snapped into catches on the existing array, completing the work.288

Meanwhile, cosmonauts Kizim and Solovyev simultaneously duplicated the work of the cosmonauts in the hydrobasin water tank simulator at Star City to provide any assistance they could with unexpected problems. Both of the spacewalks were also supervised by cosmonaut Romanenko in mission control. Communication during the EVA was augmented by the tracking ships *Volkov* and *Dobrovolskiy*, stationed in the Atlantic Ocean, and the *Gagarin* in the Mediterranean. A drawback to using the tracking ships was the need to relay the signal from the ships to the Kaliningrad mission control. The Sapphire system was used for this, which made the cosmonauts voices very distorted, but understandable if they talked very slowly. After finishing the installation, Alexandrov threw the empty array containers into space making sure they would not accidently throw them into an orbit that would decay bringing the boxes back to hit the station. Mission control was more concerned by his unplanned action because the boxes could be detected by the star trackers of the station, upsetting the guidance systems.289,290

The crew's heartbeats were faster for the first EVA than the second. They were fastest when operating the winch that deployed the array and when closing the airlock hatch.291 The Soviets said that the crew had simulated EVAs on

Earth in the hydrobasin 12 times before the mission. The crew said that the simulation was harder than the actual work.

After the first EVA, the crew immediately began preparing for the next EVA on November 3. They set up fans in their spacesuits to dry out perspiration, replaced components like air filters, which were limited to about six hours use, and tested the suits' systems. They also had to perform a medical examination before the EVA. These EVA procedures along with testing tools and maintaining the space station, eating, sleeping, and exercising took up nearly five days.

The Progress boosted the station's orbit on November 5, to 326 × 342 km and unloading was complete by November 10. On November 13, the Progress undocked. The crew continued with normal work by using the Tavria unit, to produce 35 mg of pure protein from membranes of an influenza virus that was said to supply Soviet laboratories for many months. On November 14, the Soyuz T-9 was used to lower the station's orbit from 324 × 340 km to 322 × 337 km.

The Soviets then announced the imminent return of the long-duration crew. Their return window lasted until December 14. They also announced that there would be a pause until the next long-duration crew was launched to give time to fully investigate the launch disaster of T-10A. This would also give the next crew time to practice EVA techniques to repair the Salyut's oxidizer line. The Soviets also announced that the crew of Alexandrov and Lyakhov were not supposed to fly a record breaking mission. They were supposed to be replaced by the Soyuz T-10A crew.

The return of the cosmonauts was probably a precautionary move since a new Soyuz could not be launched to give the long-duration crew a fresh return spacecraft. Rather than risk the need for a return of the cosmonauts before a new Soyuz could be launched, the Soviets decided to end the mission short of the goal of extending the previous 211-day record. The crew also had not expected or been trained for a record breaking mission. Their main goal seems to have been operating the Kosmos 1443 module and its systems.

The cosmonauts began packing the experiment results including experiment log books, holograms, processed materials, and electrotopograph results in the T-9 on November 21, and they tested the ship's engines. They also loaded trash into the orbital module of the Soyuz to be disposed of when the module is jettisoned before retrofire. Retrofire occurred as usual over the South Atlantic and was monitored by the tracking ship *Volkov*. They landed, with the capsule characteristically rolling onto its side, on November 23, at 10:58 p.m., 160 km east of Dzhezkazgan, in light fog after 149 days in space. Landing was uneventful, although this was the first time the Soyuz T used its extended lifetime as a viable manned spacecraft. The recovery forces for landing consisted of 10 aircraft, 15 helicopters, and 5 search and recovery teams along the flight path.[292]

The crew reported that they felt relatively well after landing. After 8 hours, one of the cosmonauts still experienced dizziness and nausea when sitting or standing. Both preferred maintaining a horizontal position. They lost 4.6 kg average during the flight, of which about 1 kg was gained after the first day after landing.[293] They returned to their preflight weight after 9 days. Calf volume decreased by an average of 11.8%. Both cosmonauts had an altered gait and were unable to stand perfectly still with their eyes closed during re-adaptation. They also noted dizziness and discomfort after head movements. This continued for 7 and 3 days, respectively, for the commander and flight engineer. During the flight, they were exposed to 1,755 mrad radiation, roughly 17 times the low yearly average background dose on Earth.[294]

They took about 3,000 pictures with the MKF-6M and MKS-M, and 100 with the KATE-140 during 43 sessions of more than 300 million square kilometers. Areas not only included the U.S.S.R., but also Australia, Africa, and South America in cooperation with a UNESCO program studying biological resources. They had around 20 conferences with geologists, meteorologists, and other specialists during the flight advising them on the observations they were performing. About 200 television conferences with family members and others were conducted. Eighty-eight samples of thin film coatings were obtained with the Isparitel device. They also took more than 16,000 other photographs and spectrographs in addition to holographs made by the KGA-2 holographic camera of materials processing experiment samples. Fourteen experiments with the Pion-M mass transfer experiment were performed and recorded on film and holographically. The Tavriya electrophoresis device was used to obtain 8 ampules, 35 mg total, of an anti-influenza and anti-viral preparation. They performed 19 electrophotograms of materials after exposure to space in an airlock for hours or days. They made observations with the Piramig, PCN cameras, the SKR-02M, and the RT-4M telescope. The Yelena gamma ray telescope was used for more than 300 hours measuring high-energy particles, especially over the South Atlantic Anomaly. The station's orbit had been boosted, or changed about 60 times during the mission. The Oasis plant growth unit was used to grow plants using electrical stimulation and the Svetobloc-T was also used to try growing tomatoes. Medical exams were carried out a total of 128 times. [295–298]

Kosmos 1517

Shuttle Development Flight
Launched: December 27, 1983
Landed: December 27, 1983
Altitude: 180 × 221 km @ 50.6°

This flight was the same as the Kosmos 1445 flight, but instead of landing in the Indian Ocean, the space plane landed in the Black Sea to avoid picture-taking by Australian ocean patrol aircraft as during the recovery of Kosmos 1445.

A Soviet tracking ship monitored the retrofire as it occurred over the southern Atlantic. Further details are unavailable since the recovery took place in the Black Sea.[299,300]

On January 1, Salyut 7 was in a 314 × 327-km orbit. On January 11, Salyut 7 was lowered to a 298 × 323-km orbit, and on February 1, it was lowered again to 292 × 303 km in preparation for the next 3-man Soyuz launch and long-duration mission.

Soyuz T-10B

Long-Duration Crew—237 Days
Launched: February 8, 1984, 3:07 p.m.
Landed: April 11, 1984, 5:50 p.m.
Altitude: 198 × 219 km @ 51.6°
Crew: Leonid Kizim, Vladimir Solovyev, and Oleg Atkov
Backup: Vladimir Vasyutin, Viktor Savinykh, and Valeri Polyakov
Call sign: Mayak

This crew's main task was to repair the serious engine system problem on Salyut 7, and if possible set new endurance records. Kizim and Solovyev had been training extensively in the hydrobasin at Star City for their future EVAs to repair the engine system. Solovyev was well acquainted with the Salyut systems because he had written many of the documents used on the station. Atkov was a physician trained to monitor the crew's adaptation to weightlessness and their response to various tests. Atkov also had the authority to end the mission if he observed dangerous deterioration of the crew's health.[301] The Soviets announced that a doctor would be a part of all long-duration missions in the future (they would later abandon the plan).

The Soyuz separated from its upper stage in orbit at 197 × 207 km, and maneuvered into an initial orbit at 198 × 219 km. The transfer orbit to Salyut 7 was 227 × 270 km. At 5 km distance, the crew sighted the station and activated the Salyut rendezvous system and the Soyuz docking camera. Responding to the activation of the rendezvous system, Salyut turned to point at the approaching Soyuz. After attempting to dock, Kizim backed the Soyuz away from the station because he could not see the markings on the docking target, which was in intense sunlight. The station fortunately moved into shadow shortly afterwards and Kizim approached again and docked at 5:43 p.m., February 9, at 289 × 296 km. The cosmonauts moved into the station with flashlights to turn on the station's systems and commented on the smell of burnt metal from the docking drogue. By February 17, the Salyut was completely reactivated. The Mayak's mission would follow a schedule allowing for 9 hours of experiments, one hour of preparation for the next day's work, nine hours of sleep, two hours of exercise, and three hours of personal time to eat, etc.[302]

On February 15, the crew serviced and loaded film into the MKF-6M and KATE-140 cameras. Atkov began observing crew management of the station's environment and hygiene. He was also responsible for many of the housekeeping chores and Progress unloading, which freed the commander and engineer to perform scientific experiments. A day before the launch of Progress 19, the Salyut was in orbit at 282 × 288 km.[303]

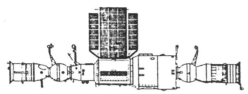

Progress 19
Launched: February 21, 1984,
9:46 a.m.
Reentry: April 1, 1984
Altitude: 186 × 245 km @ 51.6°

Progress 19 docked at 11:21 a.m., February 23, with the station in an orbit of 282 × 286 km. The crew observed the docking, but had to wait until the next day to open the cargo ship's hatch. The flight delivered 2,094 kg of supplies total including 300 kg of food, water, oxygen, mail, newspapers, and letters, 800 kg of propellant, 300 kg of film, spare parts and materials, and new Indian camera equipment to be used by the Soyuz T-11 crew. The Progress boosted the station's orbit on February 25 and 26 to 306 × 311 km.

The Progress was unloaded only as the supplies were needed, slowing the usual unloading process. Oxidizer transfer was completed by March 12, and by March 30, refueling was completed. The Progress boosted the station's orbit to 289 × 303 km, before undocking at 12:40 p.m., March 31. Retrofire was at 9:18 p.m., April 1, sending the Progress filled with trash into a destructive reentry as normal.[304]

☆ ☆ ☆ ☆ ☆

On February 24, the station was carefully oriented, and after turning off the interior lights and the solar array, the cosmonauts photographed Comet Crommelin with the French Piramig camera. On March 19, the Isparitel experiment was installed and tested in the airlock. This experiment was a prototype electron beam welder, cutter, and sprayer. The crew also began photographic sessions over India in preparation for the coming international flight.[305]

During the mission, Atkov performed a medical check of the crew about every 10 days as a part of several experiments being done to investigate aspects of adaptation to weightlessness. Some of the experiments were Biokhim, which measured blood electrolytes in flight; Optokinez, which was a vestibular eye movement test using the station's video monitor; Claznoye, which studied blood flow to the eye and movement of the blind spot; and Sport, which was an exercise program.

Soyuz T-11

Eleventh International
Crew—India

Launched: April 3, 1984,
5:09 p.m.
Landed: October 2, 1984, 1:57 p.m.
Altitude: 195 × 224 km @ 51.6°
Crew: Yuri Malyshev, Gennadiy Strekalov, and Rakesh Sharma
Backup: Anatoliy Berezovoi, Georgiy Grechko, and Ravish Malkhotra
Call sign: Beacons

The Indian international mission was similar to the French international mission of Soyuz T-6. Large-scale Indian cooperation in the space field with the Soviets began with the launch of the Aryabhata satellite on a Soviet booster in April 1975. India also cooperated with the United States and other countries to help start its own space program. Sharma and Malkhotra both arrived at Star City in September 1982 to begin training for the mission. Originally, Strekalov was the backup for Rukavishnikov, but an illness prevented Rukavishnikov from flying. Grechko was then assigned as Strekalov's backup for the mission. Sharma was eventually selected to be on the primary crew, putting Malkhotra in the backup role. Before becoming a cosmonaut, Sharma was a squadron commander with 1,600 hours jet flight time.

Soyuz T-11 separated from the booster's upper stage in orbit at 193 × 216 km and maneuvered into an initial orbit of 195 × 224 km A pair of maneuvers on the fourth and fifth orbits made the T-11's transfer orbit 222 × 271 km. The Soyuz docked at the Salyut's rear port on April 4, at 6:35 p.m., at 286 × 299 km, and the crew opened the hatch to the Salyut at 9:36 p.m.

Using the camera equipment shipped up on the last Progress flight, the cosmonauts made Earth resources observations in the Terra experiment. The Indian government was especially interested in mapping to search for water in arid regions, surveying for potential hydroelectric power plants, and searching for mineral and petroleum deposits. Salyut 7 made 9 passes over India for photography, using the MKF-6M to take about 166 photos. The KATE-140 was used to take about 200 photos and hand-held cameras were used to take more. There were also simultaneous ground surveys of the photographed areas to confirm and better analyze camera results.[306–308]

The visiting crew also experimented making thin alloy coatings in the Isparitel apparatus. In another experiment called Supercooling, a sample of silver germanium alloy was cooled with liquid helium on one side and melted by a laser on the other. It was hoped this would make the sample much stronger. Other adaptation investigations included Optokinez, Profilaktika, and Braslet. Membrana was an experiment to investigate calcium loss in methods to counter it. Vector was an experiment using an Indian designed unit to improve electrocardiogram analysis by measuring chest movement during heart beats. The Ballisto experiment also measured the accelerations of the body caused by heart pumping action, and it revealed changes in the heart's position and shape in prolonged weightlessness. Opros was a standard questionnaire to determine psychological condition, and Yoga was an attempt by Sharma at controlling weightlessness adaptation using practiced yoga exercises.[309,310]

The crew also held a joint television news conference with officials in Moscow and Prime Minister Gandhi. After completing 43 experiment sessions, the international crew returned in the T-10B spacecraft, leaving the fresh T-11 spacecraft for the long-duration crew to use. They undocked at 2:27 p.m., April 11 after loading the capsule with experiment results and exposed film. Landing was at 5:50 p.m., one minute ahead of schedule and one km from the target point, 46 km east of Arkalyk.[311–313]

☆ ☆ ☆ ☆ ☆

After the visiting crew left in the T-10B spacecraft, the long-duration crew moved the new T-11 spacecraft to the forward docking port. Atkov said the crew felt very tired after the visiting crew left the station. They knew that the others were returning to the relative luxuries of Earthly life while they still had several months left in their mission.[314]

Some typical military activities of the crew were reported in *Aviation Week* magazine to be observing Soviet military maneuvers, the release of smoke on the ground to camouflage movement; observing ABM tests; monitoring naval movement; and testing ground- and space-based laser tracking and target acquisition.

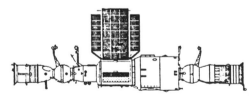

Progress 20

Launched: April 15, 1984,
      12:13 p.m.
Reentry: May 7, 1984
Altitude: 195 × 260 km @ 51.6°

Soon after separating from the booster's third stage at 183 × 247 km, Progress 20 maneuvered to 195 × 260 km and then to a transfer orbit of 236 × 269 km. The Progress docked with the station's rear docking port at 1:22 p.m., April 17, at 278 × 290 km. The flight delivered mail, air, parts, a 40-kg tool kit containing 25 tools for the engine and attitude control repairs, batteries, air regenerators, and propellant.

Progress 20 also carried a special extension on the forward end of the orbital module, which was unfolded and locked on ground command after docking. On the extension was a place for an EVA anchor to be used in repair EVAs. This enabled a cosmonaut to "stand" on the Progress orbital module while another worked attached to the Salyut.

The Progress boosted the station's orbit on April 19 to 285 × 326 km and again on April 20 to 301 × 332 km. After the Salyut engine repair EVAs, the Progress undocked on May 6, at 9:46 p.m.[315,316]

On April 24, 26, 29, and May 4, Kizim and Solovyev performed EVAs in an attempt to correct the Salyut's propulsion system problems. Atkov monitored them from inside the Salyut, which was equipped with only two EVA spacesuits. Ryumin, who saved Salyut 6 during an EVA to jettison the KRT-10 telescope, was the flight director for most of the EVAs. Communication coverage of the station's orbit during the EVAs was extended by using two tracking ships, one in the Atlantic and one in the Pacific, from 20 to 50 minutes. A special tool kit delivered by Progress 20 was used for the repairs.

The first in the series of EVAs began at 8:31 a.m., April 24, and was 4.25 hours long. During it, the crew attached a telescoping ladder-like structure to the Salyut that curved around a third of the Salyut's propulsion section. This required the cosmonauts to drive pins into the hull of the Salyut to attach the ladder to the station. They then placed a foot restraint or anchor on the ladder. An anchor was also placed on a specially built extension on Progress 20's orbital module. They also placed tool containers to the hull of the Salyut, and set up a path of hand-holds or tethers from the airlock on the transfer compartment to the work area to speed subsequent EVAs.

The next EVA was 5 hours long and replaced a valve assembly and fuel lines on the damaged half of the engine system, starting at 6:40 a.m., April 26. Kizim did most of the actual work on the ladder while Solovyev helped, attached to the Progress. It took 20 minutes for the cosmonauts to get into position to start work. First, Kizim opened an access plate on the side of the station's propulsion section near the Salyut's docking antenna. Next, he cut through an insulating covering, first poking a hole in the thin material using a pneumatic punch, and then cutting it with a tool resembling a butter knife. He then opened the filler assembly that contained the valve to be replaced. This took two hours to accomplish because at least one bolt on the filler was covered with epoxy putty and the wrench would not fit the bolt. After some hard work the bolt was removed.

They replaced the valve and installed jumper pipes in the propellant system launch feed ports to test the repair. Atkov, from inside the Salyut, pressurized sections of the propellant line with nitrogen from the Progress' supplies. Soviet mission planners had hoped that the repairs could end there, but by pressurizing individual sections of propellant line using the jumpers, they found a section of the line had to be replaced. The cosmonauts asked for more EVA time beyond the planned 4.1 hours and were allowed to continue. When they completed all the work that could possibly be done during the EVA, they replaced the insulation over the open hole in the side of the propulsion section and returned to the Salyut. The crew was not scheduled to make another EVA for a few days to allow them time to rest and take care of other routine space station needs, but they talked mission control into allowing a third EVA on April 29. This EVA, which started at 5:35 a.m. and lasted 2.75 hours, was to install a new section of line to bypass the damaged section, test its integrity by pressurizing the line with nitrogen supplied by Progress 20, and put an insulating cover over the new line.

While the commander and flight engineer dealt with EVA preparations, Atkov continued unloading Progress 20. By May 4, most of the cargo had been unloaded and water had been pumped into the Salyut tanks. Another 2.75-hour EVA started at 3:15 a.m., May 4. Its objective was to connect the secondary line to bypass the ruptured line. Although the work was completed for the time being, the repairs were not complete. To complete the repair the old propellant line needed to be closed off. A special tool to do this was being prepared on the ground. After completing the hard EVA work, the cosmonauts were given the rare time to take showers on May 8.[317-320]

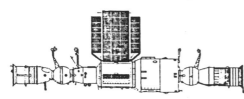

Progress 21

Launched: May 8, 1984,
2:47 a.m.
Reentry: May 26, 1984
Altitude: 190×243 km @ $51.6°$

The Progress separated from its spent upper stage in orbit at 185×237 km and maneuvered to its initial orbit of 190×243 km. The transfer orbit to the space station was 243×277 km and the Progress docked with the Salyut at 4:10 a.m., May 10, in orbit at 277×319 km.

The flight delivered propellant, food, another pair of add on solar panels, and other supplies. On May 20, the Progress boosted the station's orbit from 272×312 km to 296×347 km and again on May 25 to 334×355 km. The Progress was undocked at 1:41 p.m., May 26 and made a normal destructive reentry.[321-323]

On May 11, the cosmonauts continued to unload the Progress and began experiments measuring the atmosphere around the station and underwent medical exams. On May 18, the crew started an exercise program called Sport, which under Atkov's supervision increased the intensity of exercises and tried to shorten the time required for exercise to prevent muscle and bone mass loss.[324]

On May 19, Kizim and Solovyev went EVA again at 9:52 p.m., to add another pair of 9-square-meter gallium arsenide solar arrays to the right Salyut array (right as viewed from the rear).[325] An identical pair of the arrays had been added to the center Salyut array by the Soyuz T-9 crew. This second pair of solar panels reportedly added 1,200 watts of power (at 20 amps) to the Salyut power system. The EVA lasted 3.1 hours and was not related to the engine system work previously performed. This time, after the crew installed the first array, Atkov rotated the main solar array 180° so the crew could install the other without moving themselves and their equipment. The handle of the winch used to raise the solar arrays broke off during the EVA, but the work was completed. Solovyev then connected the new arrays to the Salyut power system by way of the connectors built into Salyut 7 before its launch more than two years earlier. The cosmonauts carefully threw the empty array containers into space to dispose of them. They cut off a piece of the left main Salyut solar array for return to Earth and analysis. They also installed a small new antenna on the Salyut for an unspecified purpose.[326–328]

Progress 22

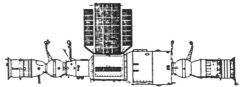

Launched: May 28, 1984,
      6:13 p.m.
Reentry: July 15, 1984
Altitude: 188 × 244 km @ 51.6°

Progress 22 maneuvered to an initial orbit at 188 × 244 km after separating from the third stage in a orbit of 179 × 209 km. The transfer orbit was 290 × 331 km and the cargo ship docked with the Salyut station at 7:47 p.m., May 30, at 334 × 358 km. The flight delivered propellant, food, mail, 45 kg of film, 40 kg of medical equipment, 25 kg of instruments, new storage batteries for the station, and supplies including air that would be needed for future EVAs.

The Progress was used to boost the station's orbit on June 22 to 365 × 383 km, then it lowered the orbit to 307 × 354 km on June 11 to adjust the time for the next launch. The Progress again raised the orbit on June 14 to 334 × 358 km. Progress 22 undocked on July 15 at 5:36 p.m. The Progress carried out a clean separation, which only used the springs in the docking collar to push the spacecraft away from the station. The Soviets theorized that short engine burns, normally made to increase the separation rate, were harming the Salyut's solar arrays when the rocket exhaust hit them.[329,330]

☆ ☆ ☆ ☆ ☆

On June 2, after loading the MKF-6M camera with new film cassettes delivered by Progress 22, the cosmonauts began a new series of Earth resources observations centered on the oceans of the world. The cosmonauts reported that they could in some cases see mountains underwater. The usual land observations were continued and the crew also began astrophysical experiments. The oceanographic studies also reportedly included experiments aimed at tracking ballistic missile submarines using a side-looking radar. The control target was a Soviet Delta type submarine in the western Pacific. Radar results from U.S. shuttle flight STS-41G showed that sub-surface waves could be detected.[331,332]

On June 8, the cosmonauts carried out the usual medical examinations and continued Earth resources work. During some of the photographic sessions, the cosmonauts used the Niva video cameras along with the traditional film cameras. Their training had included many aircraft flights simulating conditions with the same equipment they would be using on the Salyut.[333] The crew had taken around 1,800 photos using the MKF-6M, and about 500 with the KATE-140. They also conducted a practice evacuation of the station and performed medical tests using the Chibis suit.[334] On June 21, the station was in orbit at 328 × 365 km. By June 29, the crew had measured the atmosphere around the space station and refueled the Salyut with oxidizer from Progress 22.

Sub-Orbital Kosmolyot
Shuttle Development Flight
Launched: July 4, 1983
Landed: July 4, 1983
Altitude:

Nothing more is known about this flight except that it was of the same type spacecraft as Kosmos 1374. The miniature space plane was launched from Kapustin Yar by the C-1 booster into a sub-orbital flight path. *Aviation Week* magazine reported that many other sub-orbital flights in this series have apparently been flown, but the information is not publicly known.[335]

By July 6, the Salyut's propellant tanks had been refueled. By July 13, the Progress was loaded with trash for disposal to be burned up on reentry. After the undocking of the Progress, the Salyut maneuvered to adjust the station's orbit to 318 × 358 km.[336,337]

Soyuz T-12
First Female Spacewalk
Launched: July 17, 1984,
            9:41 p.m.
Landed: July 29, 1984, 4:55 p.m.
Altitude: 198 × 225 km @ 51.6°
Crew: Vladimir Dzhanibekov,
        Svetlana Savitskaya, and
        Igor Volk
Backup: Vladimir Vasyutin,
         Yekaterina Ivanova, and
         Viktor Savinykh
Call sign: Pamir

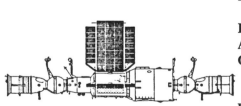

About 9 minutes after liftoff, the Soyuz separated from its booster upper stage and attained an initial orbit of 198 × 225 km. The Soyuz then entered a transfer orbit to Salyut 7 at 271 × 304 km. The Soyuz T-12 docked at 11:17 p.m., July 18, at an orbit of 333 × 356 km. The T-12 for the first time used a new communication channel that transmitted docking information to the Salyut during the approach. This information was previously only available to flight controllers. This was the first time a Soyuz with three people could reach the Salyut's operational altitude. This suggested modifications to the launch vehicle slightly increasing payload capability. After the usual 3-hour check of the docking system and shutting down the Soyuz, the crew opened the hatch and entered the Salyut.

While Savitskaya's work on the flight gained most of the attention at the time, Cosmonaut Volk was later announced as the cosmonaut in charge of shuttle flight testing. His program of research for the flight was somewhat different than usual. He exercised very little and took medical preparations to reduce the effects of adaptation to weightlessness. After landing, his flying skills were also tested. All these activities were investigations into how well cosmonauts could fly a space shuttle after a period of weightlessness in orbit. This was not clear at the time of the flight, but the description of his purpose on the mission was identical to that of the future Soyuz TM-4 mission of Soviet shuttle test pilot Levchenko.

Savitskaya's mission was to perform an EVA as a propaganda stunt that gave the Soviets another space first (NASA astronaut Kathy Sullivan would perform satellite refueling tests during an EVA on STS 41-G, in November 1984). Again, as on the Soyuz T-7 mission, Savitskaya was given exclusive use of a Soyuz spacecraft for privacy. The six people on board the Salyut complained that they were overcrowded and that some activities were not possible, like using the exercise equipment, because it would disturb the others.

Savitskaya's EVA began at 6:55 p.m., July 25. At first, Dzhanibekov stayed in the airlock and filmed Savitskaya, while she worked nearby outside the station. After placing a work station with foot restraints on the outer hull, she tested the URI multipurpose electron beam tool to cut and weld 0.5-mm titanium and stainless steel with solder of tin and lead, and sprayed a silver coating on aluminum. The URI evolved from the Isparitel and Isparitel-M experiments and was tested during weightlessness training flights. The URI comprised a portable 0.5-meter square mother unit weighing 30 kg, carried by the cosmonaut or attached to hand rails. It was connected by a cable to a hand welder that weighed 1.5 kg, and looked like a double-barrel flare pistol. One barrel was used for welding or cutting and the other for spraying. The crucible of material to be sprayed was placed in the beam. When it was vaporized by the beam, it resolidified on the nearby surfaces. The welder had four folding sample holders that each held 6 samples for experiments. The temperature of objects could also be monitored by a non-contact infrared thermometer that looked like a pistol. The welder was tested on two stainless steel and four titanium samples.[338,339]

After Savitskaya completed her work, the cosmonauts switched places and Dzhanibekov tested the device. The EVA was observed on television by mission control. Before returning to the station, they removed the Ekspozitsiya materials experiment from the Medusa fixture on the station's hull for return to Earth. After 3.5 hours they returned to the station and removed the URI samples from the URI for study on Earth.

It was later revealed that Savitskaya and Dzhanibekov were to perform the last repair tasks on the Salyut propulsion unit, giving Savitakaya credit for saving Salyut 7, in addition to being the first female EVA. However, the long-duration crew demanded that they have the honor of finishing the work they had begun. So the T-12 crew briefed the long-duration crew on the procedures and specially designed tools they would use to complete the repairs of the Salyut. They used video tapes of simulations to help illustrate the procedures.[340]

From July 19 to 26, several Electrotopograph experiments were performed in the scientific airlock using new film delivered by the Soyuz T-12. This required the airlock to be vented to space for up to 40 hours at a time, exposing it to full sunlight. Kizim was credited with greatly shortening the planned cooling period from 6 hours to only minutes by putting the station's airlock into shadow.

Among the other experiments carried out during the joint flight were more Braslet tests, the French Tsitos-3, and an experiment in which the cosmonauts mixed various substances like mortar in weightlessness to determine what could be used to plug gas and oil wells on Earth with less leakage. Tavriya was an electrophoresis device to test the separation of biological substances into four parts. Its main chamber was 90 cm long with 230 needles to draw off the different layers of material. The results were expected to be 15 to 20 times purer than obtainable on Earth. Among the preparations processed were an antifluenza vaccine and antibiotics for agricultural use. Sixty samples were obtained with the unit during the short flight. The EFO-1 photometer was also used in upper atmospheric research. The Astra-1 camera was used to measure the gases and particles emitted by and surrounding the station. Resonance experiments were conducted and samples of the station's air were taken for routine analysis on Earth.[341-343]

On July 28, the crew started packing the experiment results and tested the Soyuz T-11 main engine for the return to Earth the next day. Before undocking on July 29, the Soyuz T-11 was used to boost the station's orbit to 342 × 372 km. They landed on July 29, at 4:55 p.m., 140 km southeast of Dzhezkazgan. After landing, Volk tested his flying skills as a part of the Soviet shuttle research program. The crew returned to Star City on August 7.

Soon after the flight, at the IAF Congress, Savitskaya said that she was qualified to fly the Soyuz T. After the mission, Savitskaya said that 10 women were currently in cosmonaut training at Star City. Shatalov also said that more women would be on the third-generation stations, possibly as medical or meteorological specialists. He added that men will still probably do the heavy work like unloading transports. There were also reports that Rukavishnikov said that an all-female crew was in training.[344,345]

On July 30, the Salyut crew was given a day off and they had communications sessions with their families. The Salyut was in orbit at 343 × 387 km at the time.

On August 8, Kizim and Solovyev performed an EVA at 12:46 p.m. for 5 hours to finish the engine repairs, which included closing off the bypassed propellant line by crushing the pipe at both ends of the rupture with the special tool delivered by the T-12 crew. The tool was a hand pneumatic press with a compressed air source of 250 atmospheres and able to exert a force of 5 tons. They also cut out a 20-cm-square piece from the main solar arrays of the station for return to and examination on Earth. The tool they used to do this enabled them to cut and place the sample in a holder without touching its surface and contaminating it. Recovery in this pristine condition would allow researchers to determine how much dust and other particles were on the surface. Solar cells also degrade when exposed to ultraviolet light and micrometeors chip the surface decreasing light reception. The engine system repairs to Salyut 7 turned out to be the worst case envisioned by flight planners. Feoktistov himself thought that only four EVAs would be necessary. The total EVA time required for the repairs was 22.8 hours.[346]

Progress 23

Launched: August 14, 1984,
            10:28 a.m.
Reentry: August 29, 1984
Altitude: 186 × 250 km @ 51.6°

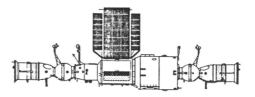

The Progress separated from the booster's third stage, and attained an orbit of 179 × 224 km. The spacecraft then maneuvered into a transfer orbit of 292 × 362 km. It docked at 12:11 p.m., August 16, at 341 × 369 km and boosted the station's orbit to 351 × 387 km. The flight delivered Siren and RS-17 X-ray telescope equipment, propellant, food, mail, air regenerators, materials, equipment, water and air to replenish that which was lost in the airlock during EVAs.

By August 20 the Salyut had been refueled with oxidizer. The crew completed maintenance work and propellant transfer on August 21. Before undocking, the Progress boosted the station's orbit to 373 × 375 km. Progress 23 undocked at 8:13 p.m., August 26, and retrofire was at 5:28 a.m., August 28.[347]

☆ ☆ ☆ ☆ ☆

After the Progress docked with the station, the repaired Salyut engine system was reactivated and functioned normally.[348] Medical checks were made on August 20, including using the Echograph ultrasound device and cardiovascular monitors. On August 24, the crew carried out experiments using the Genom electrophoresis unit to separate a DNA solution into different parts. The experiment was filmed using ultraviolet light. Seven hundred samples were taken in all during this experiment. The cosmonauts also continued filling the Progress 23 with used equipment in preparation for undocking and performed more medical tests.

When the Progress undocked on August 26, it uncovered the aft hatch, which was then opened by remote control by the cosmonauts. The Soviet/French Siren X-ray telescope had been installed in the intermediate compartment and was able to make observations through the opened hatch as the Salyut was pointed using attitude control thrusters. On August 27, the crew continued medical checks and used the Chibis vacuum suits to stress their cardiovascular systems.

On August 28 and 29, the crew participated in an Earth resources research program, which was a cooperative effort of the U.S.S.R. and some Soviet bloc countries. Photos were taken by the Salyut crew, Mi-8, AN-30, and AN-2 aircraft, and the Kosmos 1500 satellite. Research ships also took ground measurements during the comprehensive program, which was similar to international mission efforts. The cosmonauts also observed the Iran-Iraq war, reportedly seeing explosions caused by bombs or missiles. They observed ships being attacked and reported watching one ship burn for nearly the entire mission.[349,350]

In the first week of September, the Siren telescope was used to take spectrographs of the Crab Nebula and objects in Cygnus. The RS-17 telescope made similar measurements but was designed to receive high-energy X-rays from 2,000–800,000 electron volts. The telescopes were installed in the intermediate compartment and the rear docking hatch was opened during observations. The data was recorded on magnetic tape.[351] By the end of the mission the next month, the crew had performed 46 sessions operating the two telescopes.

On September 22, the crew used an infrared radiometer to remotely measure temperature at various parts of the space station. Medical examinations continued at a frequent rate in preparation for return to Earth. The Chibis suit was used and the cosmonauts cardiovascular systems were measured by various apparatus and blood samples were taken. The blood samples were spun in a centrifuge to separate the blood components for storage and later return to Earth. By September 27, the crew was using the Chibis suits often to help recondition their bodies for normal gravity, in addition to the normal two hours of exercise. Pressure in the Chibis suit was lowered to 45 mm Hg below station atmospheric pressure, which was not quite as low as Skylab astronauts used their negative pressure device in 1973–74. The usual exercise consisted of running 5 km on the treadmill, and pedaling 10 km on the ergometer.

A Soviet device called Argument was used during the flight to make images of the heart, similar to the French Echograph ultrasonic cardiovascular monitor.

The Soviet device was to be mass-produced for use in clinics and by emergency crews. Atkov reported that the supply of vitamins was being used as planned and that the Salyut medical kit had not been used during the flight. The medical kit included drugs for headaches, colds, and insomnia, as well as treatments for burns, traumas, and hemorrhages.[352–355]

On September 29, they began loading the Soyuz T-11 with experiment results, including samples of the air in the station for analysis of microflora. During reentry, the capsule experienced loads of 5–6 gravities, higher than the normal 3–4.[356] At 1:57 p.m., October 2, the crew landed 145 km northeast of Dzhezkazgan. They had been in space for a record 237 days but were in good condition (see Figure 5-14). Three hours later they were back in Baykonur and feeling fatigued. The crew's tibia volume decreased by 15% on average, and their weight remained about the same as before the flight. By the third day back on Earth, they were walking outside, but a full recovery took 3 weeks. It was noted that Atkov was 5–6 cm taller than at the start of the mission due to normal expansion of the spine in weightlessness. They had operated 100 experiments during the flight. Thirty different medical studies were performed a total of 200 times by the crew.[357,358]

Atkov and the Soviets were convinced that their research into calcium loss in flight had revealed that loss levels off after a few months. They were disappointed to learn from NASA doctors later at a conference that the Soviet measurement methods were not accurate enough to show any leveling off, if it

Figure 5-14. The Soyuz T-10B Mayak crew of Kizim, Soloyvov, and Atkov returned to Earth after 237 days in orbit. They are wearing the Soyuz T type pressure suits while they wait to go to their first medical checks before returning to Baykonur. Their capsule rolled onto its side after landing and is visible in the background. (Source: Tass from Sovfoto.)

did occur. Thus, there was still no known way to prevent continued calcium loss in weightlessness.[359]

Kosmos 1614

Shuttle Development Flight

Launched: December 19, 1984, 6:53 a.m.
Landed: December 19, 1984, 8:28 a.m.
Altitude: 173 × 223 km @ 50.7°

This was the last orbital flight of the series begun by Kosmos 1374. Again, the spacecraft landed in the Black Sea after one orbit. Five other sub-orbital flights not listed in this book have been reported by *Aviation Week* magazine, but no further information is available.[360]

☆ ☆ ☆ ☆ ☆

On December 19, 1984, a statement was issued that all was well on Salyut 7, which was in an orbit of 366 × 387 km. On December 30, the orbit was 365 × 370 km, and the next day Radio Moscow stated that manned missions to the station would continue. The Soviets intended to launch the next mission, Soyuz T-13, with the crew of Vasyutin, Savinykh, and A. Volkov to further extend the limits of manned flight to 10 months.[361] The crew would have performed EVAs to test a beam device as well.

The crew would also have been visited by the Syrian intercosmonaut crew, which started training in 1985, and an all-female crew consisting of Svetlana Savitskaya, Yekaterina Ivanova, and Yelena Dobrokvashina.[362] The female crew did not get the chance to fly to Salyut 7 after the long-duration crew returned prematurely. After that, international mission agreements with France, Afghanistan, and Bulgaria pushed back the all-female mission until it was indefinitely postponed. Even in 1987, there were still reports of up to 10 women cosmonauts training at Star City.[363] The Soviets lost control of Salyut 7 in the winter of 1984–1985, and it drifted for months totally abandoned.

On March 2, the Soviets issued a statement that Salyut 7 had completed its mission. But before the month was out, they decided to try a risky space rescue to repair the station. The repair mission was probably a result of the great success of the U.S. space shuttle and the repair of the Solar Maximum satellite. The Soviets objective would be to make Salyut 7 capable of at least making a controlled reentry. A Progress mission to stabilize the Salyut was impossible, since the Salyut must orient itself to the approaching spacecraft for an automatic docking, even when the Progress was controlled from the ground.

The likely cause of the failure was probably determined after receiving periodic weak telemetry signals from the station. At about the same time, in March, it was reported that President Reagan was considering offering a joint repair mission using the shuttle, but the offer never materialized.

Soyuz T-13

Repair and Long-Duration
Mission—169 Days

Launched: June 6, 1985, 10:40 a.m.
Reentry: September 26, 1985, 1:52 p.m.
Altitude: 198 × 222 km @ 51.6°
Crew: Vladimir Dzhanibekov and Viktor Savinykh
Backup: Leonid Popov and Aleksandr Aleksandrov
Call sign: Pamir

The mission to repair Salyut 7 was formed in March 1985. Dzhanibekov was picked for this mission because of his experience with manual rendezvous during the Soyuz T-6 flight and his expertise with the Salyut electrical system during Soyuz 27. Savinykh had been training for the 10-month mission that was originally to be launched on Soyuz T-13. He also had experience with the type of condition the station was in after his repair mission to Salyut 6 in March 1981, when a solar array failure lowered station temperatures and caused water condensation in the station.

After the rescue mission was formed, Vasyutin and A. Volkov, the rest of the original crew, continued to train for an EVA to erect a beam experiment with Grechko standing in for Savinykh, who was by then in orbit. All of the cosmonauts would eventually meet on Salyut 7 and reform the original Soyuz T-13 crew, with Dzhanibekov and Grechko as extras. On June 7, Shatalov said that the crew would conduct work "required by regulations, which will undoubtedly need to be done after ... long use of the station." [364] More simply, they would do a lot of work to make the station habitable again.

Soyuz T-13 used a special 2-day rendezvous that saved propellant and simplified approach and docking to the dead station. The transfer orbit to Salyut 7 was 298 × 334 km. The Soyuz main engine was fired to match orbits with the space station at two and again at one hour before docking. The Soyuz automatically approached the Salyut in a 356 × 358-km orbit.

The cosmonauts discovered the Salyut was in a slow roll (no more than 0.3°/second) and the Salyut's solar arrays pointed randomly. They transmitted television pictures of the station to mission control, and Flight Director Ryumin said they were alarming, but they had expected the station would be in that condition. Dzhanibekov also noted that the once green insulation blankets that covered Salyut 7's transfer compartment and its adapter section had turned gray during the station's years in orbit and exposure to the elements. Tracking ships, including the *Cosmonaut Patsayev*, were deployed in the Atlantic, off West Africa, in the Pacific and in the Mediterranean.

At 10 km distance, Dzhanibekov stopped the approach to provide the docking computer with information about the station's attitude. At the time, the Salyut was turned with its side toward the approaching Soyuz. Normally, the station would automatically point toward the Soyuz. The Soyuz automatically continued the approach to 3 km distance, at a rate of 12, and later 6 meters/second when Dzhanibekov took control. The crew had trained for approaches from as far as 30 km out. The distance depended on the accuracy of the booster. While making the final approach for docking, Dzhanibekov used specially installed control sticks near the right porthole of the Soyuz capsule.[365-367]

At 2 km, the crew used a new optical guidance system, hand-held laser range finder and a night vision instrument, to see and measure distance to the station. While Dzhanibekov flew the Soyuz he instructed Savinykh on what data to enter into the computer. Savinykh then reported to Dzhanibekov the results so he could make the necessary maneuvers. Because of these duties, the other cosmonauts had named Savinykh the "human computer." At 200 meters from the station, Dzhanibekov stopped for 10 minutes because the sun was behind the station, making visibility of the docking port poor. While Dzhanibekov waited for more favorable lighting conditions, he circled the station, inspecting it for any unexpected damage. After that, Dzhanibekov again lined up with the station's forward docking port, and in a roll matching the station, performed a normal docking at 12:50 a.m., June 8, at 356 × 375 km, 5 minutes after sunset.

The crew's first task was to confirm the lack of power on the Salyut docking collar connections. They took great caution opening the pressure equalization valve to take a sample of the station's air. It was possible that an electrical fire in the Salyut would have left a poisonous atmosphere in the station (similar precautions were taken by the Skylab 2 crew in 1973, before entering the crippled Skylab space station). Even though the air tested as good, the cosmonauts wore respirators as they opened the hatch. They later described the station as dark, completely silent, and very cold. Metallic surfaces were covered with frost, the air was stagnant, and there were icicles everywhere. A U.S. source reported that one of the cosmonauts was coughing often during the initial work.[368,369]

The temperature in the Salyut was estimated at − 10°C. The station's thermometers only went down to 0°C, so one of the cosmonauts spit on the wall and timed how long it took to freeze. Mission control used this unusual data to estimate the temperature.[370] At first they worked in arctic coats and gloves with oxygen masks and flashlights. The cosmonauts took off the station's porthole covers, which are always installed between manned missions, so they could work during the daylight passes for around 40 minutes of each orbit, before returning to the warm Soyuz to report on progress to Deputy Flight Director Solovyev.

They said that their feet got painfully cold, and after working without any ventilation they got headaches, felt sleepy, and limp from the carbon dioxide build-up. Air pressure in the station had fallen to 714 mm Hg because of the cold, but it was not dangerously low. Mission Control had instructed the crew that only one cosmonaut should enter the station at a time, presumably so that if one of the crew were overcome by carbon dioxide the other could retrieve him.[371,372]

They found all eight main space station batteries dead and two of these were ruined. They were prepared to end the mission if all the batteries were unusable and develop a way to deliver more of the heavy batteries on the next Soyuz flight. They next set up a pipe from the Soyuz ventilation system to one of the Salyut air regenerators to provide some protection from carbon dioxide build-up (otherwise, levels would be dangerous within 24 hours). The objective of the mission at this point was to repair the station so it could at least be safely deorbited.[373]

On June 9, Dzhanibekov, an expert on the Salyut electrical system, determined that the sensor which switches the solar array power between the batteries and the station's power bus had failed, so that the batteries were not recharged. Another automatic system turned on the recharging system every orbit but the faulty sensor kept shutting it off. When the batteries ran down, the power to the station dropped below the minimum needed. Occasionally, the solar arrays were able to power the systems left on by the last crew and charge the batteries enough for the radios to send the weak telemetry signals that were received by mission control. This would probably have been during a continuous sunlight period in early 1985. Normally, this failure would have been noticed by mission control and the problem averted, but a faulty telemetry radio did not report the sensor failure.[374]

To restore power to the Salyut, Dzhanibekov cut the cables to the solar arrays and Salyut electrical buses, to bypass the control relays. Then he connected the combined arrays to each battery one at a time using scavenged wire from other station equipment. While charging the first battery, the Soyuz was used to orient the Salyut's solar arrays for maximum sunlight. After a few hours, the first one was fully charged to 28.6 volts and the telemetry system was turned on. The rest of the batteries were recharged in the same way. A new sensor control unit was installed and the solar arrays and batteries were then reconnected normally. After a 24-hour work day, the crew retired to the Soyuz.[375,376]

The next day, the Salyut's lights and air heaters were turned on. Normally, the electric heaters were used only when the station was unmanned to keep temperature within normal limits for the station's equipment. A crew normally dissipates enough heat in the well insulated Salyut to maintain a comfortable temperature. Main heaters in the walls were not activated until the water covering the interior evaporated. As the station warmed, humidity levels soon reached 90%. The crew found several valves in the environmental control system were damaged by freezing water and needed to be replaced. They made a list of equipment to be delivered by the next Progress. At this time, they were still sleeping in the Soyuz orbital module and wore overalls, gloves, and down hats, but they complained that their feet froze in their flight boots.

On June 11, they had their first hot meal since the launch. The station was in orbit at 356 × 375 km. On June 13, the station's attitude control system was reactivated and successfully tested. There was rejoicing in mission control since this would allow the badly needed Progress mission to dock. The solar array pointing system was reactivated and damaged electronics units were replaced.[377]

By June 16, the station's temperature finally increased to above freezing, allowing water recovery by the environmental control system. They could not increase the temperature too fast or they would be inundated by water from all the ice melting. The environmental control system, water recovery system, air filters, and food warmers were activated at this time. The crew had been using water from the Soyuz, which had supplies for 8 days, and water that was found frozen in portable Salyut containers. The combined water supply would have run out about June 22, and Dzhanibekov later said that they were prepared to drain coolant water out of their pressure suits to drink if necessary. The Salyut's water tanks had also begun to thaw by this time, supplying the crew with 200 liters in one tank, and almost 200 in the other water tank, which the cosmonauts were warming by turning that side of the Salyut into the sun often. But, they had no hot water since the water heater had frozen and broken. To warm milk they instead used a powerful and hot television light. The main heating panels were activated, but the crew turned them off because they had become used to 16°C (58°F).[378–381]

After completing the initial repair work, Dzhanibekov said that there were still some doubts about Salyut 7's reliability. The Soviets officially had blamed the loss of the station on Salyut's faulty radio equipment, but the crew replaced the radio command unit. By June 19, the crew was participating in the Kursk-85 Earth observation exercise, which combined Salyut, Meteor-Priroda, and Kosmos Radar satellites and ground observations by Priroda and cosmonauts Kovalenok and Farkas of Hungary.

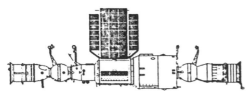

Progress 24

Launched: June 21, 1985,
4:40 p.m.
Reentry: July 15, 1985
Altitude: 185 × 237 km @ 51.6°

Progress 24 used transfer orbits of 193 × 270 km, 209 × 409 km, and 350 × 409 km to reach Salyut 7. It docked on June 23, at 6:54 p.m., at 355 × 358 km. The

Progress refueled the thawed Salyut propulsion system and delivered other supplies needed to repair systems in the station to allow automatic operations to resume.

The flight also delivered new type Salyut spacesuits (the old ones were not usable after being frozen), a new water heater, mail, food, 280 kg of water, 3 new batteries, 2 add-on solar arrays, oxygen, linen, warm shoes, toiletries, 40 kg of replacement parts, 30 kg of film, 30 kg of medical equipment, video tapes and photos of family, and medical supplies. The supplies' weight totaled 2,000 kg. Progress 23 undocked at 4:28 p.m., July 15.[382–386]

☆ ☆ ☆ ☆ ☆

On June 25, the Pamir crew installed the new water heater. On June 27, they performed medical checks and on the 29th, they tested the television cameras in the station. By July 2, the water tanks were refilled from the Progress. The crew also finished a series of experiments with the Astra-1 mass spectrometer studying the atmosphere around the space station. By July 5, 3 new batteries were installed, repairs to the water recovery system were completed, the station's air was replenished, and refueling was begun.[387]

On the night of July 7, Savinykh woke up with very irritated eyes, soon remembering that he must have looked out the unfiltered porthole too long that day. Fortunately, he soon recovered from the mild ultraviolet burn. By July 9, the crew was filling the Progress with trash and useless equipment, the station was being refueled, and the crew began a series of Earth resources photography sessions. On July 10, they replaced 8 ventilation fans and inventoried equipment on the station.[388,389] Progress 24 undocked and made a destructive reentry to make way for Kosmos 1669.

Kosmos 1669

First Use of Platform

Launched: July 19, 1985,
　　　　7:05 p.m.
Reentry: August 30, 1985
Altitude: 188 × 246 km @ 51.6°

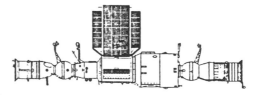

The Kosmos 1669 platform was the first flight of a "free-flyer" that could dock periodically to a station and be serviced and replenished by cosmonauts. The main purpose of the platform was to fly free of disturbances caused by the space station's crew. The platform was reportedly a modified Progress transport with solar arrays to provide power and extended duration flight when in free-flying mode. Apparently, this flight only tested some of the guidance and control systems developed for the platform because the Soviets still referred to the spacecraft as a transport, and it performed a standard Progress flight plan. Some confusion was caused when the Soviet media referred to the flight as

Progress 25. This was an obvious mistake by the reporter and did not indicate a Kosmos name change for a flight in trouble. This was further confirmed by the release of the transfer orbit data, which was not normally released for a Progress mission.

The transfer orbits to Salyut 7 were 234 × 355 km and 298 × 358 km. The platform docked on July 21, at 7:05 p.m., in orbit at 354 × 358 km. The crew soon began unloading equipment including the Comet and Medusa experiments, the Mariya X-ray noise telescope, 2 add-on solar panels, water, biological samples, and propellant. By August 27, the crew began filling Kosmos 1669 with trash and the transfer of propellant and water to the Salyut's tanks was being completed. Kosmos 1669 boosted the station's orbit on August 29, before undocking at 1:50 a.m. It then made another approach and docked testing its systems much like Progress 1 did during its flight.[390] The platform was not used in its free-flying mode during this flight. It deorbited and made a destructive reentry like a standard Progress flight.[391-393]

☆ ☆ ☆ ☆ ☆

On July 29, the Salyut crew performed medicals and continued the Kursk-85, Svetobloc-T and Rost experiments. August 1, was officially a day of rest for the cosmonauts, but they spent the time preparing for the next day's EVA.

On August 2, the cosmonauts performed a 5-hour EVA, starting at 11:15 a.m., to place the French Comet experiment on the exterior of the Salyut, and to install the last pair of gallium arsenide solar panels to the left Salyut array. They also installed a small prototype solar battery to the array. They waited for 20 minutes to come into communication range before beginning work. To install the pair of arrays, they first attached one, and then had the Salyut array rotated 180° by ground controllers to attach the next. All the Salyut arrays now had two add-on arrays installed. The add-on arrays were all gallium arsenide and measured 4.5 × 1.2 meters. The Comet experiment was used primarily to gather dust from Comet Giacobini-Zinner. The experiment had several chambers that could be opened and closed by the cosmonauts from inside the Salyut. They also installed Medusa materials experiment cartridges on the station's hull.[394,395]

The EVA also tested the new semi-rigid space suits delivered by Kosmos 1669. The suits had new shoulder joints that expanded the reach envelope, better peripheral vision, better monitoring systems, and illuminated controls. The suits had lights on the helmets and arms. The commander's suit had red stripes and the engineer's had blue stripes. Despite the new joints, the cosmonauts complained of uncomfortable pinching of their arms and their fingers tingled for two days after the EVA. The Soviets claim that suiting up and pre-breathing required only 20 minutes. Also at this time, the humidity in the Salyut began to reach normal levels after very high conditions during the repair operations.[396,397]

Experiments conducted by the crew in the first weeks of August included growing flax in the Magnetogravistat, photographic studies with the KATE-140 and MKF-6M, biological processing with the Svetobloc-T, and operation of the Mariya X-ray noise telescope. On August 16, the crew made observations of atmospheric pollution over industrial areas. They also took photographs of agricultural land simultaneously with aircraft to provide a more detailed study of its condition. By August 18, the crew had taken around 2,000 photos with the MKF-6M and 600 with the KATE-140. On August 20, the crew tested the performance of the Salyut solar panels and the additional panels by varying the angle of sunlight hitting them.[398]

Kosmos 1669's attitude control system was used in a test to control the station's attitude, but the results was said to be disappointing. By August 27, the crew began filling Kosmos 1669 with trash and the transfer of propellant and water to the Salyut's tanks was being completed. Kosmos 1669 boosted the station's orbit on August 29, before undocking at 1:50 a.m. It then made another approach and docked testing its systems much like the first Progress flight.[399]

On September 3, the crew exposed test materials to vacuum in the scientific airlock and observed them with the Electrotopograph process. The cosmonauts also performed the Biryuza and Analiz crystal growth experiments, took atmospheric measurements with the Astra-1 spectrograph, and did another Electrotopograph experiment over the next week. They began tests on September 14, of their cardiovascular systems using the Chibis suits and observed growth of cotton in the Oasis.[400]

Soyuz T-14

First Partial Crew
Exchange—56 (216) Days
Launched: September 17, 1985,
     4:39 p.m.
Landed: November 21, 1985, 1:31 p.m.
Altitude: 196 × 223 km @ 51.6°
Crew: Vladimir Vasyutin, Georgiy Grechko, and Aleksandr Volkov
Backup: Aleksandr Victorenko, Gennadiy Strekalov, and Yevgeny Saley
Call sign: Cheget

The crew of Vasyutin, Savinykh, and Volkov started training in September 1984 for a scheduled launch in March 1985, but the Salyut failure occurred in October 1984. When the Soviets launched the Soyuz T-13 repair mission, they had removed Vasyutin and Volkov from the planned crew, and teamed Dzhanibekov with Savinykh for the repairs and check out of the Salyut. The T-14 flight then brought the others from the original crew (Vasyutin and Volkov) to the station, in exchange for Dzhanibekov to reform the original crew.

The Soyuz T-14 was launched into clear skies and after separating from the booster's upper stage it maneuvered into an initial orbit at 196×223 km. The transfer orbit to Salyut 7 was 272×326 km and the T-14 docked at 6:15 p.m., September 18, at an altitude of 338×353 km. The T-14 crew opened the Salyut's aft hatch and entered the station at 9:24 p.m. The flight carried new equipment to the station, including a new spectrometer, the SKIF, which photographed the layers of the atmosphere and developed the pictures to let the cosmonauts see the results immediately.[401] It also carried the EFU-Robot pilot scale electrophoresis experiment. Its operation was fully automatic and provided pure preparations like an anti-influenza standard for use by the Soviet Health Service, food industry, and agriculture.

The cosmonauts operated the Svetobloc-T experiment to produce a synthetic gel for use in electrophoretic purification of materials on Earth, performed Resonance tests, and Dzhanibekov used the Chibis suit to prepare himself for return to Earth. The crew also participated in the Black Sea 85 program of Earth resources research combining photography from the Salyut, satellites, aircraft and measurements on the ground.[402]

The cosmonauts also operated instruments for atmospheric research and Savinykh described the work in his diary. He and Grechko were in position at a porthole to photograph the various layers of the atmosphere as the sun set on the horizon. At the same time, Dzhanibekov tracked the sun's movement from a porthole while Vasyutin had to stop work with the EFU-robot unit to record the exact time of the event. The experiment lasted less than a minute, but it took at least an hour to prepare the cameras and instruments, and position and instruct everyone properly. This illustrates the complexity and time needed to carry out the experiments on the station. Savinykh also said that the space veterans, Dzhanibekov and Grechko, were sleeping only 3–4 hours a night during the joint mission. The spaceflight would probably be the last for each cosmonaut—Dzhanibekov had made 5 flights to Salyut 6 and 7, and Grechko had already flown two long-duration missions to Salyut 4 and 6.[403]

On September 20, the crew experimented with a device that tested using acupuncture to relieve the uncomfortable symptoms of space adaptation sickness.[404] Experiments continued over the next few days, until it was time for the crews to split up. Dzhanibekov, from the T-13 crew, and Grechko from the T-14 crew would return to Earth in the T-13 spacecraft. Cosmonaut Savinykh, of the T-13 crew, was joined by Vasyutin and Volkov from the T-14 crew to continue manning the station. The Soyuz was loaded with the results of the T-13 flight including film, material retrieved from the Medusa fixture, ampules of biological substances, and flight logs.

Soyuz T-13 undocked at 7:58 a.m., September 25. The T-13 broke with usual landing practice by remaining in orbit for 30 hours of maneuvers, while the Salyut docking aids were deactivated, simulating the docking of the T-13 on June 8 (see Figure 5-15). The Soyuz was allowed to drift away from the station and then maneuvered 3 times to close to within 5 km of the station. The cosmonauts then took over, approached to less than a kilometer, and kept station with the Salyut. A docking could have easily been accomplished again from that distance and the exercise was declared a success. The crew then proceeded with normal landing procedures. The retrofire maneuver was 201 seconds in duration. The Soviets reported that the capsule experienced 4 gravities of deceleration during reentry.[405,406]

Figure 5-15. Soyuz T-14 is shown docked to the Salyut 7 in gravity gradient flight, as the Soyuz T-13 maneuvers around the station. Gallium arsenide solar panels are attached to all three main solar arrays to increase power generation capabilities. Salyut 7 had orange stripes painted on its white exterior to aid visibility during rendezvous and docking. (Source: Tass from Sovfoto.)

The Soyuz capsule's parachute opened at 9,500 meters altitude, lowering the capsule to 2 meters when the landing rockets fired, braking the capsule's fall. Soyuz T-13 landed at 1:52 p.m., September 26, at a new landing zone 220 km northeast of Dzhezkazgan in clear skies and light winds. The landing zone was a new one, and the T-13 was the first manned craft to use it. Dzhanibekov was assisted out of the capsule by doctors.[407]

After the T-13 departed, the new Salyut 7 crew was given the weekend off before starting on what was planned to be a six-month flight plan. This would end with Savinykh accumulating more than 10 months in space, accomplishing the original Soyuz T-13 mission plan.

Kosmos 1686

Star Module

Launched: September 27, 1985,
              12:41 p.m.
Reentry:
Altitude: 172 × 302 km @ 51.6°

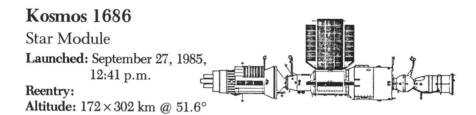

Kosmos 1686 was the fourth Star module type spacecraft. It differed greatly from its predecessors because it did not have a reentry module. The Star module was equipped with scientific equipment, telescopes for astrophysical research, and additional propellant instead of a return capsule. There were rumors that the module also had some military reconnaissance capability and that Volkov (a military cosmonaut) was to do classified work during the mission.[408] The Soviets said that the cosmonauts would test its equipment and structural elements and perfect methods for controlling complexes of large size and mass.

The Soviets referred to this as the "module" or "multi-purpose orbital module" version of the Star module. The next generation space stations with multiple docking ports would use the same type modules. Kosmos 1686 also tested systems to be used on the Mir modules, including new digital control systems as opposed to Salyut's analog control systems. The module was 15 meters long and 4.15 meters in diameter at its widest point. The module's two solar panels spanned 16 meters.[409]

The transfer orbits to reach Salyut 7 were 284 × 318 km, and 290 × 336 km. The Star module docked at the forward port of Salyut 7 at 1:16 p.m., October 2, at 336 × 352 km on automatic controls, monitored by mission control and the cosmonauts. The Soyuz, Salyut, Kosmos complex was 35 meters long. Total power output of the station complex was 8–10 kW.[410,411] The flight delivered 4,500 kg of freight and experiments, including more than 1,000 kg of large equipment like the deployable beam experiment to be used by the Soyuz T-15 crew, and the Kristallizator materials processing unit.[412,413] The module carried 3,000 kg of propellant and almost doubled the interior volume of the station.[414]

On October 8 the crew began activating the module's systems. On October 10, the first stage of the Comet experiment was over as the cosmonauts closed a remote control lid on the experiment. They also operated the Mariya high-energy particle observation experiment, the Pion materials experiment, and watched the development of cotton and flax in the Oasis. They reported on October 13 that the module was still not fully unloaded. The next day they took part in a goodwill broadcast with astronauts Slayton and Stafford and U.S. Congressman Bill Nelson. In the same month, at the IAF conference in Sweden, the Soviets announced that the next space station launched would have multiple docking ports and be expanded by adding 4–6 modules to the ports.

In the middle of October, the cosmonauts tested increasing the intensity of exercise versus time consuming, repetitive exercise. The crew still exercised 2–2.5 hours a day. They also started controlling the station's attitude with the Star module, which could control the station's attitude for a week at a time without ground instructions. On October 20, the crew opened a new chamber of the Comet experiment to collect more comet dust, and they also observed noctilucent clouds. On October 25, the crew performed routine maintenance work; the station's orbit was 357 × 375 km.[415]

In later October, the crew was having difficulty in keeping up with the work schedule because Vasyutin was not working. He had lost his appetite and was obviously sick. Vasyutin was staying in bed and doing nothing all day. On October 28, Volkov and Savinykh finally convinced Vasyutin to talk to mission control about his problem. They were told to wait for his condition to change and continue working.[416]

In early November, the cosmonauts took pictures of the Tajikistan area, which was the epicenter of a recent major earthquake. On November 5, *Pravda* reported that one of the tasks for the day's work was to investigate possible buried ancient irrigation systems in central Asia. This would most probably have used a small synthetic aperture radar on Salyut 7 that also had been used to try tracking submarines. On November 12 the crew checked the Salyut's solar panels and began using instruments delivered by the module to study atmospheric phenomenon and charged particles.[417]

On November 13, at 10:11 p.m., the crew began scrambling radio transmissions. They informed mission controllers that Vasyutin's condition had not improved. They were apparently informed to begin preparing for a landing during the next regular landing window. The first landing window opened November 17, but the lighting conditions for recovery of the capsule were poor until November 21. The crew then continued normal activities, evaluating the station complex and the Salyut's solar arrays.[418,419]

Savinykh later explained the emergency in his diary of the flight, which was partially printed in *Pravda* on December 29. He said that at first Vasyutin seemed uneasy. Then he had trouble sleeping and lost his appetite. Savinykh and Volkov tried to cheer him up, but the illness grew worse. Vasyutin wished to continue the mission, hoping the illness would go away, but by the end of the mission, he had become a bundle of nerves. He also said that he thought that the problem was in his frame of mind. On November 15, the crew performed routine medical tests using the Chibis suit. On November 17, the crew was told by mission control to prepare for landing within 3 days. After this Vasyutin's condition improved a bit, and the crew leisurely began mothballing the station, including putting covers on the portholes.[420,421]

On November 21, the crew returned to Earth, with the press initially reporting that mission commander Vasyutin was suffering from appendicitis. Savinykh was given command of the mission and the Soviets said that either Savinykh or Volkov could have assumed the duty. They landed in light snow and cloudy skies, 180 km southeast of Dzhezkazgan at 1:31 p.m.

As usual, the recovery operations were shown live on Soviet television. The reporter said that stretchers were ready (as always) but went unused and that Vasyutin was first out of the capsule, followed by Volkov and then Savinykh. Vasyutin consented to a short interview during which he said, "I'm feeling all right, the way I should after a landing. I am very happy to see people. It has been just the three of us for so long, it is nice to see so many people." When asked if the mission was a success he answered, "Well, not quite." Quickly one of the other cosmonauts (probably Savinykh) said, "Well, what can I say, the flight is over, much has been accomplished. I'd like to keep on working, but we can't stay up there forever."[422]

The Soviets reported later that after landing, Vasyutin was examined by a doctor and flown to a hospital in Moscow, a flight of several hours' duration. The fact that the cosmonauts did not return at first opportunity on November 17 conflicts with the explanation that Vasyutin had appendicitis and Savinykh's description of the illness, and Vasyutin's apparent health at the post-landing interview supported the rumors that the illness was mental.

The Soviets reported that Vasyutin was suffering from 40°C (104°F) fever and inflammation for three weeks and was released from the hospital on December 20. Later, Savinykh said in an interview that Vasyutin had suffered from a cold, and that the mission would have lasted beyond January 1, with he and Vasyutin performing the beam erecting EVA, which was later done by the T-15 crew. He also noted that there were ten female cosmonauts in training at Star City at the time. Still later, it was quietly reported that Vasyutin had suffered from a prostate infection. The Soviets never commented officially on the illness according to the wishes of Vasyutin and medical ethics. On December 30, all three of the cosmonauts attended the usual awards ceremony in Moscow, indicating that the mission ended due to reasons beyond the crew's control. During the shortened mission, the crew still managed to perform 400 experiments using 45 different instruments and photographed 16 million km^2 of Earth.[423,424]

Whatever medical problem caused the return, it was totally unexpected because the Soviets had promised Congressman Bill Nelson that Salyut 7 would be manned during Nelson's shuttle flight (STS 61-C), then planned for late December 1985. The crew left several items on the station that would be returned by the next crew to Salyut 7. The Soyuz T-15 crew would be the next crew to operate the station, but this would take place after they had set up the next space station, Mir.

References

1. Kidger, Neville, "Salyut 6 Mission Report: Part 4," *Spaceflight*, Vol. 22, No. 11–12, Nov.–Dec. 1980, p. 343.

2. "Soviet Union Takes Lead In Manned Space Operations," *Aviation Week & ST*, March 9, 1987, p. 129.

3. Foreign Broadcast Information Service, U.S.S.R., Space, JPRS-USP-84-001, Jan. 1984, Joint Publications Research Service, p. 17.

4. Ibid. JPRS-USP-86-004, April 1986, p. 16.

5. Johnson, Nicholas L., *Handbook of Soviet Manned Spaceflight*, American Astronautical Society: San Diego, 1980, p. 381.

6. Peebles, Curtis, "The Soviet Space Shuttle," *Spaceflight*, Vol. 26, No. 5, May 1984.

7. Hall, Rex, "The Soviet Cosmonaut Team, 1978–1987," *Journal of the British Interplanetary Society*, Vol. 41, p. 111.

8. Oberg, James E., *New Race for Space*, Stackpole Books, 1984, pp. 120–122.

9. U.S. Congress, Office of Technology Assessment, *Salyut, Soviet Steps Toward Permanent Human Presence in Space, A Technical Memorandum*, Washington D.C., Dec. 1983, p. 39.

10. Lenorovitz, Jeffrey M., "Soviets Planning Manned Shuttle Mission for 1989," *Aviation Week & ST*, Jan. 16, 1989, p. 36.

11. Peebles, op. cit.

12. Congressional Research Service, The Library of Congress, *Soviet Space Programs 1976–80, Manned Space Programs and Life Sciences, Part 2*, Washington: Government Printing Office, 1984, p. 458.

13. Oberg, James E., personal communication, 1989.

14. Oberg, op. cit. p. 131.

15. Johnson, Nicholas L., *Soviet Military Strategy in Space*, Janes Publishing Co., 1987, pp. 140, 158.
16. *Aerospace America*, American Institute of Aeronautics and Astronautics, Nov. 1987, p. 25.
17. Borrowman, Gerald, "The Soviet Space Shuttle,"*Spaceflight*, Vol. 25, No. 4, April 1983, p. 149.
18. Christy, Robert D., "Orbits of Soviet Spacecraft at 51.6° Inclination," *Spaceflight*, Vol. 22, No. 2, Feb. 1980, p. 80.
19. The Kettering Group, "Observations of 1977-66A, Cosmos 929," *Spaceflight*, Vol. 20, No. 9–10, Sept.–Oct. 1978, pp. 353–355.
20. Oberg, James E., *Uncovering Soviet Disasters*, Random House: New York, 1988, p. 195.
21. Clark, Phillip S., "Soviet Launch Failures," *Journal of the British Interplanetary Society*, Vol. 40, No. 10, Nov. 1987, p. 529.
22. Williams, Trevor, "Soviet Re-entry Tests: A Winged Vehicle?"*Spaceflight*, Vol. 22, No. 5, May 1980, p. 213.
23. Kidger, Neville, "Salyut 6 Space Station,"*Spaceflight*, Vol. 21, No. 4, April 1979, p. 178.
24. Johnson, Nicholas L., *Handbook of Soviet Manned Spaceflight*, op. cit. p. 334.
25. Hooper, Gordon R., "Missions to Salyut 6," *Spaceflight*, Vol. 20, No. 3, March 1978, p. 118.
26. Congressional Research Service, The Library of Congress, *Soviet Space Programs 1976–80, Manned Space Programs and Life Sciences, Part 2*, Washington: Government Printing Office, 1984, p. 615.
27. Johnson, *Handbook of Soviet Manned Spaceflight*, op. cit. p. 334.
28. Hooper, op. cit. Vol. 20, No. 3, March 1978, p. 120.
29. Hooper, op. cit. Vol. 20, No. 12, Dec. 1978, p. 437.
30. *U.S.S.R. Space Life Sciences Digests*, NASA CR-3922(7), Issue 6, p. 114.
31. Johnson, *Handbook of Soviet Manned Spaceflight*, op. cit. p. 337.
32. Hooper, op. cit. Vol. 20, No. 6, June 1978, pp. 231–232.
33. Foreign Broadcast Information Service, U.S.S.R., Space, JPRS-USP-87-003, April 1987, Joint Publications Research Service, p. 20.
34. Oberg, James E., *Red Star in Orbit*, New York: Random House, 1981, p. 167.
35. Hooper, op. cit. Vol. 20, No. 11, Nov. 1978, p. 378.
36. Hooper, op. cit. Vol. 20, No. 6, June 1978, p. 233.
37. Hooper, op. cit. Vol. 20, No. 11, Nov. 1978, p. 371.
38. Ibid. pp. 371–372.
39. Johnson, *Handbook of Soviet Manned Spaceflight*, op. cit. p. 339.
40. Hooper, op. cit. Vol. 21, No. 7, p. 321.
41. Hooper, op. cit. Vol. 20, No. 11, pp. 371–374.

42. Hooper, op. cit. Vol. 21, No. 5, p. 222.
43. *U.S.S.R. Space Life Sciences Digests*, NASA CR-3922(7), Issue 6, p. 114.
44. Hempel, Wilhelm, "MKF-6M Multi-spectral Camera in Space," *Spaceflight*, Vol. 21, No. 3, March 1979, p. 110–113.
45. Foreign Broadcast Information Service, U.S.S.R., Space, JPRS-USP-84-003, June 1984, Joint Publications Research Service, p. 45.
46. Johnson, *Handbook of Soviet Manned Spaceflight*, op. cit. p. 342.
47. Hooper, op. cit. Vol. 20, No. 11, Nov. 1978, pp. 376–379.
48. Foreign Broadcast Information Service, U.S.S.R., Space, JPRS-83430, May 1983, Joint Publications Research Service, p. 37.
49. Hempel, Wilhelm, "The Splav-01 Furnace," *Spaceflight*, Vol. 21, No. 2, Feb. 1979, p. 57.
50. Kidger, op. cit. Vol. 22, No. 4, p. 152.
51. Kidger, op. cit. Vol. 22, No. 11–12, Nov.–Dec. 1980, p. 343.
52. Hooper, op. cit. Vol. 20, No. 12, Dec. 1978, pp. 430–432.
53. Powell, Joel, "Salyut 6 Medical Monitoring Techniques," *Spaceflight*, Vol. 23, No. 9, Nov. 1981, p. 317.
54. Hooper, op. cit. Vol. 20, No. 12, Dec. 1978, pp. 432–433.
55. Oberg, *Red Star in Orbit*, op. cit. p. 178.
56. Hooper, op. cit. Vol. 20, No. 12, Dec. 1978, pp. 433–437.
57. *U.S.S.R. Space Life Sciences Digests*, NASA CR-3922(7), Issue 6, pp. 115–116.
58. Congressional Research Service, The Library of Congress, *Soviet Space Programs 1981–87, Part 1*, Washington: Government Printing Office, May 1988, p. 110.
59. Johnson, *Handbook of Soviet Manned Spaceflight*, op. cit. p. 347.
60. Hooper, op. cit. Vol. 20, No. 12, Dec. 1978, p. 435.
61. Oberg, *Red Star in Orbit*, op. cit. p. 169.
62. Williams, op. cit. pp. 240–241.
63. King-Hele, D. B., Walker, D. M. C., Pilkington, J. A., Winterbottom, A. N., Hiller, H., and Perry, G. E., *The R.A.E. Table of Earth Satellites 1957–1986*, New York: Stockton Press, 1987, p. 528.
64. Johnson, *Handbook of Soviet Manned Spaceflight*, op. cit. pp. 134–135.
65. Powell, op. cit. p. 318.
66. Hooper, op. cit. Vol. 21, No. 3, March 1979, pp. 127–128.
67. Ibid. pp. 127–129.
68. Kidger, op. cit. Vol. 22, No. 2, Feb. 1980, p. 60.
69. Hooper, op. cit. Vol. 21, No. 5, May 1979, p. 218.
70. Foreign Broadcast Information Service, U.S.S.R., Space, JPRS-83430, May 1983, Joint Publications Research Service, p. 41.
71. Powell, op. cit. p. 318.
72. Hooper, op. cit. Vol. 21, No. 3, March 1979, p. 133.

73. Hooper, op. cit. Vol. 21, No. 5, May 1979, p. 220.

74. Hooper, op. cit. Vol. 21, No. 5, May 1979, pp. 216–220.

75. Oberg, James E. and Oberg, Alcestis R., *Pioneering Space*, New York: McGraw-Hill, 1986, p. 81.

76. Oberg, *Red Star in Orbit*, op. cit. p. 206.

77. Johnson, *Handbook of Soviet Manned Spaceflight*, op. cit. p. 352.

78. Oberg, *Red Star in Orbit*, op. cit. p. 203.

79. Hooper, op. cit. Vol. 21, No. 7, July 1979, p. 324.

80. Kidger, op. cit. Vol. 22, No. 4, April 1980, p. 154.

81. "Progress Docking Extends Intensive Salyut Activities," *Aviation Week and Space Technology*, McGraw-Hill, New York, July 9, 1979, p. 20.

82. Johnson, *Handbook of Soviet Manned Spaceflight*, op. cit. p. 353.

83. Hooper, op. cit. Vol. 21, No. 5, May 1979, p. 222.

84. Hooper, op. cit. Vol. 21, No. 5, May 1979, pp. 221–222.

85. Foreign Broadcast Information Service, U.S.S.R., Space, JPRS-83430, May 1983, Joint Publications Research Service, p. 41.

86. Hooper, op. cit. Vol. 21, No. 7, July 1979, pp. 318–322.

87. Johnson, *Handbook of Soviet Manned Spaceflight*, op. cit. p. 355.

88. Hooper, op. cit. Vol. 21, No. 7, July 1979, pp. 322–323.

89. Hooper, op. cit. Vol. 21, No. 8–9, Aug.–Sept. 1979, p. 359.

90. Oberg and Oberg, *Pioneering Space*, op. cit. p. 10.

91. Hooper, op. cit. Vol. 21, No. 8–9, Aug.–Sept. 1979, p. 361.

92. Ibid. pp. 360–361.

93. Powell, op. cit. p. 318.

94. Kidger, op. cit. Vol. 22, No. 2, Feb. 1980, p. 50.

95. Hooper, op. cit. Vol. 21, No. 8–9, Aug.–Sept. 1979, p. 362.

96. Johnson, *Handbook of Soviet Manned Spaceflight*, op. cit. p. 360.

97. Oberg, *Red Star in Orbit*, op. cit. p. 210.

98. Kidger, op. cit. Vol. 22, No. 2, Feb. 1980, p. 50.

99. Hooper, op. cit. Vol. 21, No. 8–9, Aug.–Sept. 1979, p. 363.

100. Christy, op. cit. p. 80.

101. Oberg and Oberg, *Pioneering Space*, op. cit. pp. 5, 203.

102. Kidger, op. cit. Vol. 22, No. 2, Feb. 1980, pp. 51–52.

103. Kidger, op. cit. Vol. 22, No. 4, April 1980, pp. 146–151.

104. Kidger, op. cit. Vol. 22, No. 2, Feb. 1980, pp. 51–57.

105. Ibid.

106. Ibid.

107. Kidger, op. cit. Vol. 22, No. 3, March 1980, p. 110.

108. Kidger, op. cit. Vol. 22, No. 2, Feb. 1980, p. 58.

109. Congressional Research Service, The Library of Congress, *Soviet Space Programs 1976–80, Manned Space Programs and Life Sciences, Part 2,* Washington: Government Printing Office, 1984, p. 628.

110. Foreign Broadcast Information Service, U.S.S.R., Space, JPRS-83994, July 1983, Joint Publications Research Service, p. 14.
111. Oberg, *Red Star in Orbit*, op. cit. p. 196.
112. Foreign Broadcast Information Service, U.S.S.R., Space, JPRS-83994, July 1983, Joint Publications Research Service, pp. 16–17.
113. Ibid. JPRS-83994, July 1983, Joint Publications Research Service, pp. 18–19.
114. Kidger, op. cit. Vol. 22, No. 2, Feb. 1980, pp. 58–60.
115. Foreign Broadcast Information Service, U.S.S.R., Space, JPRS-83994, July 1983, Joint Publications Research Service, p. 14.
116. Kidger, op. cit. Vol. 22, No. 3, March 1980, p. 109.
117. Foreign Broadcast Information Service, U.S.S.R., Space, JPRS-83430, May 1983, Joint Publications Research Service, p. 42.
118. Kidger, op. cit. Vol. 22, No. 4, April 1980, p. 150.
119. Congressional Research Service, The Library of Congress, *Soviet Space Programs 1976–80, Manned Space Programs and Life Sciences, Part 2*, Washington: Government Printing Office, 1984, p. 579.
120. Kidger, op. cit. Vol. 22, No. 3, March 1980, p. 110.
121. Peebles, op. cit.
122. Williams, op. cit. pp. 213–214.
123. Kidger, op. cit. Vol. 22, No. 3, March 1980, p. 111.
124. Kidger, op. cit. Vol. 22, No. 4, April 1980, p. 154.
125. *Technology Review*, July 1988, p. 39 (photo).
126. "Salyut 6 Boosted to Record High Orbit," *Aviation Week and Space Technology*, McGraw-Hill, New York, July 16, 1979, p. 23.
127. Johnson, *Handbook of Soviet Manned Spaceflight*, op. cit. pp. 370–371.
128. Kidger, op. cit. Vol. 22, No. 3, March 1980, p. 112.
129. Johnson, *Handbook of Soviet Manned Spaceflight*, op. cit. p. 371.
130. Kidger, op. cit. Vol. 22, No. 3, March 1980, pp. 112–113.
131. Oberg, *Red Star in Orbit*, op. cit. pp. 216–217.
132. Kidger, op. cit. Vol. 23, No. 8, Oct. 1981, p. 267.
133. Kidger, op. cit. Vol. 22, No. 11–12, Nov.–Dec. 1980, p. 343.
134. Kidger, op. cit. Vol. 22, No. 4, April 1980, p. 145.
135. Oberg, *New Race for Space*, op. cit. pp. 70, 72.
136. Kidger, op. cit. Vol. 22, No. 4, April 1980, pp. 148, 153.
137. Kidger, op. cit. Vol. 23, No. 8, Oct. 1981, p. 268.
138. Kidger, op. cit. Vol. 22, No. 3, March 1980, p. 114.
139. *U.S.S.R. Space Life Sciences Digests*, NASA CR-3922(7), Issue 6, p. 114.
140. Kidger, op. cit. Vol. 22, No. 11–12, Nov.–Dec. 1980, p. 344.
141. Congressional Research Service, The Library of Congress, *Soviet Space Programs 1976–80, Manned Space Programs and Life Sciences, Part 2*, Washington: Government Printing Office Washington, 1984, p. 534.

142. Kidger, op. cit. Vol. 22, No. 11–12, Nov.–Dec. 1980, p. 345.

143. Ibid. p. 344.

144. Oberg, *Red Star in Orbit*, op. cit. p. 5.

145. Kidger, op. cit. Vol. 22, No. 11–12, Nov.–Dec. 1980, p. 345.

146. Kidger, op. cit. Vol. 23, No. 8, Oct. 1981, p. 269.

147. Foreign Broadcast Information Service, JPRS-USP-86-005, Sept. 12, 1986, Joint Publications Research Service, p. 141.

148. Oberg and Oberg, *Pioneering Space*, op. cit. p. 56.

149. Kidger, op. cit. Vol. 23, No. 8, Oct. 1981, p. 269.

150. Kidger, op. cit. Vol. 22, No. 11–12, Nov.–Dec. 1980, p. 345.

151. Kidger, op. cit. Vol. 23, No. 2, Feb. 1981, p. 44.

152. Oberg, *New Race for Space*, op. cit. p. 95.

153. Kidger, op. cit. Vol. 23, No. 2, Feb. 1981, p. 45.

154. Kidger, op. cit. Vol. 23, No. 3, March 1981, p. 74.

155. Kidger, op. cit. Vol. 23, No. 2, Feb. 1981, p. 45.

156. Kidger, op. cit. Vol. 23, No. 3, March 1981, p. 74.

157. Oberg, *Red Star in Orbit*, op. cit. p. 199.

158. Kidger, op. cit. Vol. 23, No. 3, March 1981, p. 76.

159. Foreign Broadcast Information Service, JPRS-USP-86-005, Sept. 12, 1986, Joint Publications Research Service, p. 141.

160. Kidger, op. cit. Vol. 23, No. 3, March 1981, p. 77.

161. Kidger, op. cit. Vol. 23, No. 8, Oct. 1981, p. 269.

162. Kidger, op. cit. Vol. 23, No. 3, March 1981, p. 77.

163. Kidger, op. cit. Vol. 23, No. 7, Aug.–Sept. 1981, p. 214.

164. Oberg, *Red Star in Orbit*, op. cit. p. 199.

165. Kidger, op. cit. Vol. 23, No. 7, Aug.–Sept. 1981, p. 216.

166. Foreign Broadcast Information Service, JPRS-USP-86-005, Sept. 12, 1986, Joint Publications Research Service, p. 141.

167. Kidger, op. cit. Vol. 23, No. 7, Aug.–Sept. 1981, p. 216.

168. Oberg, *Red Star in Orbit*, op. cit. p. 184.

169. Kidger, op. cit. Vol. 23, No. 7, Aug.–Sept. 1981, p. 217.

170. Foreign Broadcast Information Service, JPRS-USP-86-005, Sept. 12, 1986, Joint Publications Research Service, p. 141.

171. Borisenko, I. and Romanov, A., *Where All Roads to Space Begin*, Moscow: Progress Publishers, 1982, p. 60.

172. Kidger, op. cit. Vol. 23, No. 8, Oct. 1981, p. 266.

173. Ibid. pp. 269–270.

174. Oberg and Oberg, *Pioneering Space*, op. cit. p. 202.

175. Kidger, Neville, "The Flight of Soyuz T-3," *Spaceflight*, Vol. 24, No. 1, Jan. 1982, p. 29.

176. Congressional Research Service, The Library of Congress, *Soviet Space Programs 1976–80, Manned Space Programs and Life Sciences, Part 2,* Washington: Government Printing Office, 1984, p. 449.

177. Kidger, "The Flight of Soyuz T-3," op. cit. p. 29.

178. Congressional Research Service, The Library of Congress, *Soviet Space Programs 1976–80, Manned Space Programs and Life Sciences, Part 2,* Government Printing Office: Washington, 1984, p. 534.

179. Kidger, "Salyut Mission Report—Part 10," op. cit. p. 74.

180. Ibid.

181. "Soviet Scene," *Spaceflight,* Vol. 29, No. 1, Jan. 1987, p. 16.

182. Foreign Broadcast Information Service, U.S.S.R., Space, JPRS-83612, June 1983, Joint Publications Research Service, p. 15.

183. Kidger, "Salyut Mission Report—Part 10," op. cit. p. 74.

184. Ibid. pp. 75, 78.

185. Kidger, "Salyut Mission Report—Part 11," op. cit. p. 174.

186. "Soviet Scene," *Spaceflight,* Vol. 29, No. 1, Jan. 1987, p. 17.

187. Foreign Broadcast Information Service, U.S.S.R., Space, JPRS-83430, May 1983, Joint Publications Research Service, p. 44.

188. Kidger, "Salyut Mission Report—Part 11," op. cit. p. 177.

189. Foreign Broadcast Information Service, U.S.S.R., Space, JPRS-83612, June 1983, Joint Publications Research Service, pp. 15–16.

190. Ibid. p. 3.

191. Kidger, "Salyut Mission Report—Part 11," op. cit. p. 178.

192. Ibid.

193. Kidger, "Salyut Mission Report—Part 12," op. cit. p. 28.

194. Kidger, "Salyut Mission Report—Part 11," op. cit. p. 176.

195. Foreign Broadcast Information Service, U.S.S.R., Space, JPRS-82970, Fcb. 1983, Joint Publications Research Service, p. 19.

196. Kidger, "Salyut Mission Report—Part 12," op. cit. p. 28.

197. Foreign Broadcast Information Service, U.S.S.R., Space, JPRS-USP-84-004, Aug. 22, 1984, Joint Publications Research Service, p. 63.

198. Ibid. JPRS-82771, Jan. 1983, Joint Publications Research Service, p. 5.

199. Ibid. JPRS-USP-84-004, Aug. 22, 1984, Joint Publications Research Service, p. 56.

200. Ibid. pp. 57–58.

201. Kidger, "Salyut Mission Report—Part 12," op. cit. p. 29.

202. Ibid.

203. Foreign Broadcast Information Service, U.S.S.R., Space, JPRS-82771, Jan. 1983, Joint Publications Research Service, pp. 3–6.

204. Kidger, "Salyut Mission Report—Part 12," op. cit. p. 29.

205. Congressional Research Service, The Library of Congress, *Soviet Space Programs 1981–87, Part 1*, Washington: Government Printing Office, May 1988, p. 273.

206. Turnill, Reginald, (ed.), *Janes Spaceflight Directory 1986*, London: Janes, 1986, p. 224.

207. Peebles, op. cit.

208. Foreign Broadcast Information Service, U.S.S.R., Space, JPRS-82970, Feb. 1983, Joint Publications Research Service, p. 25.

209. Kidger, "Salyut Mission Report—Part 13," op. cit. p. 123.

210. Foreign Broadcast Information Service, U.S.S.R., Space, JPRS-USP-84-001, Jan. 1984, Joint Publications Research Service, p. 27.

211. Oberg, *New Race for Space*, op. cit. p. 95.

212. Foreign Broadcast Information Service, U.S.S.R., Space, JPRS-82970, Feb. 1983, Joint Publications Research Service, p. 29.

213. Kidger, "Salyut Mission Report—Part 13," op. cit. p. 124.

214. Foreign Broadcast Information Service, U.S.S.R., Space, JPRS-82771, Jan. 1983, Joint Publications Research Service, p. 7.

215. Ibid. JPRS-82970, Feb. 1983, Joint Publications Research Service, p. 29.

216. Ibid. JPRS-USP-84-001, Jan. 1984, Joint Publications Research Service, pp. 6–7.

217. Kidger, "Salyut Mission Report—Part 13," op. cit. p. 125.

218. Foreign Broadcast Information Service, U.S.S.R., Space, JPRS-83612, June 1983, Joint Publications Research Service, p. 10.

219. Ibid. JPRS-USP-84-001, Jan. 1984, Joint Publications Research Service, p. 12.

220. Ibid. JPRS-82970, Feb. 1983, Joint Publications Research Service, p. 32.

221. Kidger, "Salyut Mission Report—Part 13," op. cit. p. 125.

222. Foreign Broadcast Information Service, U.S.S.R., Space, JPRS-82970, Feb. 1983, Joint Publications Research Service, p. 37.

223. Ibid. JPRS-83430, May 1983, Joint Publications Research Service, p. 14.

224. Ibid. JPRS-USP-84-004, Aug. 22, 1984, Joint Publications Research Service, p. 59.

225. Lebedev, Valentin, "Diary of a Cosmonaut," *Final Frontier*, Feb. 1989, p. 44.

226. Oberg and Oberg, *Pioneering Space*, op. cit. pp. 190, 195.

227. Foreign Broadcast Information Service, U.S.S.R., Space, JPRS-82771, Jan. 1983, Joint Publications Research Service, p. 32.

228. Ibid. p. 33.

229. Ibid. p. 34.

230. Ibid. pp. 33–35.

231. Ibid. JPRS-USP-84-002, March 1984, Joint Publications Research Service, p. 2.

232. Lebedev, op. cit. p. 45.
233. Foreign Broadcast Information Service, U.S.S.R., Space, JPRS-83612, June 1983, Joint Publications Research Service, p. 12.
234. Ibid. JPRS-USP-84-004, Aug. 22, 1984, Joint Publications Research Service, pp. 59–63.
235. Ibid. JPRS-USP-84-003, June 1984, Joint Publications Research Service, p. 42.
236. Oberg, *New Race for Space*, op. cit. p. 79.
237. Foreign Broadcast Information Service, U.S.S.R., Space, JPRS-USP-84-004, Aug. 22, 1984, Joint Publications Research Service, pp. 52–53, 66.
238. Ibid. JPRS-USP-85-001, Feb. 4, 1985, Joint Publications Research Service, p. 13.
239. Ibid. JPRS-USP-84-004, Aug. 22, 1984, Joint Publications Research Service, pp. 53–62.
240. Ibid. JPRS-84161, Aug. 1983, Joint Publications Research Service, p. 5.
241. Ibid. JPRS-83612, June 1983, Joint Publications Research Service, p. 2.
242. Ibid. JPRS-USP-86-004, April 1986, Joint Publications Research Service, p. 25.
243. Ibid. JPRS-84946, Dec. 1983, Joint Publications Research Service, p. 6.
244. Kidger, "Salyut Mission Report," *Spaceflight*, Vol. 26, No. 5, May 1984, p. 230.
245. Foreign Broadcast Information Service, U.S.S.R., Space, JPRS-84161, Aug. 1983, Joint Publications Research Service, p. 6.
246. Congressional Research Service, The Library of Congress, *Soviet Space Programs 1981–87, Part 1*, Washington: Government Printing Office, May 1988, p. 274.
247. Turnill, op. cit. p. 224.
248. Kidger, "Salyut Mission Report," op. cit. Vol. 26, No. 3, p. 137.
249. Foreign Broadcast Information Service, U.S.S.R., Space, JPRS-83994, July 1983, Joint Publications Research Service, p. 3.
250. Kidger, "Salyut Mission Report," op. cit. Vol. 26, No. 3, p. 137.
251. Oberg, *New Race for Space*, op. cit. pp. 96, 97.
252. Foreign Broadcast Information Service, U.S.S.R., Space, JPRS-83612, June 1983, Joint Publications Research Service, p. 5.
253. Oberg, *New Race for Space*, op. cit. p. 116.
254. Foreign Broadcast Information Service, U.S.S.R., Space, JPRS-84946, Dec. 1983, Joint Publications Research Service, p. 6.
255. Turnill, op. cit. p. 217.
256. *U.S.S.R. Space Life Sciences Digests*, NASA CR-3922(7), Issue 6, p. 1.
257. Kidger, "Salyut Mission Report," op. cit. Vol. 26, No. 3, p. 138.

258. Congressional Research Service, The Library of Congress, *Soviet Space Programs 1976–80, Manned Space Programs and Life Sciences, Part 2,* Washington: Government Printing Office, 1984, p. 469.

259. Foreign Broadcast Information Service, U.S.S.R., Space, JPRS-USP-85-001, Feb. 4, 1985, Joint Publications Research Service, p. 16.

260. Ibid. JPRS-84946, Dec. 1983, Joint Publications Research Service, p. 21.

261. Ibid. p. 14.

262. Ibid. pp. 19–20.

263. Turnill, op. cit. p. 217.

264. Foreign Broadcast Information Service, U.S.S.R., Space, JPRS-USP-85-001, Feb. 4, 1985, Joint Publications Research Service, p. 23.

265. Ibid. JPRS-USP-84-004, Aug. 22, 1984, Joint Publications Research Service, p. 61.

266. *U.S.S.R. Space Life Sciences Digests,* NASA CR-3922(7), Issue 6, p. 4.

267. Kidger, "Salyut Mission Report," op. cit. Vol. 26, No. 3, p. 140.

268. Oberg, *New Race for Space,* op. cit. p. 103.

269. Foreign Broadcast Information Service, U.S.S.R., Space, JPRS-USP-86-004, April 1986, Joint Publications Research Service, p. 25.

270. Turnill, op. cit. p. 218.

271. Kidger, "Salyut Mission Report," op. cit. Vol. 26, No. 3, p. 140.

272. Kidger, "Salyut Mission Report," op. cit. Vol. 26, No. 5, p. 230.

273. Turnill, op. cit. p. 218.

274. Oberg, *New Race for Space,* op. cit. p. 109.

275. *U.S.S.R. Space Life Sciences Digests,* NASA CR-3922(7), Issue 6, p. 4.

276. Oberg, *New Race for Space,* op. cit. p. 116.

277. Furniss, Tim, "Soviets Open 1988 Space Account," *Flight International,* Jan. 26, 1988, p. 26.

278. Foreign Broadcast Information Service, JPRS-USP-87-005, Aug. 1987, Joint Publications Research Service, p. 93.

279. Ibid. JPRS-USP-87-006, Nov. 24, 1987, Joint Publications Research Service, p. 49.

280. Kidger, "Salyut Mission Report," op. cit. Vol. 26, No. 5, p. 231.

281. Oberg, James E., *Uncovering Soviet Disasters,* Random House: New York, 1988, pp. 188–190.

282. Turnill, op. cit. p. 219.

283. Oberg, *New Race for Space,* op. cit. p. 108.

284. Kidger, "Salyut Mission Report," op. cit. Vol. 26, No. 5, pp. 231–232.

285. Ibid. p. 232.

286. Ibid. p. 233.

287. Foreign Broadcast Information Service, U.S.S.R., Space, JPRS-USP-84-005, Oct. 26, 1984, Joint Publications Research Service, p. 2.

288. Ibid. JPRS-USP-84-002, March 1984, Joint Publications Research Service, p. 17.
289. Ibid. p. 13.
290. Kidger, "Salyut Mission Report," op. cit. Vol. 26, No. 5, p. 233.
291. *U.S.S.R. Space Life Sciences Digests*, NASA CR-3922(7), Issue 6, p. 3.
292. Foreign Broadcast Information Service, U.S.S.R., Space, JPRS-USP-84-002, March 1984, Joint Publications Research Service, pp. 14–21.
293. Kidger, "Salyut Mission Report," op. cit. Vol. 26, No. 5, p. 233.
294. *U.S.S.R. Space Life Sciences Digests*, NASA CR-3922(7), Issue 6, pp. 1–4.
295. Turnill, op. cit. p. 219.
296. Foreign Broadcast Information Service, U.S.S.R., Space, JPRS-USP-85-001, Feb. 4, 1985, Joint Publications Research Service, pp. 12–13, 16, 22.
297. Congressional Research Service, The Library of Congress, *Soviet Space Programs 1976–80, Manned Space Programs and Life Sciences, Part 2*, Government Printing Office: Washington, 1984, p. 469.
298. Kidger, "Salyut Mission Report," op. cit. Vol. 26, No. 5, p. 234.
299. Turnill, op. cit. p. 224.
300. Peebles, op. cit.
301. Kidger, "Salyut Mission Report," op. cit. Vol. 27, No. 7, p. 327.
302. Foreign Broadcast Information Service, U.S.S.R., Space, JPRS-USP-84-003, June 1984, Joint Publications Research Service, pp. 15, 16, 22.
303. Ibid. pp. 19, 26.
304. Ibid. pp. 23–26.
305. Ibid. pp. 23–26.
306. Nikitin, S. A., "Soyuz T-11, The Spaceflight of the Soviet Indian Crew," NASA TM-77615, March 1985, pp. 5, 7, 10.
307. Congressional Research Service, The Library of Congress, *Soviet Space Programs 1981–87, Part 1*, Government Printing Office: Washington, May 1988, p. 47.
308. Foreign Broadcast Information Service, U.S.S.R., Space, JPRS-USP-84-001, Jan. 1984, Joint Publications Research Service, p. 28.
309. Ibid. JPRS-USP-84-004, Aug. 22, 1984, Joint Publications Research Service, pp. 7, 17, 26.
310. Congressional Research Service, The Library of Congress, *Soviet Space Programs 1981–87, Part 1*, Government Printing Office: Washington, May 1988, p. 43.
311. Foreign Broadcast Information Service, U.S.S.R., Space, JPRS-USP-84-004, Aug. 22, 1984, Joint Publications Research Service, pp. 9, 23, 27.
312. Nikitin, op. cit. p. 10.
313. Turnill, op. cit. p. 221.
314. *U.S.S.R. Space Life Sciences Digests*, NASA CR-3922(8), Issue 7, p. 101.

315. Foreign Broadcast Information Service, U.S.S.R., Space, JPRS-USP-84-004, Aug. 22, 1984, Joint Publications Research Service, pp. 34, 38.
316. Kidger, "Salyut Mission Report," op. cit. Vol. 26, No. 12, p. 462.
317. Ibid.
318. Foreign Broadcast Information Service, U.S.S.R., Space, JPRS-USP-86-001, Jan. 13, 1986, Joint Publications Research Service, p. 67.
319. Turnill, op. cit. p. 220.
320. Foreign Broadcast Information Service, U.S.S.R., Space, JPRS-USP-84-004, Aug. 22, 1984, Joint Publications Research Service, pp. 31–36.
321. Ibid. JPRS-USP-84-005, Oct. 26, 1984, Joint Publications Research Service, p. 4.
322. Turnill, op. cit. p. 220.
323. Kidger, "Salyut Mission Report," op. cit. Vol. 26, No. 12, p. 463.
324. Foreign Broadcast Information Service, U.S.S.R., Space, JPRS-USP-84-004, Aug. 22, 1984, Joint Publications Research Service, pp. 41–42.
325. Congressional Research Service, The Library of Congress, *Soviet Space Programs 1981–87, Part 1*, Government Printing Office: Washington, May 1988, p. 56.
326. Foreign Broadcast Information Service, U.S.S.R., Space, JPRS-USP-84-004, Aug. 22, 1984, Joint Publications Research Service, p. 44.
327. Ibid. JPRS-USP-84-005, Oct. 26, 1984, Joint Publications Research Service, pp. 2–3.
328. Kidger, "Salyut Mission Report," op. cit. Vol. 26, No. 12, p. 463.
329. Foreign Broadcast Information Service, U.S.S.R., Space, JPRS-USP-84-005, Oct. 26, 1984, Joint Publications Research Service, pp. 6, 7, 18.
330. Ibid. JPRS-USP-84-006, Nov. 14, 1984, Joint Publications Research Service, p. 13.
331. Turnill, op. cit. p. 220.
332. Burrows, William E., *Deep Black: Space Espionage and National Security*, Random House: New York, 1986, p. 313.
333. Foreign Broadcast Information Service, U.S.S.R., Space, JPRS-USP-84-005, Oct. 26, 1984, Joint Publications Research Service, p. 16.
334. Kidger, "Salyut Mission Report," op. cit. Vol. 26, No. 12, p. 464.
335. Bozlee, Art and Vick, C. P., "The Soviets' Mystery Craft," *L5 News*, May 1986, p. 12.
336. Foreign Broadcast Information Service, U.S.S.R., Space, JPRS-USP-84-005, Oct. 26, 1984, Joint Publications Research Service, pp. 14, 15, 17.
337. Kidger, "Salyut Mission Report," op. cit. Vol. 26, No. 12, p. 464.
338. Turnill, op. cit. p. 221.
339. Kidger, "Salyut Mission Report," op. cit. Vol. 26, No. 12, p. 466.
340. Foreign Broadcast Information Service, U.S.S.R., Space, JPRS-USP-84-006, Nov. 14, 1984, Joint Publications Research Service, pp. 9, 13.

341. Ibid. JPRS-USP-84-005, Oct. 26, 1984, Joint Publications Research Service, pp. 24–26, 32, 37, 40, 41.
342. Congressional Research Service, The Library of Congress, *Soviet Space Programs 1981–87, Part 1*, Government Printing Office: Washington, May 1988, p. 39.
343. Kidger, "Salyut Mission Report," op. cit. Vol. 26, No. 12, p. 466.
344. Turnill, op. cit. p. 221.
345. Foreign Broadcast Information Service, U.S.S.R., Space, JPRS-USP-84-006, Nov. 14, 1984, Joint Publications Research Service, p. 1.
346. Congressional Research Service, The Library of Congress, *Soviet Space Programs 1981–87, Part 1*, Government Printing Office: Washington, May 1988, p. 57.
347. Foreign Broadcast Information Service, U.S.S.R., Space, JPRS-USP-84-006, Nov. 14, 1984, Joint Publications Research Service, pp. 12–20.
348. Turnill, op. cit. p. 220.
349. Foreign Broadcast Information Service, U.S.S.R., Space, JPRS-USP-84-006, Nov. 14, 1984, Joint Publications Research Service, pp. 14–21.
350. *U.S.S.R. Space Life Sciences Digests*, NASA CR-3922(8), Issue 7, p. 103.
351. Foreign Broadcast Information Service, U.S.S.R., Space, JPRS-USP-86-001, Jan. 13, 1986, Joint Publications Research Service, pp. 61–62.
352. *U.S.S.R. Space Life Sciences Digests*, NASA CR-3922(5), Issue 5, pp. 2, 3, 6.
353. Oberg and Oberg, *Pioneering Space*, op. cit. p. 93.
354. *U.S.S.R. Space Life Sciences Digests*, NASA CR-3922(8), Issue 7, p. 102.
355. Foreign Broadcast Information Service, U.S.S.R., Space, JPRS-USP-84-006, Nov. 14, 1984, Joint Publications Research Service, p. 30.
356. Turnill, op. cit. p. 220.
357. Foreign Broadcast Information Service, JPRS-USP-86-006, Nov. 1986, Joint Publications Research Service, p. 126.
358. Turnill, op. cit. p. 221.
359. Oberg, *New Race for Space*, op. cit. p. 72.
360. Bozlee and Vick, op. cit. pp. 10–12.
361. Clark, Phillip S., "Soviet Space Activity, 1985–1986," *Journal of the British Interplanetary Society*, Vol. 40, No. 5, May 1987, p. 203.
362. *Spaceflight*, Vol. 31, No. 2, Feb. 1989, p. 58.
363. "Soviet Scene," *Spaceflight*, Vol. 29, No. 1, Jan. 1987, p. 17.
364. Kidger, "Salyut Mission Report," op. cit. Vol. 27, pp. 420, 469.
365. Foreign Broadcast Information Service, U.S.S.R., Space, JPRS-USP-85-005, Sept. 30, 1985, Joint Publications Research Service, p. 2.
366. Congressional Research Service, The Library of Congress, *Soviet Space Programs 1981–87, Part 1*, Government Printing Office: Washington, May 1988, p. 63.

367. Kidger, "Salyut Mission Report," op. cit. Vol. 27, p. 469.
368. Foreign Broadcast Information Service, U.S.S.R., Space, JPRS-USP-86-001, Jan. 13, 1986, Joint Publications Research Service, p. 8.
369. Kidger, "Salyut Mission Report," op. cit. Vol. 27, p. 420.
370. Canby, T. Y., "Are the Soviets Ahead in Space," *National Geographic*, Oct. 1986, p. 433.
371. Congressional Research Service, The Library of Congress, *Soviet Space Programs 1981–87, Part 1*, Government Printing Office: Washington, May 1988, p. 64.
372. Foreign Broadcast Information Service, U.S.S.R., Space, JPRS-USP-86-001, Jan. 13, 1986, Joint Publications Research Service, p. 8.
373. Kidger, "Salyut Mission Report," op. cit. Vol. 27, p. 470.
374. Foreign Broadcast Information Service, U.S.S.R., Space, JPRS-USP-86-001, Jan. 13, 1986, Joint Publications Research Service, pp. 4–5.
375. Canby, op. cit. p. 433.
376. Foreign Broadcast Information Service, U.S.S.R., Space, JPRS-USP-86-001, Jan. 13, 1986, Joint Publications Research Service, p. 5.
377. Ibid. JPRS-USP-86-005, Sept. 1986, Joint Publications Research Service, p. 70.
378. Ibid. JPRS-USP-86-001, Jan. 13, 1986, Joint Publications Research Service, pp. 5–6.
379. Canby, op. cit. p. 422 .
380. Foreign Broadcast Information Service, U.S.S.R., Space, JPRS-USP-86-005, Sept. 1986, Joint Publications Research Service, p. 70.
381. Kidger, "Salyut Mission Report," op. cit. Vol. 27, p. 470.
382. Foreign Broadcast Information Service, U.S.S.R., Space, JPRS-USP-85-005, Sept. 30, 1985, Joint Publications Research Service, pp. 10–19.
383. Ibid. JPRS-USP-86-001, Jan. 13, 1986, Joint Publications Research Service, pp. 1–16.
384. Ibid. JPRS-USP-86-005, Sept. 1986, Joint Publications Research Service, p. 72.
385. Turnill, op. cit. p. 223.
386. Kidger, "Salyut Mission Report," op. cit. Vol. 27, p. 421.
387. Ibid. p. 470.
388. Foreign Broadcast Information Service, U.S.S.R., Space, JPRS-USP-86-005, Sept. 1986, Joint Publications Research Service, p. 73.
389. Ibid. JPRS-USP-85-005, Sept. 30, 1985, Joint Publications Research Service, p. 17.
390. Ibid. JPRS-USP-86-001, Jan. 13, 1986, Joint Publications Research Service, pp. 31, 41.
391. Kidger, "Salyut Mission Report," op. cit. Vol. 27, p. 471.
392. Turnill, op. cit. p. 223.

393. Foreign Broadcast Information Service, U.S.S.R., Space, JPRS-USP-86-001, Jan. 13, 1986, Joint Publications Research Service, p. 30.
394. Ibid. JPRS-USP-86-005, Sept. 1986, Joint Publications Research Service, p. 74.
395. Kidger, "Salyut Mission Report," op. cit. Vol. 27, p. 471.
396. Foreign Broadcast Information Service, U.S.S.R., Space, JPRS-USP-86-001, Jan. 13, 1986, Joint Publications Research Service, pp. 16–19.
397. Ibid. JPRS-USP-86-005, Sept. 1986, Joint Publications Research Service, p. 74.
398. Ibid. JPRS-USP-86-001, Jan. 13, 1986, Joint Publications Research Service, pp. 20–28.
399. Ibid. pp. 30, 31, 41.
400. Ibid. pp. 32–35.
401. Turnill, op. cit. p. 224.
402. Foreign Broadcast Information Service, U.S.S.R., Space, JPRS-USP-86-001, Jan. 13, 1986, Joint Publications Research Service, pp. 42–45, 48.
403. Ibid. JPRS-USP-86-005, Sept. 1986, Joint Publications Research Service, p. 75.
404. Turnill, op. cit. p. 224.
405. Foreign Broadcast Information Service, U.S.S.R., Space, JPRS-USP-86-001, Jan. 13, 1986, Joint Publications Research Service, pp. 49–50.
406. Turnill, op. cit. p. 223.
407. Foreign Broadcast Information Service, U.S.S.R., Space, JPRS-USP-86-001, Jan. 13, 1986, Joint Publications Research Service, pp. 51–52.
408. Clark, "Soviet Space Activity, 1985–1986," op. cit. p. 206.
409. Foreign Broadcast Information Service, U.S.S.R., Space, JPRS-USP-86-002, Feb. 1986, Joint Publications Research Service, p. 4.
410. Clark, "Soviet Space Activity, 1985–1986," op. cit. p. 206.
411. Foreign Broadcast Information Service, U.S.S.R., Space, JPRS-USP-86-002, Feb. 1986, Joint Publications Research Service, pp. 2, 4.
412. "Soviet Scene," *Spaceflight*, Vol. 29, No. 1, Jan. 1987, p. 17.
413. Congressional Research Service, The Library of Congress, *Soviet Space Programs 1981–87, Part 1*, Government Printing Office: Washington, May 1988, p. 73.
414. Foreign Broadcast Information Service, U.S.S.R., Space, JPRS-USP-86-004, April 1986, Joint Publications Research Service, p. 26.
415. "Soviet Scene," *Spaceflight*, Vol. 28, No. 3, March 1986, pp. 111, 112.
416. Foreign Broadcast Information Service, U.S.S.R., Space, JPRS-USP-86-005, Sept. 1986, Joint Publications Research Service, p. 76.
417. Ibid. JPRS-USP-86-002, Feb. 1986, Joint Publications Research Service, pp. 10–11.
418. "Soviet Scene," *Spaceflight*, Vol. 28, No. 3, March 1986, p. 112.

419. Ibid. Vol. 28, No. 2, Feb. 1986, p. 63.
420. Foreign Broadcast Information Service, U.S.S.R., Space, JPRS-USP-86-005, Sept. 1986, Joint Publications Research Service, p. 76.
421. "Soviet Scene," *Spaceflight*, Vol. 28, No. 3, March 1986, p. 112.
422. Translated excerpts from Soviet TV, "Channel 3 Moscow," Public Broadcasting System, Oct. 8, 1986.
423. Turnill, op. cit. p. 224.
424. "Soviet Scene," *Spaceflight*, Vol. 29, No. 1, Jan. 1987, p. 17.

Chapter 6

Mars Precursors

During the flights of Salyut 6 and 7, the Soviets had begun experimenting with prototype production facilities for semiconductor materials and pharmaceuticals. The results of those experiments showed the possibility of economic benefits by producing those materials in space in large quantities. Soon after the launch of Salyut 6, the Soviets began proposals and planning for the next generation of space stations. By 1980, many scientists in the Soviet Union reportedly advocated docking two Salyut's together by 1986, and increasing crew size to allow specialists to fly and obtain better data.[1] Even before Salyut 7 seemed to reach the end of its usable life, the Soviets planned to expand their experiments in materials processing and began planning large-scale production in space. To support this large-scale production required the development of a third generation of space station and new spacecraft, just as the second-generation Salyut required new spacecraft to support it in the 1970s.

This idea of docking Salyuts together was taken to its logical conclusion with the design of the Mir space station. Mir's construction was already a major effort facing the Soviets by 1983.[2] At the IAF conference in Sweden, in October 1985, the Soviets announced that Mir would have multiple docking ports and be expanded by adding 4–6 modules to the ports.[3] This chapter covers the years 1986 to 1989, when the Soviets began operations of the Mir space station and flight testing of the Energia heavy booster and space shuttle.

Aleksey Leonov, then deputy head of cosmonaut training, said that Mir was a third-generation station that would take materials processing from research and development to full-scale production. To help carry out this mission, specialized modules were also developed to allow expanded research in many fields and allow larger space station crews. A new version of the Soyuz transport also was developed to allow more supplies to be sent to the space station and more results returned to Earth. Simultaneous with furthering efforts in materials processing, the Soviets continued extending the limits of human endurance in space with future manned planetary missions becoming a possibility.

Third-Generation Space Station

Mir means peace in Russian, but Mir also translates into commune or village, and world or universe. The station was named Mir by the Soviets to try and give their manned space program a better image, as a part of a larger public relations blitz directed at the nations of the West, particularly the United States. The Soviets had to do this because their manned space program's activities include significant work in military reconnaissance and ABM tests and research. Space station crews, in general, don't spend a disproportionate amount of time doing military work; it's just that the rest of the Soviet space program is so largely devoted to military work that the manned program was the only area to try and show a different face. Military activities carried out by cosmonaut crews reportedly were anti-submarine warfare research into using plankton biolumi-nescence in the wake of submerged craft, calibrating and testing targeting systems for ground-based lasers, and experimenting with tracking of ballistic missiles and satellites.[4]

The Mir space station was planned to operate for several years, and was the first space station equipped with several docking ports. Soviet plans as of 1986 were to launch four specialized modules to Mir over four years for additional research and habitation space. This schedule later proved to be optimistic and the modules were delayed by many months. Some of the modules were planned to operate in a free-flying mode and be serviced occasionally by space station crews. The Soviets also began planning to phase out the Progress spacecraft and replace them with a Star module type spacecraft for space station refueling and return of experiment results and processed materials (see Figure 6-1).

Space Shuttle Program

During this period, the Soviet shuttle program was also finally entering the flight stage. With a shuttle operating, the Soviets could return large amounts of material from space and reduce the number of unmanned boosters needed to supply a space station. The shuttle program was officially announced in 1987 with the launch of the Energia heavy booster. The Soviets even offered to allow NASA to use their runway at Baykonur for emergency shuttle landings and stated that the runway was primarily built as a shuttle recovery runway.[5] Interestingly, during the development of the NASA space shuttle, one of the military require-ments was that the shuttle be capable of landing up to 2,100 km away from its flight path to avoid an emergency landing in a communist country. Components of the Soviet shuttle's launch vehicle were tested on the first test flight of the Energia booster. The Soviets later stated that the construction of the Soviet shuttle had cost $10 billion dollars and took 10 years.[6] This agrees very well with the reports of failures in the first Soviet shuttle program in 1978, which led to

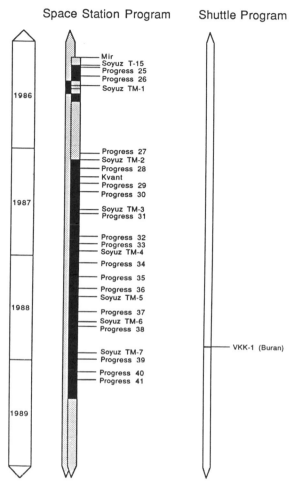

Figure 6-1. The space station and shuttle programs were continued from 1986 to 1989. All known flights related to these programs are shown here and are described in this chapter. The gray portions show space station spacecraft lifetimes and black portions show manned mission durations.

the complete change to using the NASA shuttle orbiter design in the second shuttle program. On the second test of the Energia booster, an unmanned Soviet shuttle was carried successfully into orbit for a short flight (see Figure 6-2).

The shuttle's mission was primarily to build and resupply space stations that would follow the current Mir complex. The larger Mir 2 permanent space station is planned for launch in pieces by the Energia booster and Soviet shuttle around 1994. Initial design of Mir 2 began in 1988.[7] It will initially house 12 people

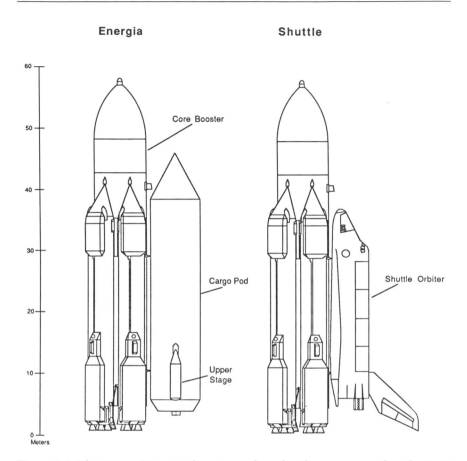

Figure 6-2. The K-type (Energia) booster can launch either a cargo pod or the Soviet shuttle orbiter. The Energia may be modified in the future to be fully recoverable by adding wings to the core stage. The Energia may also be enlarged by using more strap-on boosters.

and could eventually become Kosmograd, a "space city" with a population of at least 100.

Return to the Moon

By 1995, the Soviets plan to have the large station in Earth orbit for large-scale manufacturing of materials impossible to make on Earth. From there, they will be able to launch manned missions to the planets, and be able to support a lunar base.[8] These initial systems may also be expanded in the future to be more capable. The Soviets already anticipate modifying the Energia booster to lift twice as much weight into orbit as its first flights.[9] The same Soviet space infrastructure that can launch a manned Mars mission can support a large lunar effort, massive space manufacturing capabilities, and military defense and offense.

The development of a transportation system like the Energia/Shuttle is not an end in itself—it is required by some overriding need. Precisely what this program is has not been revealed. The Soviets plainly say it was developed for the space station program and flight to the moon and planets, but these are not scheduled for many years to come. Dr. Boris Gubanov, the chief designer of the Energia space rocket system said that the shuttle was not the ultimate goal or destination of the Soviet space program.[10] It was only one of the elements of the extensive program of developing Soviet aerospace systems, or space infrastructure. The payloads that the Energia and shuttle will be launching in the near future are of great interest to many observers.

The possible long-term uses for the Energia/Shuttle are more obvious. Using lunar materials, the Soviets can build large structures like solar power satellites and produce rocket propellant for use in Earth orbit at a fraction of the cost of using Earth supplied materials. Perhaps they will sell their products and services to other countries that now plan manned space activities in the next century. Glavkosmos, the official Soviet agency for selling space services like boosters and experiments, has already sold Soviet space services to Western countries.

Before the turn of the century, the Soviets hope to use the Energia to launch large mirrors into orbit to light northern regions of the U.S.S.R. in winter and launch power relay satellites as prelude to solar power satellites[11] to supply remote areas of Siberia. Feoktistov commented that construction of large solar power satellites is still not possible and launch systems cost must still be reduced tens or hundreds of times.[12] But, use of lunar material would greatly reduce this limitation.

Mars Missions

The Soviets are now extending the duration of space flights on the Mir station so that when the permanent station is built, it can be used to launch the first manned missions to the planets. The Soviets stated in 1987 that after a successful 8-month mission, they felt confident to fly a 12-month mission by extrapolation. The Soviets had also stated that when cosmonauts have flown for 30 months in weightlessness, the decision will be made about manned flight to the planets.[13] By that time, the new permanent space station could support the launch of a manned interplanetary spacecraft made of the well tested Mir type components. Initial Soviet design concept studies for a nuclear-powered electric-drive Mars transport capable of carrying four crewman were already underway in 1988 (see Figure 6-3).[14] The current design philosophy is an evolutionary approach

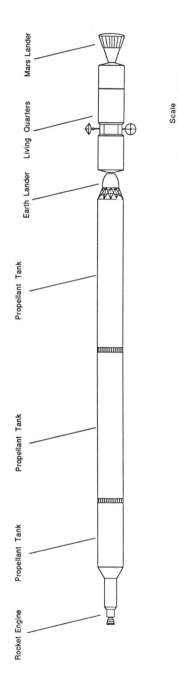

Figure 6-3. Soviet design studies for manned missions to Mars after the year 2000 are underway. The drawing is based on Soviet drawings for nuclear or conventionally propelled spacecraft.

beginning with the design of unmanned Mars probes that could be modified for manned use after initial unmanned tests. Using a Mir type station with 5 additional modules, a crew of 3 could be supported for about 40 months— enough time for a journey to Mars, a short landing, and the trip back. The time that the mission would take is totally dependent on the size of the spacecraft, the type of boosters used, and the trajectory used to reach Mars. The time of the flight must be within the limits the human body can withstand after returning from weightlessness. The only other option is to create ·artificial gravity, which increases the weight, size, cost and complexity of the spacecraft.

The problem of bone decalcification or demineralization in weightlessness is caused by the same factors that cause disuse osteoporosis on Earth, i.e., the lack of muscular exercise. In weightlessness, a lot of effort is still required to move about using one's arms, but the lower body loses strength as the muscles deteriorate from disuse. The bone mass and muscle mass decrease unless counteracted by deliberate use of the legs as in running on a treadmill. Research is underway in the U.S. and U.S.S.R. into drugs to help counter the deterioration, which also can help people on Earth suffering from disuse osteoporosis due to physical disabilities.[15] Despite the problem, the Soviets have been successful so far in using exercise techniques to prevent serious deterioration in cosmonauts.

The Soviets are currently planning to perform a Mars mission in weightlessness, even though the technique of creating artificial gravity by rotating a spacecraft has been successfully tested on a small scales. During the flight of Kosmos 936 in 1977, rats were placed in a centrifuge rotating to simulate gravity of almost Earth normal for 19 days. The results showed that there was no skeletal or muscle degradation during the flight.[16] Despite this great advantage, to build a spacecraft to rotate and carry humans would require more structural weight and complicate the mission. In the Soviets' opinion, the effects of weightlessness can be adequately countered with less cost by exercise as has been demonstrated on Soviet space stations. Shortly before the launch of the shuttle Buran in 1988, the Soviets announced that the Energia and shuttle would be used to return men to the moon early in the next century, with hopes to establish an international base there. But, they noted that flight to Mars would have priority over the return to the moon.[17] The Soviets publicly do not plan to make a manned Mars landing until about 2010, after several unmanned sample return missions. Although a Mars mission would have great propaganda and scientific value, it would more importantly be an indicator of the Soviets' substantial Earth orbit capabilities.

International Reactions

This realization of the growing Soviet capabilities is slowly beginning to have its effects in the U.S. The United States, Europe, and Japan realize the manufacturing potential of a space station and hope that planned U.S. space station will be heavily involved in producing materials impossible to make on Earth. One U.S. company is also planning a small man-tended space station devoted to materials processing. The fact that so many countries are now planning Earth orbit space systems is an indicator that near Earth space will be the most important area in the 1990s and into the next century.

As of 1988, the European Space Agency was planning two different space shuttle systems for use in the 1990s and beyond the Hermes and Sanger. The Japanese also are building an unmanned shuttle and even the Chinese are interested in building a space shuttle. The same countries are also interested in launching space stations of their own. Of all the countries today planning space stations, the Soviets are well ahead in experience and development. Only the United States has the capability to match Soviet efforts if it begins soon. The planned Mir 2 and the U.S. space station Freedom will be built at about the same time, if no more delays occur in the programs.

In 1987, General Secretary Gorbachev proposed a joint U.S.-U.S.S.R. mission to Mars, but the U.S. administration was uninterested in the offer. Mars may be an attractive target for exploration and future space development, but the next step has long been believed to be the development of the moon and the exploitation of its easily available resources for use on and around Earth. The Soviets also recognize this, but say that trips to Mars may come before the moon because of forces other than scientific interests.

Even the Soviets are unsure about how to proceed with their space program with so many options available with the Energia, shuttle, and space stations. General Shatalov stated that progress in formulating a clear program for the future had been slowed by insufficient progress in materials processing experiments and experiment analysis.[18] But, by building an Earth orbit space infrastructure of large space stations, space shuttles, and heavy-lift boosters, the Soviets are determined to find out what is possible.

Mir

Salyut Station with Multiple Docking Ports
Launched: February 20, 1986, 12:28 a.m.
Altitude: 169 × 277 km @ 51.62°

Mir's development began in 1983, apparently when the status of the new Energia booster was still in doubt requiring the design to parallel the old Salyut's.[19] Some details about Mir were released at the 1985 IAF Congress, including the fact that it would have multiple docking ports and would be launched in 1986 (see Figure 6-4).[20] The launch of Mir was shown on television in Western countries, 12 hours after the fact, and Mir's differences from previous Salyuts were greatly exaggerated by the U.S. press.

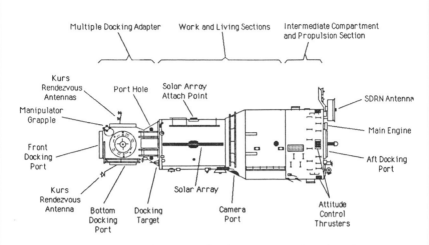

Figure 6-4. Mir was an evolutionary development from the Salyut design. New features included a multiple docking adapter and large new solar arrays. Mir was equipped only as a living quarters when launched and did not carry any experiments other than medical equipment. All experiments were delivered to the station by later flights.

Mir Space Station

The Mir station was an evolutionary modification of the previous Salyut stations. Mir was the same dimensions, 13.13 meters long and 4.15 meters in diameter, weighing about 20,000 kg at launch. The station was designed to operate for at least 10 years and be expanded by adding five Star modules to the station. To expand the station five docking drogues on a multiple docking adapter were put in the place of the transfer compartment on a normal Salyut. The forward port also was equipped with the new Kurs docking system, which eliminated the need for the space station to be orientated toward the approaching spacecraft. A new manipulator system was also designed to be used on Star modules to move the modules to a side port after docking to the forward port.

Mir's other modifications included two newly designed gallium arsenide solar arrays. Each array's total area was 76 square meters; they spanned 29.73

(continued on next page)

(continued from page 305)

meters and produced an estimated 9 kW of power. On the aft end of the station was a small dish antenna for communication through geostationary Luch satellites to ground stations.

Mir was controlled by seven computers of the Strela system. A digital data bus provided connection from the computers to the station's systems and experiments. The computers could be programmed from the ground to operate the station and experiments at least a few days in advance. Mir also was equipped with a new computer called the EVM, which was capable of maintaining the station's orientation indefinitely without human intervention, as was needed with the old Delta system. The space station controls included a new optical sight and a new portable orientation control stick. The environmental system eliminated lithium hydroxide canisters and used the Vozduyk system to reject carbon dioxide directly to space.

Most of the station's volume consisted of the living section. The galley and folding table were similar to Salyut equipment with built-in food heaters for a crew of two. Mir was also equipped with a treadmill and an ergometer, which was normally stowed beneath the floor behind the table. Mir had no Salyut-type telescope housing, and instead of a Salyut's equipment racks Mir had only living space for a crew of two. Each cosmonaut had a separate closet-like compartment off the living section for sleep and privacy. The sleeping compartments each had a folding chair, mirror, porthole, and sleeping bag. There was only one scientific and trash airlock mounted hidden in the floor.

Mir was launched by an up-rated Proton booster (SL-13), which had been flown successfully twice in 1985 (see Figure 6-5). The launch of Mir was specially planned to enable the next Soyuz mission to visit both Salyut 7 and Mir. To simplify the maneuvers needed for the upcoming Soyuz T-15 transfer between the Mir and Salyut stations, Mir was put into the same orbital plane as Salyut 7, necessitating the launch as Salyut 7 passed over the launch site. The launch window lasted 5 seconds as Salyut 7 passed over the Cosmodrome.[21] The launch was observed in mission control by cosmonauts Makarov, Kubasov, Ivanchenkov, Aksenov, Strekalov, and Savinykh.[22]

After separating from the Proton third stage, Mir soon maneuvered to an orbit of 172×301 km and then 324×340 km, and slowly approached Salyut 7. This led some Soviet space analysts to believe that it would be docked with Salyut 7, however the Soviets had other plans. The Proton booster's third stage orbited for 5 days before reentering, in the early morning over the Atlantic Ocean about 1,100 km east of Newfoundland (48°N, 45°W). The destruction

Figure 6-5. Here a Proton (D-1e) booster launches a Vega space probe. The Proton (D-1h) booster, which launched Mir, was very similar. (Source: Tass/Sovfoto.)

of the stage was observed by the pilots of an airliner in the area. The stage broke up at an estimated 11.3-km altitude.[23]

Mir required installation of new computer systems over the next few months at the Kaliningrad mission control center to handle the increase in telemetry and commands the new station required. On February 24, mission control continued a series of control tests on structure and systems of the station.[24] By March 3, the power, environmental systems and docking systems were tested and checked. The station's systems were all in order and ready to receive a crew.

Soyuz T-15

First Station-to-Station Transfer—125 Days

Launched: March 13, 1986, 3:33 p.m.
Landed: July 16, 1986, 4:34 p.m.
Altitude: 193×238 km @ $51.6°$
Crew: Leonid Kizim and Vladimir Solovyev
Backup: Aleksandr Victorenko and Aleksandr Aleksandrov
Call sign: Mayak

The T-15 crew arrived at Baykonur March 3, as the testing of Mir by mission control was completed. The launch was announced a day in advance and shown live on Soviet television and U.S. networks, although without advanced notice.

The Soviets provided American-like coverage to most of the launch highlights. This was at least the fourth Soviet launch shown live, contrary to numerous Western news reports (the first was ASTP, the second Soyuz T-11, and the third a Vega probe).[25] The launch was from the primary A-2 launch pad.

For this complex mission, there were four controller teams at Kaliningrad, one for Salyut 7, one for Mir, one for the T-15, and one for upcoming Progress flights.[26] The mission's objectives were to check out and prepare Mir for operations, and complete the Soyuz T-14's mission at Salyut 7.

Soyuz T-15's transfer orbits were 239 × 289 km and 292 × 331 km. A Progress type approach was used, which required two days to rendezvous with the station, but it saved propellant that would be used later. Mir had been in an orbit close to the Salyut 7 complex since the middle of February. Kizim approached Mir's aft docking port, which still used the old docking system that was compatible with the Soyuz T and Progress. Then he flew around the station to dock at the forward port, which had the new Kurs docking system for use with the soon to be launched Soyuz TM and Star modules. It was impossible to make an automatic docking at the forward port in a Soyuz T. The Soyuz T-15 also carried a laser range finder that was first used on the Soyuz T-13 flight. At 2.2 km, the range finder was used by the crew as they approached Mir.[27]

By the time Kizim had maneuvered into position to dock, Mir was turned for best lighting conditions for docking. Kizim docked with Mir manually from 60 meters distance, 12 minutes ahead of schedule at 4:38 p.m., March 15, at 332 × 354 km. No television coverage of the approach or docking was provided, but the cosmonauts resumed the television coverage while touring the facilities of the station (see Figure 6-6).[28]

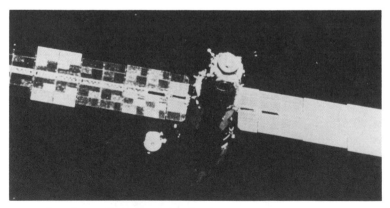

Figure 6-6. This photo of Mir was taken by the U.S. space surveillance cameras soon after it was launched. The cameras are used to monitor all Soviet spacecraft in orbit and have been used to observe cosmonauts during EVAs. (Source: U.S. Department of Defense.)

The cosmonauts began activating life-support systems, installing experiments, and testing instruments while waiting for more to be delivered on Progress 25. This mission did not include much scientific research, because no equipment was launched on Mir for that purpose. The crew would only install equipment as it was delivered by Progress transports. On March 16, the station was in orbit at 332×354 km. The cosmonauts had finished activating the environmental control system by March 18.[29]

Progress 25

Launched: March 19, 1986, 1:08 p.m.
Reentry: April 21, 1986
Altitude: 184×249 km @ $51.6°$

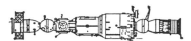

The first maneuver after separating from the A-2 booster's upper stage put Progress 25 into an orbit of 189×268 km. The final transfer orbit to Mir was 227×295 km. The Progress docked March 21, at 2:16 p.m., to Mir's aft port. This flight carried supplies for 20 days including propellant, air, tools, film, mail, food, 200 kg of water and equipment to the T-15 cosmonauts on Mir. The crew had some difficulty in unloading the cargo because some fasteners were over tightened. The crew commented on the need of additional wrenches for this purpose. The cosmonauts then began the first refueling of Mir. On March 26 and April 18, the Progress was used to boost the station's orbit to 336×360 km. The Progress undocked on April 20, at 11:34 p.m. and reentered normally.[30,31]

☆ ☆ ☆ ☆ ☆

The crew's major duty during the first weeks of the mission was to test Mir's 7 computers and put the Luch communication system in order. On March 29, the video communication channels were tested, and the next day, telemetry and voice channels were tested. The Soviet's Space Data Relay Network (SDRN) system, using Kosmos 1700 stationed roughly over Baykonur, increased telemetry volume ten times over that previously available.[32,33] Salyut 7 had required 100 commands per orbit compared to Mir's 300, an estimated 1,000 with 4 modules attached. The Soviets said that the system was not in continuous use yet. To test the SDRN system before launch, Mir was rolled out of the Proton assembly building and its dish antenna pointed at Kosmos 1700.[34] The communications link was tested successfully. Similar tests of the U.S. TDRSS system with shuttle orbiters at the Kennedy Space Center were routine. Later, it was reported that the Luch SDRN system was much more difficult to use than expected, and it was very time consuming to align the space station so the antenna could face the Luch satellite. In addition, the Luch satellites were not very reliable, and launch failures also occurred.

On April 1, the cosmonauts performed Resonance tests to determine the dynamic stability of the station complex, and continued unloading and installing equipment delivered by the Progress. On April 4, it was announced that the first stage of the mission to Mir station was ending. On April 15, they tested the backup radios and temperature control radiators. On April 18, they performed more Resonance tests and changed the station's orbit with Progress 25's engine, to 336 × 360 km.[35] When not checking the station's systems, the cosmonauts observed Earth, tended plant growth experiments, or performed other small experiments.

Progress 26

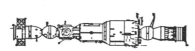

Launched: April 23, 1986, 11:40 p.m.
Reentry: June 23, 1986
Altitude: 192 × 264 km @ 51.6°

Progress 26 required extensive orbit adjustments before docking in an unusual 4-day rendezvous. The first orbit correction resulted in an orbit of 190 × 274 km, then another maneuver resulted in a transfer orbit of 205 × 265 km. A day later, another transfer orbit was attained of 255 × 304 km. The next day the orbit was brought to near Mir's, at 338 × 349 km, and then the next day it docked with Mir at 1:26 a.m., April 27, at 335 × 345 km.

The supply flight carried propellant, food, water, mail, and equipment to continue the outfitting of the space station. The Progress was also used to provide the first rendezvous maneuver for the Soyuz flight from Mir to Salyut 7. Propellant transfer to Mir was performed under ground control while the crew was away on Salyut 7. The fuel transfer was completed by June 19, and the oxidizer was transferred by June 20. The Progress 25 and 26 missions completed the refueling of Mir, which had used most of its original propellant to reach its final orbit after launch.

On June 20, the Progress changed the station's orbit in preparations for the T-15 redocking after the Salyut 7 phase of the Mayak mission. Progress 26 undocked on June 22, at 10:25 p.m., and retrofire was at 10:40 p.m. the next day.[36,37]

☆ ☆ ☆ ☆ ☆

On April 13, Mir was one minute ahead of Salyut 7 and in a similar orbit. On May Day, the crew was given the day off and video of the festivities in Red Square was provided on their monitor. On May 4, they started deactivating the station and packing 500 kg of equipment and supplies to be taken to Salyut 7 in the Soyuz's orbital module, including the crew's plants.[38–40]

On May 5, Progress 26 lowered the orbit of Mir to 309 × 345 km, and the complex began closing on Salyut 7. The T-15 did not have enough propellant for the planned trip to Salyut 7 and back, so the Progress made up for that by pushing Mir into a different orbit as well. The Soyuz T-15 undocked from Mir at 4:12 p.m., May 5. They performed no immediate engine burns near the station for fear of getting the station's windows dirty and harming instruments and the solar arrays.[41]

The Soyuz T-15 approach to Salyut 7 used techniques similar to the Soyuz T-13 approach. Salyut 7 was 4.15 minutes and 3,000 km ahead of Mir at the time. After 4 orbits, Soyuz T-15 adjusted its orbit with two maneuvers to 311 × 343 km and 307 × 342 km. It then performed two more maneuvers to rendezvous with Salyut 7, which was being maneuvered by the Kosmos 1686 Star module to the correct docking attitude under ground control (see Figure 6-7). During the approach, the crew used the new hand-held laser range finder to determine distance from the station. This equipment was flown on all flights since Soyuz T-13 and a similar device was made standard equipment on Soyuz TM transport. The automatic system was used from 5 km distance. Kizim then took manual control to dock at Salyut 7's aft port at 8:58 p.m., May 6.[42,43]

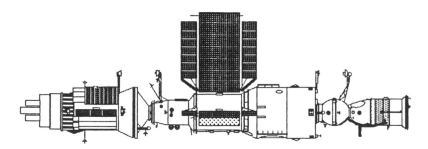

Figure 6-7. After a stay on the Mir station, the Soyuz T-15 crew flew to the Salyut 7 station to use its experiments and retrieve some usable equipment for use on Mir.

The crew reactivated Salyut 7 on May 8, and began maintenance work on the radios, life support and temperature control systems. Salyut 7's orbit was then 358 × 360 km. On May 13, the activation was complete and the cosmonauts started their first investigations by using the Astra mass spectrometer to investigate the thin cloud of gases around and emitted by the Salyut. Similar studies were carried out on previous U.S. shuttle missions.

The crew then began to prepare for EVA tests and experiments. During a press conference, the cosmonauts said that they would close down Salyut 7 during their 6-month mission. They also said that they had trained extensively for EVA, and that after they completed their mission, another long-duration crew would be launched to Mir.

Soyuz TM-1

First Flight of Modified
Soyuz T—Unmanned

Launched: May 21, 1986, 12:22 p.m.
Landed: May 30, 1986, 10:49 a.m.
Altitude: 195 × 224 km @ 51.6°
Crew: none

The launch of this flight was announced in mid-March. This was the first flight of the modified Soyuz T spacecraft. The Soviets announced that the Soyuz TM would be used to deliver future crews to space stations of the modular type, meaning Mir (see Figure 6-8).

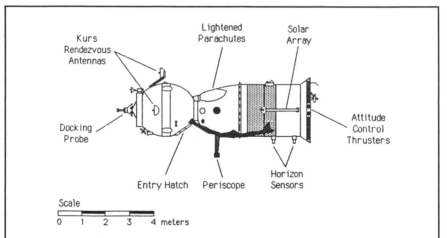

Figure 6-8. The Soyuz TM was a modified version of the Soyuz T. Improvements included the new Kurs docking system and more redundant systems. Payload capacity was also improved slightly by reducing the weight of the spacecraft's systems.

Soyuz TM Spacecraft

The Soyuz TM retained the same basic Soyuz T design, but several changes and redesigned systems were developed to increase the utility and safety of the spacecraft. The Soyuz TM was equipped with the new Kurs (course) rendezvous system and new computers to enable the Soyuz to dock with a space station while the station was at any relative attitude. The Kurs system

(continued on next page)

also did not require the station's radar transponders to be on to dock, making it capable of docking with an unresponding station. The Kurs system makes contact with the station at a range of 200 km and docking lock-on begins at 20 to 30 km distance.

A new feature was also the display of closing rate information, as detected by the docking radar, for the cosmonauts. The TM also had new lighter, 140-kg, 1,000-square-meter parachutes, which increased the launch payload by 250 kg and the landing payload by about 90 kg. The new parachutes also took up less volume leaving more space inside the reentry module. New Rassvet system radios provided communication through the new SDRN network by relaying the signal from the Soyuz TM to Mir for transmission to the SDRN satellites. Separate voice channels for each cosmonaut were also provided. An improved landing altimeter radar, improved inertial guidance units, acceleration sensors, and computer systems were installed. The launch abort tractor motor was lightened and propellant tanks in the instrument section were improved. A new non-cooled main engine nozzle was also used. The Soyuz TM had triple redundant electrical systems and redundant hydraulic systems.

Just as with the Soyuz T, the Soyuz TM also introduced new pressure suits to be worn during launch and landing. The new suits provided 307 mm Hg pressure in normal mode, and 200 mm Hg in backup mode, when easier movement was needed. The Soviets claim that the suit could be donned and activated in about 30 seconds. The suit weighed only about 10 kg and could be used in emergencies for crew transfers in open space, although this would require umbilicals or the use of a small backpack to provide oxygen. Normally, the suit was provided cabin air, but if the cabin decompresses, the suit is isolated from the cabin and provided with oxygen automatically.

The Soyuz used a transfer orbit of about 200 × 240 km, and then 195 × 291 km, and docked after what had become a normal 2-day approach, at 340 × 377 km on May 23, at 2:12 p.m. The 2-day approach saved propellant and simplified final rendezvous maneuvers. At the time of docking, Mir was 20 minutes ahead of the Salyut 7 complex in roughly the same orbit, but Mir was slowly drifting back to its original position behind Salyut 7.[44]

The Soyuz TM was equipped with the Kurs rendezvous system and other new features (see box). Soyuz TM-1 undocked to clear the forward port for the returning T-15 crew at 1:23 p.m., and landed at 10:49 a.m. while the station was in orbit at 332 × 343 km.[45,46]

The cosmonauts on Salyut 7 performed an EVA on May 28 to retrieve the French Comet experiment, and also test the URS beam, which was delivered by the Kosmos 1686 in 1985. The EVA was originally intended to be performed by the Soyuz T-14 crew, but had to be postponed after the medical emergency ended that mission. The URS beam was preassembled in a cylinder one meter tall, and deployed into a 15-meter-long square beam, similar to a beam built on the STS-61A shuttle mission six months before. The Soviets described the beam as lattice and pin frames, which could be welded together once the beam was deployed; however, the cosmonauts only welded sample pieces of the beam with an improved URI portable welding unit like that tested on Soyuz T-12. The beam was made of an aluminum-titanium alloy. It could be extended to varying lengths up to 15 meters and measured 40 by 40 cm square. The beam container weighed 150 kg, but the beam itself weighed only 20 kg. The canister was attached to a fixture, which was placed on the Salyut's forward docking unit. On the end of the beam was a square platform, on which experiments were placed.[47,48]

The first EVA started at 9:43 a.m., lasted 3.8 hours, and was shown live on Soviet television. As with the rest of the mission, the Soviets announced the EVA in advance and made all television coverage available to the West. This was also the second use of new EVA spacesuits, which were similar to the old Salyut suits. The major new features were improved controls, flexibility, and lights attached to the sleeves and helmet.

After the beam was deployed and measurements taken, the beam was folded back into its container. The Soviets noted that applications for the beam were joining space stations and making large telescope mirrors, solar panels, and spacecraft hangers. During the EVA, the cosmonauts also retrieved space exposure cassettes, which were on the hull of Salyut 7, including Medusa (bipolymers), Spiral (cables), Istok (connectors and bolts), and Resurs (structural metals). For the next 2 days, the cosmonauts caught up on documentation and cleaned the spacesuits.[49,50]

On May 31, they performed another EVA at 8:57 a.m., for 5 hours, and continued tests of the URS beam. The beam was deployed to 12 meters length, and was monitored by a seismic detector, which measured the vibrations of the beam. Information from the detector was sent using a low power (3 mW) laser data link to a receiver placed in a Salyut porthole. The results of these tests would take a reported 6 months to analyze. Kizim climbed out on the beam commenting that it rocked a little. The crew then folded the beam and dismantled the beam unit. Fifteen minutes of the EVA were covered on Soviet television. They also placed an experiment on the hull to test the strength of aluminum/magnesium alloy exposed to space under stress.[51,52]

During the EVA, they also took measurements of the cloud of particles and gas surrounding the Salyut. The cosmonauts also retrieved a piece of an experimental solar array that Dzhanibekov and Savinykh had installed during their EVA on August 2, 1985. The sample would be returned to Earth to study the effects of exposure to the space environment over a long period. The Soviets noted that the team of Kizim and Solovyev had accumulated

31.67 hours of EVA time, during 8 EVAs, including the Soyuz T-10B mission.[53]

After the EVA, the cosmonauts began intense observations of the Kiev area after the accident at the Chernobyl nuclear power plant. They supplied television pictures of the weather systems in the area. Scientists on the ground would use this and satellite data to try to control the local weather by cloud seeding to prevent heavy rains from carrying contaminated soil into the rivers and reservoirs supplying Kiev. During the early part of June, the cosmonauts continued routine work including tending plants in the Biogravistat, Oasis, and Svetbloc; measuring Earth's atmosphere; operating the Kristallizator materials processing unit; photographing Earth; and continuing routine physical exercises.[54]

On June 14, Mir was 2 minutes ahead of Salyut 7 in a similar orbit. On June 22, Progress 26 undocked from Mir and reentered the next day. It had been used to change Mir's orbit to set up the redocking of the Soyuz T-15.

At 6:58 p.m., June 25, at 356 × 359 km, the cosmonauts undocked from Salyut 7 after completing the Soyuz T-14 crew's aborted mission. They had packed 400 kg of Salyut 7's equipment and experiments on the Soyuz, some for installation on Mir, including the French Comet experiment, spacesuits, biopolymer cassettes, film, spectrograms, the KATE-140 mapping camera, spectrometers, the Pion-M, the EFU-Robot electrophoresis unit, the video complex (cameras and recorders), the French-made Echograph, PCN and other apparatuses.[55–57]

At the time of undocking, Mir was 3,000 km ahead of and 25–30 km above Salyut 7. After undocking, the Soyuz T-15 went into a 325 × 338-km transfer orbit to return to the Mir station. The Soyuz had about 420 kg of propellant at the time for the transfer, of which 250 kg would be needed for retrofire to return to Earth.[58,59]

On June 24, Mir maneuvered to lower its orbit by 30 km. On June 26, the Soyuz was closing on Mir, and at 200 meters distance, the cosmonauts activated the Igla rendezvous system, which maneuvered the Soyuz to within 50 meters of Mir's rear docking port. Kizim then maneuvered around the station and docked manually to the forward port at 11:46 p.m., at 332 × 366 km (see Figure 6-9). On July 1, they installed the experimental Strela information system, which displayed systems status, station documentation, and uplinked information from mission control. The crew also participated in the Geoeks-86 experiment, which combined Earth observations from Mir, aircraft, and Kosmos 1602.[60]

On July 14, the cosmonauts gave a interview in which their return was announced. Mir had been cleaned and prepared for the next long-duration crew and the cosmonauts looked well. They undocked at 1:07 p.m., July 16; the Soyuz orbital module separated at 2:23 p.m., followed by retrofire at 4:05 p.m. They landed in clear sky at 4:34 p.m., July 16, 55 km northeast

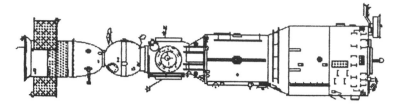

Figure 6-9. The Soyuz T-15 crew returned to the Mir station with equipment scavenged from Salyut 7. After a short stay, the crew returned to Earth because of delays in the launch of the Kvant experiment module for Mir.

of Arkalyk. Kizim had lost .5 kg of weight during the flight, while Solovyev had lost 5 kg.[61]

The Soviets had announced that the landing would be shown live on television, but this was not true. Instead, the Soviets provided live pictures of a commentator at mission control and 2 hours later showed pictures of the cosmonauts after recovery. The crew had operated 175 instruments during more than 2,500 sessions during their stay in Salyut 7. More than 3,000 photographs were taken, mostly of Earth to be used by more than 500 organizations in the U.S.S.R.[62] The mission accomplished the check-out of the newly launched Mir station and also closed-out Salyut 7. Unfortunately for the Soviets, the Kvant module to be launched to Mir was significantly delayed and the T-15 crew could not wait for it to continue their mission on the sparsely equipped Mir.[63]

Between August 19 and 22, the Salyut 7 complex was boosted using both Salyut 7's and Kosmos 1686's engines to a 474 × 492-km orbit, a record Salyut altitude. This was to keep the Salyut out of the way of Mir operations and preserve it for a future inspection mission. Later, the Soviets said that they hoped to send a Soviet shuttle crew to Salyut 7 before the end of the 1990s to access micrometeoroid damage, assess systems performance over the long period, and perhaps retrieve pieces of the station for analysis on Earth. Vladimir Shatalov later stated that Salyut 7 will either be returned to Earth by Soviet shuttle or sunk in the ocean after descending from orbit.[64,65]

Until then, ground controllers would maintain temperature in the station between five and 20°C, and observe the station's systems endurance, hermetic qualities, and performance of the thermal control system. By September 30, Salyut 7's orbit was 473 × 475 km and Mir was at 325 × 359 km. The Soviets reported that the mass of Salyut 7 and its equipment had almost doubled since its launch because of the accumulation of equipment delivered by Progress and Kosmos spacecraft (that would make the weight about 37,000 kg plus 20,000 kg for Kosmos 1686). The orbital lifetime of Salyut 7 was estimated at no less than 8 years.[66,67] By December 19, 1986, Mir's orbit was 321 × 349 km.

Progress 27

Launched: January 16, 1987, 9:06 p.m.
Reentry: February 25, 1987
Altitude: 114 × 164 km @ 51.6°

Progress 27 was launched to supply the next long-duration crew when they arrived at Mir, the next month. The transfer orbit to Mir was about 185 × 260 km. Progress 27 docked to Mir at 10:27 a.m., on January 18, in orbit at 312 × 343 km. It then boosted the station's orbit to 328 × 369 km on February 11. The main purpose of the Progress 27 was to complete the outfitting of Mir by delivering experiments and equipment like the 41-kg Pion-M heat and mass transfer experiment to study capillary action.[68,69]

On January 18, the Progress boosted the station's orbit to 328 × 363 km. The Progress was mostly unloaded by the Soyuz TM-2 crew by February 13, and refueled the station by February 19, and boosted its orbit twice before undocking at 2:29 p.m., February 23. Retrofire was at 6:17 p.m., February 25.[70–72]

Soyuz TM-2

First Mir Long Duration
Mission—11 Months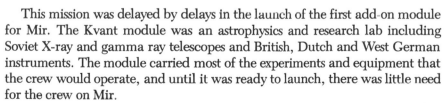
Launched: February 6, 1987, 12:38 a.m.
Landed: July 30, 1987, 4:04 a.m.
Altitude: 176 × 179 km @ 51.6°
Crew: Yuri Romanenko and Aleksandr Laviekin
Backup: Vladimir Titov and Aleksandr Serebrov
Call sign: Tamyr

This mission was delayed by delays in the launch of the first add-on module for Mir. The Kvant module was an astrophysics and research lab including Soviet X-ray and gamma ray telescopes and British, Dutch and West German instruments. The module carried most of the experiments and equipment that the crew would operate, and until it was ready to launch, there was little need for the crew on Mir.

Pictures released before the Soyuz TM-2 launch showed 3 Soyuz and 2 boosters being assembled in the Soyuz assembly building at Baykonur. The Soviet efforts were apparently not hampered by the extreme cold wave hitting Europe and the U.S.S.R. at the time. In a possibly related flight, a Proton booster carrying what may have been a Luch communications satellite failed on January 30. The Kosmos 1700 Luch satellite, which was tested and used by the T-15 crew had failed and drifted out of its orbital slot, in the last half of 1986.[73] The Proton may have been launching its replacement.

The original crew for the Soyuz TM-2 mission was Titov and Serebrov. Serebrov was disqualified because of his medical condition in the months before the launch, forcing the less experienced backup crew of Romanenko and Laviekin to replace them.[74] This was Laviekin's first space flight, and he had not trained as a backup until the Soyuz TM-2 mission. The crew arrived at Baykonur on January 23 to prepare for the launch.

Soyuz TM-2 was launched on time, about 3 hours after Mir passed over the Baykonur Cosmodrome. The transfer orbit to Mir was 269 × 308 km.

Two hours before docking, Mir was turned to provide best lighting conditions for docking. At 130 km distance, the Kurs docking rendezvous system was activated and the TM-2 approached the rear of the station. Progress 27 occupied the rear port at the time and the Soyuz automatically moved past the station and docked at the forward port at 2:28 a.m., February 8. The crew opened Mir's forward hatch 90 minutes later and entered the station. The Soyuz T-15 crew had left a television camera pointed at the hatch so that mission control would have a view of their entrance.[75,76]

On February 9, the station's orbit was 328 × 369 km. The cosmonauts began unloading Progress 27 the next day and used the spacecraft to boost the station's orbit again on February 11. The crew's main mission was to prepare Mir for permanently manned operations. A major task to accomplish in the mission was for the cosmonauts to reprogram Mir's computers for use with the new Kvant module, which would be added, and update other programs. Since the last crew's visit, Mir had been operating with only one of its computers functioning. On February 9 the station was in orbit at 328 × 369 km.[77]

As usual for the mission, the crew followed a normal work day referenced to Moscow time, waking at about 8:00 a.m. and ending the day at about 11:00 p.m. The Soviets reported that Laviekin adapted poorly to weightlessness. On February 12, the Soviets reported that the temperature in Mir was 22°C and the pressure was 830 mm Hg, after the crew had changed the temperature control settings and activated the water regeneration system. The crew had also finished inventorying the consumables on the station.[78,79]

On February 15, they reported seeing several waves on the ocean that were 200 to 300 km wide. The next day they underwent routine medical checks. On February 18, the crew installed additional components to Mir's electrical system, transferred water from Progress 27, and prepared the station for refueling, which began the next day. The cosmonauts requested that more entertainment video tapes be sent on the next Progress and installed new medical equipment delivered by the Progress. On February 20, Progress 27 boosted the station's orbit before undocking on February 23. On February 24, the cosmonauts donned spacesuits for an unknown experiment, but apparently did not exit the station. By February 26, the crew started the first Earth resources work with the new Sever camera, which took pictures looking to the side of the station's flight path. This enabled a greater opportunity to see relief in the photos. This work went on for the next week along with experiments using the Pion-M. The Soviets reported strange results from previous tests of the Pion-M that prompted the new series of material processing experiments. Particles of silica aerogel (a glass) in suspension formed saucer shapes 5 mm in diameter, fluoroplastics formed tree shapes, and glass pellets formed in clumps that could not be easily broken up.[80,81]

Progress 28

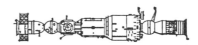

Launched: March 3, 1987, 2:14 p.m.
Reentry: March 27, 1987, 8:07 a.m.
Altitude: 176 × 221 km @ 51.6°

Progress 28 used a transfer orbit of 191 × 272 km to reach Mir and docked at 3:43 p.m., on March 5, at 344 × 369 km. The flight delivered mail, food, the 136-kg Korund-1M semiconductor crystal growth unit, a new KATE-140 mapping camera, spectrometers and the Gamma 1 medical equipment. On March 9, the Progress was used to boost the station's orbit to 355 × 386 km. On March 13, water was transferred to Mir. By March 20, the crew completed unloading the Progress and began filling the transport with trash. By March 24, the station had been refueled. After boosting the station's orbit twice, the Progress undocked at 8:07 a.m. March 27 to clear the way for the Kvant module. Retrofire was at 5:59 a.m., March 28, from an orbit of 257 × 339 km.[82–84]

☆ ☆ ☆ ☆ ☆

On March 3, the crew performed more Earth resources work, performed the Kolosok experiment investigating aerosol structure in weightlessness, and performed routine medical examinations using the Gamma-1 unit. The crew also prepared semi-conductor production facilities, and by March 13 had installed the Korund-1M furnace, KATE-140, and automation units for the station's electrical system. The crew also performed another medical check using the Gamma-1 and an ultrasonic cardiograph while pedaling the ergometer. The space station's orbit was about 355 × 386 km.[85] By March 13, the crew finished installing the Korund. The Korund furnace was a modified version of the Kristall and Splav flown on earlier stations, and it was described as a pilot scale unit weighing 136 kg. It could process 1 kg of samples placed in a turntable that moved the different samples into the furnace at programmed intervals. The Korund could operate 6–150 hours maintaining temperatures from 20–1,270°C with .5°C accuracy at a time. The furnace's electric heating elements consumed 1,000 watts power. The crystal ampules were 25 mm in diameter. The processed materials made in the Korund would be used in Soviet industry for infrared and laser equipment. The Soviets claimed that the yield of usable crystals grown in space was ten times that of crystals grown on Earth. By the 1990s the Soviets expect to be able to process individual 35-kg samples for super large integrated circuits, solar panels, lasers and infrared imaging devices. The cosmonauts were scheduled to perform 48 experiments using the Korund to process ten different materials, including producing cadmium selenium and indium antimony semiconductors for use in the Soviet electronics industry.[86–89]

After installing the Korund, the cosmonauts found it did not operate correctly. They soon realized that the furnace was producing a lot of hot air, which in weightlessness remained around the furnace and its adjacent computer control box. It apparently got too hot for the electronic components in the control box to function properly. To fix the problem, they installed a heat reflecting material between the furnace and its control panel and installed a cooling duct and a fan to blow air onto the control box. The unit then worked properly. By March 17, the crew had finished the first round of experiments with the Korund and began another week of Earth resources work and Resonance experiments. About March 23, the crew photographed the Caucasus region for spring water run-off calculations. On March 27, the cosmonauts performed experiments using the Pion-M. The next day's work continued with the Korund, and the cosmonauts damp-mopped the station's walls.[90]

Kvant

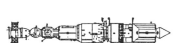

Aft Module

Launched: March 31, 1987, 4:06 a.m.
Propulsion Module Reentry: August 25, 1988
Altitude: 177×320 km @ $51.6°$

This was the first flight of the new short space station module. Dzhanibekov said in an interview, in 1986, that the module delivery had slipped behind schedule and was initially planned for launch in the middle of 1986. Kvant was designed to be used with Mir mainly to provide experiment space and gyroscopic control of the station's orientation (see Figure 6-10). By using electricity provided by solar arrays, the gyros can substitute for using attitude control thrusters and save what would be a large amount of propellant that would have to be delivered by many Progress flights.

The module was launched 1° out of plane with Mir, in an approach that was very much like that made by Star modules. By April 2, the propulsion section had boosted the module to a 297×364-km orbit. On April 5, the spacecraft approached Mir. The cosmonauts were suited up, and in the Soyuz TM in case of a collision.[91] The spacecraft started its approach at 17 km distance using the old Igla docking system. At 500 meters distance, the Kvant's forward docking camera was activated and the docking probe extended. When Kvant was only 200 meters from the station and preparing for final docking maneuvers,

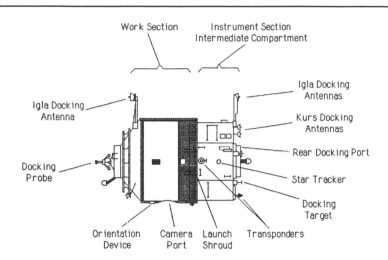

Work Section   Instrument Section
Intermediate Compartment

Igla Docking
Antenna

Igla Docking
Antennas

Kurs Docking
Antennas

Docking
Probe

Rear Docking Port

Star Tracker

Docking
Target

Orientation   Camera   Launch   Transponders
Device   Port   Shroud

Figure 6-10. The Kvant module was the first module launched to Mir. It was delivered by a propulsion module that was discarded after docking. Kvant carried telescopes and astrophysical experiments clustered around its rear docking port. Refueling connections were carried around Kvant to enable refueling of Mir Progress transports.

Kvant Module

Kvant was the first Mir expansion module. The module was docked at launch with a maneuvering module, called the Functional Auxiliary Block, which would deliver the module to the station. The "Block" weighed 9,600 kg, was 8.7 meters long, and would act as a space tug to deliver the module to the space station. The Kvant module was 4.15 meters in diameter, 5.8 meters long, weighed 11,000 kg and consisted of a large diameter work section and an intermediate compartment. The combined spacecraft weighed 20,600 kg at launch. Kvant carried 1,500 kg of internal systems and instruments and 2,500 kg of cargo. The expected lifetime of the module was 5 years.

The module had 40 cubic meters habitable volume, and with the interior was divided into an instrument and living section by panels. The module contained the electron water electrolysis device used for converting humidity into oxygen for addition to the station's atmosphere. The module was connected to Mir's air ventilation system through flexible pipes laid through the hatches. Mir was configured to allow these pipes to be run to all modules that would be added later, in addition to the Soyuz transports docked to the forward and aft ports.

(continued on next page)

(continued from page 321)

The module's equipment also included six magnetically suspended gyroscopes called Gyrodins, each weighing 165 kg. The gyroscopes were used to control the station's attitude by converting electrical energy from the space station's solar array into torques by spinning the gyroscopes. The Kvant module also carried the Roentgen experiment package around its intermediate section. Instruments carried there included the Soviet Pulsar-1 X-ray telescope, the British TTM X-ray wide-angle telescope, the ESA Sirene 2 spectrometer, and the West German Phoswich X-ray telescope. Also in the experiment cluster was the 800-kg Svetlana experimental electrophoresis production plant.

Flight Director Ryumin radioed to the cosmonauts that Kvant had lost its lock-on to Mir's docking transponders. He requested that they take a look at the situation. Romanenko reported that he saw Kvant drifting by the station. The module was rotating slightly as it passed within 10 meters of Mir.[92,93]

Over the next few days, mission controllers investigated the problem and the crew returned to their regular work schedule. The next day Ryumin said that the flight controllers may have been over cautious about the docking and that the problem was understood. During the analysis, the module had drifted 400 km away from the station. At 12:00 a.m., April 9, the cosmonauts donned their pressure suits again and closed themselves off in the Soyuz for another docking attempt. Kvant had been brought back near Mir by ground commanded maneuvers and at 22 km the Igla automatic docking system was activated. At 4:15 a.m., the Igla achieved lock-on to Mir's docking transponder signal. At a distance of 1 km, relative velocity between Mir and Kvant was 2.5 meters per second. At 26 meters, relative velocity was .32 meters per second, and Kvant soft docked to Mir's aft port at 4:36 a.m., in a 344 × 363-km orbit. Ryumin commented that the propulsion module had enough propellant for several docking attempts.[94,95]

As normal, the docking probe of the module was reeled in to pull together the module's docking collar and the station's docking collar. Just as Mir passed out of radio communication with the ground, the probe stopped centimeters short of full retraction. This meant there was still no air-tight seal to the station or electrical connections. When Mir entered the next communications window, Flight Director Ryumin asked Romanenko to examine the aft docking port through the portholes in Mir's intermediate compartment. The cosmonauts also checked using Mir's external television camera, but saw nothing unusual. The Soviets reported that Kvant and Mir were still separated by centimeters for unknown reasons. Since the module was relatively stable and not wobbling on the docking probe, the Soviets chose to leave the module soft docked and perform an emergency EVA to inspect the docking port. The cosmonauts were scheduled to make an EVA in early May to install an add-on solar array to Mir, which was delivered on Kvant, and had been well trained in EVA procedures. Along with the additional 340-kg solar array, Kvant carried 2,500 kg of supplies to Mir.[96,97]

On April 11, the cosmonauts prepared all day for the emergency EVA to determine the cause of the incomplete module docking. At 11:41 p.m., the cosmonauts opened one of Mir's lateral docking ports and began moving to the aft end of the station to examine the docking collar. Cosmonauts in Star City had already worked out any procedures needed for the inspection in the hydrobasin water tank simulator. The hydrobasin contained a mock-up of Salyut 6-7 type station that had been modified to look like Mir by adding a multiple docking adapter and other minor parts. The mock-up could be rotated 360° to allow easy access to any portion of the station.[98]

Part of the 3-hour 40-minute EVA was televised live in the U.S.S.R. (until a Phil Donahue special preempted it). The crew was to use the portable television camera to show mission control the docking unit but it failed and mission control had to rely on the cosmonauts' description of the situation. Shortly into the EVA, Laviekin thought his suit pressure was falling, but Romanenko reminded him it was normal procedure. The suit's pressure was lowered during work to enable easier movement. When they reached the rear docking port, Laviekin found that a white bag had been caught in between the Kvant probe and the Mir's drogue, blocking the docking collars of the two spacecraft from being drawn together. On command of mission control, Kvant unreeled its docking probe, which pushed the module away from Mir several centimeters. The cosmonauts took turns trying to pull the plastic out from between the drogue and the probe where the plastic was tightly compressed. They finally pulled the bag out and the docking probe was reeled in to hard dock the module and make an airtight seal, electrical, and refueling connections. The bag had been inadvertently caught between the Mir drogue and the probe of Progress 28 when the cosmonauts loaded trash into the Progress before it undocked.[99] One of the first things the crew did after the EVA was to replace Mir's control computer's memory with information for control of the expanded space station. The next day they opened the hatch to the module and began unloading equipment. Kvant carried 1,500 kg of internal systems and instruments, and 2,500 kg of cargo for the space station. The expected lifetime of the module was 5 years, meaning that Mir itself was intended for operations until at least 1992. Kvant's propulsion module was undocked from the rear of the module at 12:18 a.m., April 13, and it was maneuvered to a higher storage orbit. On April 17, the crew underwent another routine medical check, but most of their time for the next week was devoted to testing control of the expanded station complex. They also installed a new component for Mir's computers that was delivered by Kvant.[100,101]

Progress 29

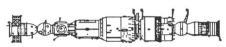

First Quadruple Docking

Launched: April 21, 1987, 7:14 p.m.
Reentry: May 12, 1987
Altitude: 194×218 km @ $51.6°$

After separating from the A-2 booster's upper stage, the Progress boosted its orbit to 194×257 km. Progress 29 docked at 9:05 p.m., April 23, to Kvant's rear port, at 343×363 km. This was the first time that four separately launched objects had been docked together. Progress 29 carried propellant, 250 kg of food, 140 kg of film, medical equipment, mail, newspapers, 170 kg of water, 138 kg of scientific equipment, 275 kg of replacement parts for Mir, personal hygiene supplies, and some apparatuses for use in the upcoming Syrian international mission. Total propellant carried to the station was 750 kg and the total boxed cargo amounted to 1,200 kg. The Progress refueled Mir by passing the propellant through pipes that went around the Kvant module and into its docking collar. Refueling was completed by May 1, and unloading of supplies was completed by May 6. Progress 29 undocked on May 11 at 7:10 a.m. and performed retrofire at 6:59 a.m., May 12.[102,103]

☆ ☆ ☆ ☆ ☆

On April 27, unloading of the Progress was completed, and the crew started preparations for an EVA to install the new solar array on Mir, which was delivered in Kvant. They inspected equipment and reviewed procedures to be used. At this time, the EVA was scheduled for the middle of May. On May 1, the crew was given the day off and watched televised coverage of May Day ceremonies in Red Square, and participated themselves in the ceremonies as usual by reading prepared statements.[104] By May 6, the crew was still activating the Kvant's systems and preparing the Roentgen experiments for future use. At this time, the cosmonauts' work schedule for the next months was changed because they were falling behind the original schedule. The planned EVA to install the new solar array had to be postponed.[105] On May 8, the crew activated the Kvant's electron water electrolysis system. This system provided oxygen for the space station and reduced the amount of air that would need to be delivered on Progress flights. Kvant also carried a new system called Vozdukh, which filtered carbon dioxide from the air and dumped it overboard, reducing the need for air-regeneration canisters on Mir.

Since the docking of the Kvant module, mission controllers experimented with controlling the space station's attitude with the module's gyroscopes. These initial experiments were scheduled to end about May 12. The cosmonauts were not scheduled to begin using the Kvant's Roentgen experiment cluster until June.[106]

Progress 30

Launched: May 19, 1987, 8:02 a.m.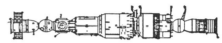
Reentry: July 19, 1987
Altitude: 178 × 218 km @ 51.6°

Soon after orbital insertion and booster separation, the Progress boosted its orbit to 192 × 265 km. The resupply spacecraft docked at 9:53 a.m., May 21, at 343 × 366 km. The Progress carried food, water, equipment, instruments, and mail to the space station. The station was refilled with fuel by July 3, and with oxidizer by July 14. On July 8, and again on July 10, the Progress boosted the station's orbit. Progress 30 undocked July 19, at 4:20 a.m. to uncover the aft docking port hatch enabling Glazar telescope observations, and ultimately making way for the Soyuz TM-3.[107]

☆ ☆ ☆ ☆ ☆

On May 22, the cosmonauts used the Yantar apparatus to vaporize and deposit copper and copper-silver alloy coatings onto polymer film. They continued unloading the Progress until May 26. They also continued experimenting with the Kvant's gyroscopes to control the station's attitude, and tested a new high-precision star tracker that would be used to update the gyroscopes' positions.[108] On May 28, the station was suffering a severe shortage of power due to the demand of various experiments. The addition of the Kvant module, and the delay in the scheduled EVA to erect a new solar panel to help power the Kvant meant that the station was limited in what experiments could be operated. A semiconductor processing experiment caused the problem to surface because the electric furnace required too much power from the electrical system. The station was also in a period of reduced sunlight due to the season and orbit, which also reduced the power normally available.

On June 2, the crew set up the Rost and Fiton plant growth units to continue experiments on growing higher order plants in space. On June 5, the cosmonauts installed an electronic photometer instrument and a multi-zonal spectrometer for Earth observation. The Mir space station was in a 345 × 383-km orbit at the time. On June 9, the crew made the first observations with the Roentgen experiments of the SN1987a supernova in the large Magellanic cloud, performed routine medical checks, and started preparing equipment for and checking the station's spacesuits for an EVA.[109]

During the first observations of the supernova, an X-ray source interfered with the normal operation of the X-ray detectors. The crew had to manually point the telescopes off center from the supernova to avoid the interference.

On June 13 at 8:55 p.m., the cosmonauts started the first EVA to install the new solar array delivered by Kvant to the top of Mir. The EVA was delayed from early May due to the problems encountered docking the Kvant module and the cosmonauts' falling behind in other work. The solar array components were too large to fit into Mir's multiple docking adapter airlock, along with the two cosmonauts. For more room, the cosmonauts used the Soyuz orbital module, which was docked to the forward port, as an extension of the airlock. They closed the hatch to the Soyuz reentry module and depressurized the airlock and orbital module at the same time. The Soyuz orbital module had not been used as an airlock since the Soyuz 6 flight in 1969. The Soviets stated at this time that a special spacious airlock with tools and equipment would be attached to Mir in the future.[110,111]

During the 1-hour 54-minute EVA, they installed the first section of beam and solar arrays to each side of the unextended beam. The beam portion of the solar array was said to be similar to the beam tested by the Soyuz T-15 crew on Salyut 7 in 1986. The cosmonauts chose to perform the EVA without the use of tethers. This was probably the first EVA performed by anyone, excluding U.S. manned maneuvering unit flights and moon walks, while not attached to the spacecraft by tethers.[112,113] On June 16, 7:30 p.m., they completed the work during a 3-hour 15-minute EVA by adding another beam and another pair of solar panels on top of the assembly installed 3 days before. During the second EVA, the cosmonauts used tethers, tying them to the station as normal. They then connected the solar arrays to Mir's electrical system. The combined assembly was then deployed, the upper beam with its arrays, and then the lower beam and its arrays, to a length of 10.6 meters. The 2 similar sections of beams and 4 solar array packages then made up the full length of the solar panel. Before returning to the station's interior, the cosmonauts placed cassettes of materials samples onto Mir's exterior for space exposure tests, as had been done on previous Salyut missions.[114]

After the EVA, the cosmonauts connected the new solar array to Mir's power system inside the station. Total area of the array was 22 square meters, which provided 2.4 kW power, bringing Mir's total power generating capability to 11.4 kW. The Soviets commented that as the power needs of the station grew in the future, they would add more solar arrays on other modules that would be docked to Mir. The solar array could also be retracted to replace failed components.

On June 19, the crew underwent medical examinations using the Gamma-1 instrument and they continued work using the Roentgen experiments. On June 23, the new solar array was switched into Mir's power system. On June 26, they replaced units of the Strela information display system expanding its capabilities, continued experiments with the Yantar, Kround, Biostoykost, Roentgen, and underwent medical checks. By June 30, the cosmonauts had activated the Glazar telescope in the Roentgen package, and the orbit of the Mir space station was 341 × 365 km.[115] On July 3, the crew installed a new star tracker and continued Earth resources work. Testing the star tracker system, the station was pointed with an accuracy of one arc minute, which was 10 times better than expected. On July 7, the crew was continuing photography sessions over the Urals and Far East and refueled Mir's propellant tanks. The Progress was used the next day to boost the station's orbit. On July 10, the Svetoblok electrophoresis unit and the Yantar electron beam were used, the Kristallizator, a Czechoslovakian crystallization experiment was installed, routine medical checks were made, and the orbit of the station was boosted using the Progress.[116,117]

Kristallizator was a microprocessor controlled semiconductor furnace capable of maintaining constant heating and pressures on the crystallization front. Temperatures produced by the units' 5 heating elements could reach 1,000°C and experiments could last from several hours to several days. Biological instruments Rost, Fiton, and Svetoblok were operated in conjunction with plant growth experiments, and Earth resources sessions were held on July 21 as the crew awaited the Soyuz TM-3 launch. [118,119]

Soyuz TM-3

Twelfth International
Crew—Syria

Launched: July 22, 1987, 5:59 a.m.
Landed: December 29, 1987,
        12:16 p.m.
Altitude: 190 × 193 km @ 51.6°
Crew: Aleksandr Victorenko, Aleksandr Aleksandrov, and Mohammed Faris
Backup: Anatoliy Solovyev, Viktor Savinykh, and Munir Habib
Call Sign: Knights

The Syrian international mission was first planned in 1985, and the crew trained for the mission for about 18 months. This was much longer than normal because the Syrian crew was to have visited the Salyut 7 space station during the flight of the Soyuz T-14 crew. After the Soyuz T-14 mission was unexpectedly aborted, the international mission was put off until the next convenient long-duration mission. The crew and backup crew were announced on December 17, 1986. The international mission was the latest in a series of international missions begun in 1978. Faris became the second Arab to fly in space (Saudi Arabian Prince Sultan Al-Saud flew on U.S. STS-51G in June 1985).[120] On July 19, the crew assembled in the Cosmonaut Hotel at Baykonur for a formal meeting approving the launch. A press conference was then held and the flight was officially announced. The assembled Soyuz TM-3 and booster were rolled out of the assembly building on July 20 for the short trip to the launch pad.

The launch was observed by a delegation from Syria and General Gherman Titov, the pilot of Vostok 2. The Soyuz was put into transfer orbits of 249 × 304 km, and then 236 × 365 km, with maneuvers on the fourth and fifth orbits to rendezvous with Mir two days later. The orbit was raised to 304 × 365 km the next day in preparation for docking.[121]

Docking to the aft port of Kvant module occurred at 7:31 a.m., July 24, and was shown live on Soviet television. The crew had to use a crow bar to open the hatch because of slightly lower air pressure in the station. They finally entered the station at 9:00 a.m.[122]

During the flight, the crew performed some Syrian planned experiments including atmospheric research, cardiovascular monitoring, materials processing, and photography of Syria using the station's MKF-6M camera during the station's three passes over that country. Most experiments were standard types, and were like those done on previous international missions. Biological experiments using the Svetlana electrophoresis unit were performed with interferon, anti-influenza, and agricultural antibiotic substances. The Soyuz TM-3 also carried the Ruchey electrophoresis unit for preparing antibiotics, which was installed in the Kvant module with the Svetlana. Atmospheric measurements were taken using the Bosra device, jointly developed by the Syrians and Soviets. The Kristallizator was also used in two joint Syrian-Soviet experiments producing an aluminum/nickel alloy and gallium/antimonide. In the Palmyra experiment, a material similar to bone tissue was produced.[123,124]

The crews reportedly worked ahead of schedule, but they also wished for more time. Faris was said to have adjusted to weightlessness by the second day of the mission. Victorenko also said his 10 years of training helped him to adjust quickly to weightlessness. Romanenko commented that there was sometimes confusion with 3 people on board with Aleksandr as a first name, referring to Aleksandrov, Laviekin, and Victorenko. On July 25, a press conference was held between the crew on Mir, and President Assad in Syria. On July 28, the crew gave a press conference for journalists covering the flight.[125,126]

During the flight, Aleksandrov replaced Laviekin as the flight engineer of the long-duration crew. This had probably been planned when the Soyuz TM-3 crew was first announced in the spring of 1987. It was unusual that an experienced long-duration mission cosmonaut would be assigned as flight engineer to an unimportant visiting flight, instead of another long-duration mission. It was also suggested that Aleksandrov was an expert on operations of the Kvant module, and was to remain on the station replacing Laviekin, who was an expert at Mir's systems. The Soviets' explanation of Laviekin's replacement was that he had developed a heart irregularity indicated by electrocardiogram changes. Although they said that the replacement was a last-minute development, they also said that Aleksandrov had been trained in the Mir simulator for a month in anticipation of a long mission.

Deputy Flight Director Blagov said Laviekin's heart problem could have been serious, but Shatalov said that there were no conclusive evidence about Laviekin's health. His return to Earth was for more detailed medical checks only. At the same time, Blagov said that it was now planned to rotate crews on a continuous basis if necessary, to permanently man the space station.

Victorenko, Faris, and Laviekin left the station in the Soyuz TM-2 spacecraft on July 29, leaving the new Soyuz TM-3 for the long-duration crew to use in December. The Soyuz must be exchanged every six months to ensure spacecraft reliability for return of cosmonauts to Earth. The Soyuz was loaded with exposed film, biological specimens, processed material samples, and other experiment results.[127]

They closed hatches between Mir and Soyuz TM-2 at 9:08 p.m. and the spacecraft undocked at 12:34 a.m., July 30. After 2 more orbits, the orbital module was jettisoned before retrofire. The Soyuz TM-2 capsule landed 140 km northeast of Arkalyk at 4:04 a.m. July 30. The capsule landed 2 km from a small village in high winds, shortly after local dawn. Two hours later the cosmonauts arrived back at Baykonur. The landing was originally scheduled for 3:30 a.m., but had been postponed because of flooding in the primary landing site.[128,129]

After the flight, Laviekin said that he was disappointed at having to return to Earth, but that he couldn't argue with the doctors. He was not in the best of condition after the long flight, and had to be helped off the airplane after returning to Baykonur. He used gravity-suit leggings, as many cosmonauts previously had, to counter the unaccustomed blood pressure in the legs. Laviekin had also commented that the designers of the equipment used on Mir should pay more attention to reliability, and that minor equipment malfunctions due to defects cost cosmonauts valuable time to fix.[130,131] The Soviets also made comments that the Kiev type cameras often jammed in weightlessness, and the crystals of wrist watches built by Moscow Watch Plant 1 often came out (probably due to pressure differences), even though these products are considered to be of the highest Soviet quality.[132] The Soviets said that Laviekin's condition did not differ from previous returning cosmonauts, and that the irregularity only occurred during physically stressful activities in orbit. He was considered to be in very good condition after more than 5 months in orbit, and post-landing examinations could not detect any heart irregularities. The Soviet doctors commented that if it hadn't been for the measurements from orbit, there would be no evidence of any physical problems. Laviekin was subsequently cleared to make future spaceflights.[133] The international cosmonaut, Faris, commented that he was tired after the flight, because he didn't sleep much during the mission, trying to experience as much as he could during the short time available.

The Soviets announced in August that an Afghan cosmonaut would fly to Mir soon. Other international missions planned at the time were a Bulgarian mission set for June 22, 1988, to make up for the Soyuz 33 failure in 1979, and a month-long French flight in November 1988.

After the departure of the Soyuz TM-2, the crew of Romanenko and Aleksandrov resumed regular operations of the space station. The crew's call sign, Tamyr, usually chosen by the mission commander, remained the same as before the crew exchange.

The Soyuz TM-3 needed to be moved around to the forward docking port to allow the next Progress flight to be received at the aft refueling port on Kvant. At 3:28 a.m., July 31, Romanenko and Aleksandrov undocked the Soyuz TM-3 from the Kvant module's aft port. The space station complex was then commanded to activate its forward docking system, causing the station to turn to face the Soyuz. Romanenko then redocked the spacecraft to the forward port at 3:48 a.m.[134,135]

The crew then returned to normal activities, including supernova observations with Roentgen instruments and tending to 3-month-old cedar trees that were delivered to Mir by the Soyuz TM-3 crew.[136]

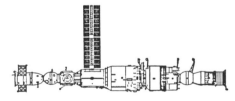

Progress 31

Launched: August 4, 1987,
12:44 a.m.
Reentry: September 23, 1987
Altitude: 190 × 239 km @ 51.6°

After separating from its upper stage, the Progress maneuvered to 193 × 269 km before maneuvering into a transfer orbit to Mir. The supply ship docked August 6 at 2:28 a.m. Unloading of the food, equipment, mail, and water supplies began the next day. On August 11, water was transferred. Refueling of Mir with propellant began on September 15 and the cosmonauts began refilling the Progress with used equipment and other garbage. The Progress boosted the station's orbit before undocking September 22 at 3:58 a.m. Retrofire was at 4:22 a.m. the next day, and the supply ship was destroyed during reentry as normal less than an hour after its successor was launched.[137]

☆ ☆ ☆ ☆ ☆

After the arrival of the Progress, the cosmonauts returned to the routine work of operating the various experiments and carried out routine maintenance work while unloading the Progress. On August 12, the Korund was operated. On

August 17, they began another series of Earth resources sessions, photographing land masses and observing atmospheric pollution.[138]

On August 30, the cosmonauts performed an emergency evacuation drill. This included donning pressure suits and closing the hatches to the Soyuz. On September 2 the crew completed installation of a hydraulic pump for the temperature control system.[139]

During early September, the crew also tested the station's solar array's generating capacity and checked radiation exposure of the space station. On September 4, the cosmonauts made routine medical checks. The cosmonauts' main duties consisted of operating the telescopes in the Kvant's Roentgen experiment module. By late September, the telescopes of the Roentgen complex had been operated 300 times, of those, 115 were of the supernova. A program of Earth resources was also carried out in cooperation with Soviet bloc countries studying pollution in the air, water, and soil using the familiar methods of simultaneous observation from the space station, Earth resource satellites, aircraft, and teams of investigators on the ground.[140]

September 12 was a day of rest for the crew and included special video transmissions from mission control of requested programs and the usual family communications. The crew also ate 3 onions harvested from their plant growth experiments. Other plants grown by the crew included wheat and radish.[141]

Refueling of Mir by Progress 31 began on September 15, and the cosmonauts began refilling the Progress with used equipment and other garbage. The Progress was used to boost the station's orbit before undocking September 22. Also during September, the tracking ship *Korolev* took up a routine position in the North Atlantic, off Nova Scotia, and remained on station for the next 5 months relaying communications between Mir and Kaliningrad mission control.[142]

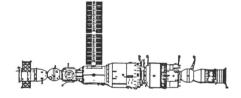

Progress 32

Launched: September 24, 1987,
          3:44 a.m.
Reentry: November 17, 1987
Altitude: 186 × 250 km @ 51.6°

After separating from the booster's upper stage, the Progress soon boosted its orbit to 193 × 267 km, and then again to a transfer orbit to Mir at 212 × 277 km. The spacecraft docked at 10:08 p.m., September 25 in an orbit of 297 × 355 km. The mission delivered 315 kg of food supplies including onions, lemons, and garlic, and 850 kg of propellant and other supplies. Propellant transfer was accomplished in the first week of November.[143]

The Progress was undocked November 10, at 6:09 a.m., and moved about 2.5 km away from the station. After an orbit, the Progress redocked at 8:47 a.m. to check on the stability of Mir's new solar panel that was erected in June. The procedure also tested new docking procedures that would save propellant on future Progress flights. The Progress undocked for the final time on November 17, at 10:25 p.m. and performed retrofire at 3:10 p.m. the same day.[144]

☆ ☆ ☆ ☆ ☆

The orbit of the Mir space station on September 30 was 295 × 355 km. It was reported that Romanenko's work schedule was reduced to 5.5 hours of work a day due to his tiring. Before the end of the mission, the crew's work schedule was reduced to only 4.5 hours of work a day, and up to 9 hours of sleep. The Soviets reported that even 2 days off from scheduled work could not relieve Romanenko's feeling of monotony. Romanenko reportedly became very depressed and argumentative toward the end of the mission, however he acted similarly during the final days of the Soyuz 26 mission.[145,146]

In early October, the crew used the Svetbloc-T to make polyacrylamide gel for use on Earth, and they also made observations of the upper atmosphere with the EFO-1 photometer. At the same time, the crew slowly unloaded Progress 32 of supplies. On October 4, the crew participated in a television conference commemorating the anniversary of the Sputnik 1 launch. Progress 32 had also delivered 1,000 envelopes to be postmarked on the anniversary, and sold after the crew's return to Earth. By October 20, the crew began an experiment program using the Glazar telescope. The crew also carried out requested observations to aid in weather forecasts. Work also continued on tending plant growth experiments and continuing Earth resources photography.[147]

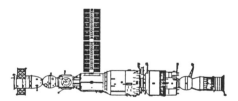

Progress 33

Launched: November 21, 1987, 2:47 a.m.

Reentry: December 19, 1987

Altitude: 187 × 249 km @ 51.6°

Progress 33 used a transfer orbit of 326 × 343 km to reach Mir's orbit and get in position for docking. During the docking, the Soviets experimented with using new procedures for orienting the station with the approaching Progress, which still used the Igla docking system that required active maneuvering by the space station. Propellant expenditures by Mir for docking a Progress were around 192 kg using the old Igla system. The new Igla procedure reduced this amount to about 82 kg.[148] The Progress docked to the Kvant port on November 23, at 4:39 a.m. Unloading of the cargo was started the same morning and was completed by December 15. Cargo included the Mariya high-energy particle detector and an experimental materials processing furnace. The Progress had been filled with trash and undocked from the station at 11:16 a.m., December 19, and made a normal destructive reentry.[149]

By November 26 the crew had started operations of the materials processing furnace delivered by the last Progress. The unit featured a heat producing lamp that was focused on the material sample by mirrors to intensify the heat. On December 3, the crew increased exercise periods to total 2.5 hours a day in preparation for returning to Earth.

Soyuz TM-4

Long-Duration Relief
Crew—12 Months
Launched: December 21, 1987,
        2:18 p.m.
Landed: June 17, 1988, 2:13 p.m.
Altitude: 168 × 243 km @ 51.6°
Crew: Vladimir Titov, Musa Manarov, and Anatoliy Levchenko
Backup: Aleksandr Volkov, Aleksandr Kaleri, and Aleksandr Shchukin
Call Sign: Ocean

The new long-duration crew's commander was announced in September 1987. The crew, and their backup crew, left Star City on December 10 to fly to Baykonur. The primary crew members were officially announced shortly before the launch. A doctor, Valeri Polyakov, was originally training with Serebrov to examine Romanenko before landing.[150] They were replaced by Manarov and Levchenko late in the training for the flight. Levchenko was such a late addition to the crew that Romanenko and Aleksandrov had never met him; however, he had been training for spaceflight since 1978. This illustrates the great separation between the civilian and military space programs in the U.S.S.R. Levchenko was a cosmonaut from the space shuttle program, which was designed by a military design organization. His assignment to the flight was to familiarize himself with spaceflight and to test his flying abilities immediately after landing.[151] In both the U.S. and U.S.S.R., there was a little concern that a shuttle pilot's physical state might degrade over a week or longer of flight, and that they might loose some flying proficiency. In preparation for the first Soviet shuttle flight, Lyakhov would test methods to avoid the problem.

The Soyuz used a transfer orbit of 255 × 296 km to reach the vicinity of the space station. At 3:11 p.m., Titov sighted Mir as the Soyuz was 35 km from the station. The Soyuz approached Mir's forward docking port and then circled around the station to dock at the aft Kvant port at 3:51 p.m., December 23, in orbit at 333 × 359 km. Two and a half hours later, after checking the docking system and shutting down the Soyuz, the crew entered the space station.[152,153]

The Soviets estimated that the in-flight exchange of crews saved a week's time that the new crew would have used reactivating a mothballed station. Although the crews were trained in the space station simulators on Earth, weeks were saved when Romanenko and Aleksandrov showed Titov and Manarov how the station's equipment was operated in real flight conditions. Among other things, the old crew demonstrated EVA equipment and practiced procedures for EVA. The new crew was scheduled to make an EVA to replace a section of the solar panels that Romanenko and Laviekin had installed.

The cosmonauts also installed the Aynur biological crystal growth unit in the Kvant module, which was delivered by the Soyuz along with fish specimens, tissue cultures, and decorative plants. The Ruchey unit was used to produce small samples of interferon. Romanenko and Aleksandrov also spent time using the Chibis suit to condition themselves for return to Earth. The cosmonauts transferred a seat from the Soyuz TM-4 to the TM-3 in preparation for landing on December 24. They also transferred biological samples, flight logs, recorded tapes, and film to the reentry module while they filled the orbital module of the Soyuz TM-3 with trash.[154]

The crews also performed an evacuation drill of the station. The simulated emergency was orchestrated on Mir's computer to give the crew realistic warnings and prompts. Mir's computer was also used for keeping a complete inventory of the items on the station. The Soviets announced that Mir crew members would soon be issued their own personal computers for use during the flight. One of the more unique pieces of equipment on Mir was a guitar, which the cosmonauts enjoyed playing often. On December 27, the crews began packing experiment results in the Soyuz TM-3 and preparing for the old crew's departure. The work became so hurried that the crew told mission control the next day to quit bothering them with questions about experiments and experiences during the mission. That same day, recovery forces 60 km northeast of Arkalyk had assembled to practice locating a landing capsule. A parachutist carrying a capsule beacon radio transmitter was used to simulate the landing capsule in flight.[155] On December 29, the TM-3 delivered to Mir by the international crew in July undocked at 8:55 p.m. carrying Romanenko, Aleksandrov, and Levchenko. Romanenko commented that none of the crew had trained together on Earth before the mission. Aleksandrov was substituted for Romanenko's original crewmate earlier in the year, and Levchenko was a shuttle cosmonaut making a training flight during the relief mission. The simplicity and standard procedures of the Soyuz transport made this kind of mission flexibility possible.

Retrofire lowered the orbit of the capsule into the atmosphere at about 11:23 p.m., and entry into the atmosphere began about 27 minutes later. The capsule was sighted shortly thereafter by the recovery helicopters and tracked until just before landing, when it was obscured by fog or blowing snow. The capsule landed at 12:16 p.m., in near blizzard conditions and 100 km/hr winds, rolling onto its side, 80 km from Arkalyk. Romanenko had spent 326 days in space during the mission. He was carried out of the capsule and placed into one of the usual divan chairs. He looked healthy and talked with the waiting reporters. Aleksandrov also was helped to the chairs while Levchenko walked without assistance. After a short time, Romanenko was put on a stretcher and taken to a nearby medical helicopter for immediate examinations. Within 30 minutes of landing, Levchenko was piloting a TU-154 to Moscow to test his flying abilities. From there, he returned to Baykonur to rejoin the crew for a press conference the next day.

The capsule returned experiment results from the last 5 months of work on the station, including 270 pictures taken in 500 periods of operation of the Roentgen telescopes and the Glazar telescope. Materials samples processed in the Kround, Earth resources photographs, and biological experiment results were also returned.[156]

During the flight of Romanenko, there were more than 1,000 scheduled work sessions in more than 70 different experiment programs. At least 700 observations using the Kvant modules telescopes were made and 95 materials processing experiments performed. The night after landing, Romanenko and Aleksandrov were flown back to Baykonur for more extensive tests. Romanenko walked off the plane on his own, with people steadying him along the way, to greet his wife. He then started more thorough medical examinations and 5 days after landing, he began to write his mission report. Officials planned to observe his recovery for at least the next year before allowing him to become eligible for flight again. After the first few weeks of examinations, Radio Moscow announced that there were no unexpected medical effects from the long mission in weightlessness.

Romanenko had lost no weight during the flight, but he did loose about 1.6 kg or 15% of his muscle volume which would be slowly regained after returning. Aleksandrov gained 2.3 kg during the mission. While Romanenko and Aleksandrov were on Mir, 34 different medical tests were performed a total of 170 times, and Romanenko ran a total of almost 1,000 km on the treadmill.[157,158]

On January 12, Romanenko returned to Star City and on the 18th, he and cosmonauts Laviekin, Aleksandrov, and Levchenko received the standard awards for their flights from President Gromeko at a presentation in Moscow. By the end of January, the Soviets said that Romanenko was completely recovered and showed no signs of his 11-month flight. At the time, Romanenko and Aleksandrov were vacationing with their families at a resort in the north Caucasus.

In August 1988, doctors discovered that Levchenko had a brain tumor that required emergency surgery. Levchenko subsequently died, and 12 days later, on August 18, another shuttle pilot, Aleksandr Shchukin, died while flying a one-seat Su-26M aerobatic stunt plane. The losses of two experienced shuttle pilots represented a serious blow to the program since the low initial flight rate planned for the shuttle required only a small number of cosmonauts to begin with.

On December 30, the day after the landing of the TM-3, Titov and Manarov undocked the Soyuz TM-4 from the rear of the station at 12:10 p.m. and waited as the station was turned, allowing Titov to dock to Mir's forward port 19 minutes later, clearing the aft port for the next Progress mission. The cosmonauts were allowed a day off from the work of preparing the station for their mission to celebrate New Year's Day by talking with their families.[159]

On January 4, the new crew started their research program for the mission. They installed new materials processing equipment and started a new program of materials processing of semiconductors. The crew was said to have completely adapted to weightlessness by January 12.[160]

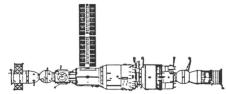

Progress 34

Launched: January 21, 1988,
1:52 a.m.
Reentry: March 4, 1988
Altitude: 185×258 km @ $51.6°$

Progress 34's transfer orbit to Mir was 242×271 km. The spacecraft docked to Mirs' aft port at 3:09 a.m., January 23, at 334×355 km. The flight delivered the usual food, propellant, water, instruments, music tapes, letters, packages, and other cargo. By February 24, refueling was completed. The Progress undocked at 6:40 a.m. on March 2, and made a destructive reentry 2 days later.[161]

☆ ☆ ☆ ☆ ☆

During the next month, the crew performed experiment programs using several different apparatuses. The Kround furnace was used to produce semi-conductor materials and the Pion was used to investigate capillary action. The Glazar ultraviolet telescope was used to study galaxies during exposures up to 8 minutes long, making magnitude 17 stars visible.

Starting on February 12, the crew received periodic briefings about their upcoming EVA and procedures that were to be used. The objective was to replace a section of Mir's additional solar panel that had been installed by the previous long-duration crew with a section containing experimental solar array devices. The replacement section was to be placed in the lower section of the 4-piece array, and would increase power by 20% over the old array section.[162]

On February 17, the crew stopped the normal experiment program to prepare for the EVA. On February 23, the crew unpacked the station's spacesuits and stowed them in the airlock or docking module. Over the next few days, the crew tested the suits' systems and prepared them for the EVA.[163] Normally, the suits' air tanks would be refilled, the carbon dioxide filters checked or replaced, and the batteries recharged. The suits would then be connected to the station's air supply and tested for air tightness and the fit adjusted for the individual cosmonaut.

On February 26, at 12:30 p.m., the cosmonauts exited the station while out of communication range of mission control. Manarov was the first out of the airlock, followed by Titov. They first installed their work equipment on the outside of the station and secured the new solar panel section. The crew had to retract the lower section of the center solar array and disconnect the power connections to the folded solar panel. The solar panel section was then removed and replaced by the new unit. After connecting the power lines to the new section and extending the lower section again, the cosmonauts moved down along the station to Progress 34, which was docked to the Kvant module. They placed several new instruments to the hull of the station, deployed a new antenna, inspected the Progress, and photographed the station's exterior before returning to the airlock. The EVA lasted 4 hours and 25 minutes. The cosmonauts had practiced the procedures for replacing the array in the hydrobasin, and in weightlessness training flights on aircraft at Star City before the flight (see Figure 6-11).[164]

Figure 6-11. Cosmonauts Titov and Manarov trained for EVA at the Hydrolab water tank weightlessness simulator. Here they are wearing spacesuit cooling garments including the temperature sensor worn on the ear. Salyut-type EVA spacesuits are shown nearby. (Source: Sovfoto.)

By the end of February, the crew had performed more than 130 sessions operating Kvant's Roentgen telescope package and 20 sessions with the ultraviolet Glazar telescope. In the beginning of March, the crew continued experiments using the Pion and the new mirrored material processing furnace. The Roentgen telescopes was also operated and Earth resource photography was conducted. The cosmonauts also complained that the air in the station had a bad smell and was dusty. By March 21, Titov had lost 1.5 kg and Manarov had gained 3.5 kg. The crew was given the day off to celebrate Manarov's birthday, and tested a new teletype unit for relaying incoming information.[165,166]

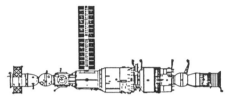

Progress 35
Launched: March 24, 1988,
12:05 a.m.
Reentry: May 5, 1988
Altitude: 184 × 262 km @ 51.6°

After upper stage cut-off, Progress 35 separated from the booster and began a series of maneuvers to raise its orbit. The upper stage remained in a low orbit that would normally decay after about a day, and the next day the upper stage reentered over parts of Texas, Oklahoma, and Arkansas at about 10:50 p.m., on March 24, local time. It was sighted by many people in the area who had no idea what it was. On the second day of flight, Progress 35 docked to the rear of the Kvant module at 1:22 a.m. The flight carried the usual cargoes including 400 kg of food, mail, propellant, and instruments. The Progress was unloaded during the next few days, but propellant was not transferred to Mir until more than a month later. On April 22, the Progress was used to boost the station's orbit to 340 × 366 km. On May 1, preparations for refueling the station were underway. During the next few days, the crew filled the Progress with trash and used equipment to be disposed of when the Progress reentered the atmosphere at the end of its mission. On May 5, at 4:36 a.m., the Progress undocked and made a destructive reentry the same day.

☆ ☆ ☆ ☆ ☆

After completing the unloading of the Progress, the cosmonauts continued their experiments. In early April, they continued using the Roentgen telescopes and making Earth observations. They made atmospheric measurements with the EFO-1 and performed crystallization studies with the Kristallizator unit. By April 7, the crew had installed units of a new, more capable Earth communications system, but they were falling behind the work schedule. By April 12, the cosmonauts had performed 330 experiment sessions, including 276 observation sessions using the station's telescopes. In late April, the crew worked with the Pion-M and continued using the Roentgen telescopes. Earth resources work was also performed, specifically including photographing areas of Cuba.

Progress 36
Launched: May 13, 1988, 3:30 a.m.
Reentry: June 5, 1988
Altitude: 185 × 246 km @ 51.6°

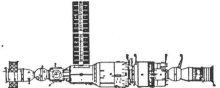

Progress 36 carried nine instruments to be used by the upcoming joint Bulgarian mission. The Progress used a transfer orbit of 223 × 334 km to reach Mir's orbit. The spacecraft approached Mir in orbit at 331 × 357 km and docked to the rear port at 5:13 a.m. May 15. The crew began unloading the Progress soon after docking. The flight delivered equipment to be used by the cosmonauts during their June EVA. On June 3, the Progress boosted the station's orbit. At 2:12 p.m., June 5, the Progress undocked to make way for the next international flight. Progress 36 drifted in orbit for several hours before approaching the deorbit point. Retrofire was performed at 11:28 p.m., sending the spacecraft into a destructive reentry.[167,168]

☆ ☆ ☆ ☆ ☆

On May 19, Radio Moscow reported that the crew started installing equipment delivered on Progress 36 that would be used during the Bulgarian mission in June. On May 23, it was reported that Titov had lost 2 kg of weight and Manarov had gained 3 kg of weight since the beginning of the mission. By May 26, the crew had finished preparing the experiments that would be used during the Bulgarian international mission. On May 30, the crew started refueling operations, operated the Svetoblok-T electrophoresis unit, and tended their plant growth experiments.

Soyuz TM-5
Thirteenth International
Crew—Bulgaria
Launch: June 7, 1988, 6:03 p.m.
Landed: September 7, 1988,
        4:50 a.m.
Altitude: 198 × 216 km @ 51.6°
Crew: Viktor Savinykh, Anatoly Solovyev, and Aleksandr Aleksandrov
Backup: Vladimir Lyakhov, Andrey Zaytsev, and Krasimir Stoyanov
Call Sign: Rodnik

This mission had earlier been scheduled for a launch on June 21. The reason for the advance was the necessity for certain lighting conditions in orbit for operating the Rozhen astronomical experiment. This advance in the schedule did not significantly affect the crew's preparation because they had been in training since January 1988. The cosmonauts boarded the Soyuz on schedule at 3:45 p.m. June 7. Launch was also on time, at 6:03 p.m., placing the Soyuz TM-5 into orbit at 198 × 216 km. A transfer orbit of 282 × 343 km was used to place the Soyuz 40 km away from Mir after 33 orbits. At that point, the Soyuz began its automatic approach to Mir. After approaching the occupied forward port the Soyuz circled around the station to the rear port. The Kurs system malfunctioned during the approach, and mission control was forced to diagnose the malfunction and make corrections before continuing the docking. The problem was solved quickly and the Soyuz docked to Kvant's rear port in orbit at 349 × 355 km, at 7:57 p.m. June 9.

More than 40 experiments were to be performed during the flight including photography of Bulgaria and other normal international mission experiments investigating materials processing, astrophysics, and human adaptation to weightlessness. Many of the experiments used the 3-kg Zora portable microcomputer to gather and store information. The computer was intended to monitor many of the flight's experiments, but some program diskettes were lost and had to be sent on the next Progress flight for use by the next international crew in August.

The crew would use new equipment delivered by previous Progress flights including the Rozhen device, which was a telescope that used digital electronic sensors for processing the images. It also displayed the pictures to the crew and were downlinked to mission control where they were displayed on a personal computer. Atmospheric measurements would be taken using a new Spektr-256 spectrometer that replaced the old Spektr-15 device. The new spectrometer measured 256 spectral wavelengths and was controlled by the Zora microcomputer. Another optical spectrometer, the Terma, was used to study luminescence in the upper atmosphere and ionosphere including aurora. Data from both the spectrometers was stored on floppy diskettes for return to Earth.

The materials processing experiments included Voal, which was an experiment program that used the material processing equipment on Mir to produce a wolfram (a raw form of tungsten) aluminum alloy. Strukura was a material processing program to produce aluminum copper iron alloy, aluminum tungsten alloy, and aluminum copper alloy. The Kilmet-Rubidium experiment developed technology to make light-weight batteries from rubidium silver and iodine. The experiment used the Kristallizator furnace, which could be operated only during the cosmonauts' sleep period when other equipment was turned off because of its large power consumption. The Ruchei experiment was to purify interferon by electrophoresis.

The life sciences experiments included Prognoz, which was a study of the cosmonauts' operational performance with applications to Earth workers; Lyulin was an experiment for testing cosmonaut reaction times; Potential was an experiment investigating the muscular and nervous system, Stratokinetika investigated body movements in weightlessness, and the Son experiment recorded electrical signals from the cosmonauts' bodies during sleep periods. The Doza-B experiment placed radiation sensors and biological samples at various places in the space station to study radiation exposure and effects on the crew. On a lighter note, the Dosug experiment assessed the effect of music, television, and computer games on crew moral.[169]

On June 10, the crew exchanged seat liners and personal items from the new Soyuz TM-5 to the old Soyuz TM-4 capsule. They also prepared experiments and participated in a communications session with Bulgarian leader Todor Zhivko. An Australian experiment investigating vaccines was also started that would continue until September. The experiment grew larger crystals than possible on Earth of a membrane protein of an influenza virus. The crystals atomic structure would be examined on Earth by three-dimensional X-ray analysis. Similar crystal experiments were planned for U.S. shuttle flights the following year.

On June 13, the cosmonauts started plant growth experiments in the Magnitogravistat studying wheat, arabidopsis, and ginseng. They also held a press conference and continued astronomy observations with the Rozhen. Before going to sleep on June 14, a new experiment was started in the Kristallizator furnace investigating aluminum copper iron alloy.[170]

On June 16, the crew began packing the first of 30 kg of experiment results including computer diskettes, film, biological specimens, and documentation into the capsule while the last experiments were being finished. The next day, the international crew boarded the Soyuz TM-4 for return to Earth. They undocked at 10:18 a.m. and drifted for two orbits (see Figure 6-12). Retrofire was at about 1:50 p.m., and 20 minutes later, the capsule entered the atmosphere. They landed 202 km southeast of Dzhezkazgan on a dry lake bed, at 2:13 p.m. Light winds blowing at 21 km/hr disrupted recovery forces shortly after blowing over a tent set up for initial medical examinations of the crew. In addition to the bothering wind, the temperature was a scorching 43°C (120°F). The capsule's hatch stuck during initial attempts to open it. Later, the recovery forces were preparing to lift the capsule onto a truck using one of the helicopters when fear of a sand storm made them abandon the effort temporarily. The cosmonauts were flown back to Dzhezkazgan by helicopter, and by jet back to Baykonur.[171]

On June 18, at 2:11 p.m., Titov and Manarov boarded the Soyuz TM-5, and moved it to the forward port 16 minutes later, clearing the aft port for the next Progress flight. The cosmonauts were exercising for two hours every day including one hour on the treadmill and one on the ergometer. Titov had lost 1.6 kg of body weight and Manarov had gained 2.7 kg. Lower leg volume had decreased by 16 and 18%, respectively.

Figure 6-12. This photo of the Mir station with the Kvant module and a Soyuz TM spacecraft was taken by a departing Soyuz TM. The add-on solar array on top of Mir gives it the appearance of an old Salyut station. (Source: Tass/Sovfoto.)

By June 21, the cosmonauts were preparing for an EVA to repair the British TTM X-ray wide-angle telescope outside the Kvant module.[172] The unit suffered periodic failures since late 1987. Although the telescope was not designed for repair in space, the Soviets had experience with similar repairs and decided to repair the telescope. The crew trained for the EVA by watching video tapes of simulations performed in the hydrobasin trainer at Star City that were delivered to Mir by a Progress.

On June 30, the crew exited the airlock to begin the EVA. The telescope's replacement 40-kg detector unit and tools were attached to a portable carrier to be carried across to the Kvant module. The detector unit was not designed to be serviced in flight, and the cosmonauts had to cut through a 20-layer-thick blanket of insulation on the side of the Kvant unpressurized section. The detector was fastened to the module's structure by small bolts and clips. The cosmonauts cut off the bolts and removed the clips. Their work was hampered because they had problems installing the normal foot restraints to provide a stable work platform and had to work without the restraints to help hold them to the station. They took turns holding each other in place as they struggled with removing the detector's clips and bolts.[173] The cosmonauts' problems continued as the bolt removal took 1.5 hours, several times longer than expected. The next obstacle was a brass clamp sealing a band holding the detector to the module. A special tool had been delivered to Mir to remove the clamp but it broke, and the cosmonauts were forced to put the insulation cover back into place and postpone the replacement until a new tool could be delivered and a new EVA scheduled. The EVA was running late anyway, and the cosmonauts returned to the airlock after 5 hours.[174,175]

Progress 37

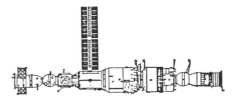

Launched: July 19, 1988, 1:13 a.m.
Reentry: August 12, 1988
Altitude: 187 × 256 km @ 51.6°

Progress 37 used a transfer orbit of 235 × 319 km to reach the Mir station two days later in orbit at 343 × 347 km. The cargo ship docked to the station's rear port on July 21, at 2:34 a.m. The spacecraft carried the usual items of food, clothes, propellant, air regenerators, water, mail, and equipment for the space station. Other items included a new computer, a new color television monitor, and new spacesuits that could be used with future manned maneuvering units. Refueling operations were started on August 7 and continued for the next two days. On August 9, the Progress boosted the station's orbit to 355 × 375 km. After being loaded with trash and used equipment, the Progress undocked at 12:32 p.m. August 12, and reentered the same day.[176]

Soyuz TM-6

Fourteenth International
Mission—Afghanistan

Launch: August 29, 1988, 8:23 a.m.
Landed: December 21, 1989,
           12:57 p.m.
Altitude: 195 × 228 km @ 51.6°
Crew: Vladimir Lyakhov, Valeri Polyakov, and Abdul Ahad Mohmand
Backup: Anatoliy Berezovoi, Gherman Arzamzov, and M. Dauran
Call Sign:

This flight had originally been scheduled for 1989, but the Soviet pledge to remove military forces from Afghanistan by February 1989 forced the international mission to be advanced or be canceled, because relations with Afghanistan after the Soviet Army withdrawal could not be predicted. After the launch of the flight, Afghan President Najibullah used the flight as a reason to declare a unilateral cease fire in the Afghan civil war for the duration of the mission.[177] The advance of the launch also meant that the crew's training was shortened and the mission may not have been as well prepared as usual.

For this mission and the previous international flight, the commander, Lyakhov, and his reserve were taken from a small group of cosmonauts trained to fly emergency evacuation missions to space stations. Polyakov was a doctor who was to check Titov and Manarov's condition before their landing. He also was participating in an unusual medical experiment in which before the flight he and his backup underwent an operation to remove a sample of their bone marrow. After his return to Earth in the spring of 1989, he would have another sample removed to help research into physiological changes caused by long-term weightlessness. This was also the first three-man crew that did not carry a flight engineer to assist the commander in flying the spacecraft. Earlier international flights with two-person crews also relied heavily on the commander to carry out all flight tasks, reducing the pilot's ability to cope with unusual flight situations.

The booster and spacecraft were rolled to the launch pad on August 27. Two days later, the Soyuz was launched and after two maneuvers at 1:20 and 1:43 p.m. was placed into a transfer orbit of 234 × 259 km. Two days later, the Kurs automatic docking system was activated at 40 km distance, which caused the station to use its gyrodyne gyroscopes to turn its rear docking port into the best lighting conditions for docking. The Soyuz then had to fly around the station to position itself 150 meters away from the rear docking port. The Soyuz approached and docked to the Kvant's rear port at 9:41 a.m., August 31 in an orbit of 339 × 366 km.[178] The crew performed 24 experiments including the usual Earth resources photographic sessions over Afghanistan using the KATE-140, MKS-M, Spektr-256, and hand-held cameras.[179] Until the flight, only 30% of Afghanistan's agricultural area had been surveyed by the Afghans. Most of the mission's experiments were routine international mission types, and most duplicated the experiments done by the last visiting crew and used the same Bulgarian-made apparatus. A new biological experiment studied calcium in plants, and a closed ecosystem in a small aquarium.

On September 1, Titov and Manarov made measurements of the upper atmosphere and performed electrophoresis experiments purifying interferon. The crews also exchanged items and seat liners from their Soyuz preparing to trade the spacecraft and leave the fresh Soyuz TM-6 for the long-duration crew. Mohmand also operated the Bulgarian Zora computer used by the last international crew. Its experiment program was not completed during the Bulgarian mission since some program diskettes were lost. The TM-6 carried replacement disks containing programs for medical tests evaluating vestibular function and its effect on space adaptation. That night Polyakov and Mohmand had their sleep monitored to access how much rest the cosmonauts were getting.

The next day, the cosmonauts continued their experiment program by tending to plants. Flax was planted in the Magnitogravistat and plant and bacteria relationships were studied in the Svetoblok-G. Later in the day a television conference was held including Afghanistan president Najibullah and the cosmonauts. The usual series of international mission experiments testing reaction time, senses, and psychological state of the cosmonauts were performed using the Zora computer. The previous day's electrophoresis experiment was finished and a new preparation of interferon was placed in the Ruchcy unit for separation. The crew also held a press conference during two televised communications periods.

On September 3, Polyakov and Mohmand conducted the Labirint experiment investigating the vestibular and visual systems. Meanwhile, Titov and Manarov continued studies of Earth's upper atmosphere and ionosphere using the Bosra device and the Bulgarian Terma photometer. By September 5, the crew began packing experiment results in the Soyuz TM-5 including film, tape recordings, computer diskettes, and biological specimens from the international mission. The long-duration crew also had space allocated for return of their experiment results in the small capsule. Among the results were crystals grown over the last three months for an Australian experiment. The crystal's atomic structure would be examined on Earth by three-dimensional X-ray analysis.

On September 6, Lyakhov and Mohmand left Dr. Polyakov with the long-duration crew and undocked at 2:55 a.m. in the old Soyuz TM-5 and began the usual landing procedures. As normal, the Soyuz orbital module was jettisoned at 3:35 a.m. to lower the weight of the spacecraft for retrofire. The Soyuz main engine ignited at 5:24 a.m. and fired for 60 seconds and then was shut down by the spacecraft's computer, which had detected an error in the attitude control system. The Soyuz had crossed from the day to the night side of the planet, confusing the infrared horizon sensors that normally detect Earth's horizon and the sun.[180] The sensors can operate in either darkness or sunlight, but not both during the same operation. Mission control was informed of the failure as the Soyuz passed over the tracking ship stationed in the South Pacific. Seven minutes later, as the spacecraft passed into Earth's shadow, the computer automatically started the engine when the infrared horizon sensors began working properly again, but Lyakhov shut it down after 6 seconds because the capsule was hundreds of miles from its hundreds of miles from its normal reentry flight path. Normal retrofire duration was 213.5 seconds, so the 60-second burn left the Soyuz in a slightly lower orbit that did not enter the atmosphere. Ten minutes later as the spacecraft flew over the Soviet Union, mission control was informed of the failure. The Soyuz computer was reprogrammed for another landing attempt.

Two orbits and three hours later, Lyakhov manually oriented the spacecraft to the correct attitude and preset the horizon sensor before starting the automatic landing sequence. The engine started, but shut down after 6 seconds. Lyakhov manually overrode the shutdown and continued the burn for about 60 seconds before realizing that he should stop the burn and wait to consult with mission control.[181] It was fortunate that he did because the 6-second firing the computer mistakenly made was part of its programming for the docking of the Bulgarian mission three months ago. If Lyakhov had continued the firing, he would have had to make a completely manual landing, something that had not yet been done by the Soyuz TM spacecraft, and something he had not been trained for. The spacecraft would not make another pass near the landing zone until the next day, so the crew and mission controllers, and Flight Director Valeriy Ryumin, settled down to sort out the problems encountered.

The cosmonauts were forced to sit in the cramped capsule in their pressure suits for a day, without the normal water, food and toilet facilities that were located in the jettisoned orbital module. There were, of course, the normal emergency rations in the capsule, primarily intended for use after landing in remote regions. But, the crew did not eat any of the rations, opting to go without food, which was probably wise because their pressure suits had no facilities for waste management. The Soyuz had enough supplies to orbit for at least two days before a serious situation would develop.

The next day, as the spacecraft's orbit brought it toward the landing area, the automatic retrofire sequence was performed successfully. Retrofire started at 4:01 a.m., and lasted 2 seconds shorter than planned. The capsule separated from the service module shortly afterwards and a few minutes later the tracking ship *Nevel*, in the south Atlantic, acquired the capsule's signal and relayed it to mission control. At 4:50 a.m. the capsule landed 160 km southeast of Dzhezkazgan, 10 km from the target. The crew was successfully recovered within 5 minutes. In the day before landing, the Western press had for the most part sensationalized the problems. The Soviets plainly explained that there were many options available for the cosmonauts' return including a manual reentry, reentry using the backup system, and even if there were engine problems, the Soyuz could deorbit using its maneuvering engines only.[182,183]

A day after the landing of Lyakhov and Mohmand in the Soyuz TM-5, the Mir crew of Titov, Manarov, and Polyakov undocked the Soyuz TM-6 from the rear of the station at 4:05 a.m. September 8, and redocked to the front of the station 20 minutes later to clear the aft port for the next Progress mission. The crew continued their program of experiments and Earth and astrophysical observations and waited for the arrival of the next Progress.

Progress 38

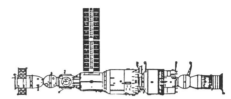

Launched: September 10, 1988,
       3:34 a.m.
Reentry: October 23, 1989
Altitude: 186 × 246 km @ 51.6°

Progress 38 used a transfer orbit of 234 × 332 km to dock to the station in orbit at 337 × 363 km at 5:22 a.m. September 12. The flight delivered the usual cargo of propellant, water, 300 kg of food including fresh fruit and vegetables, and experiments including part of the French deployable space structure experiment to be tested by the French international mission in December. Tools for the repair of the British TTM X-ray telescope were carried, and an amateur radio transceiver also was delivered for Titov and Manarov to experiment contacting amateur radio operators all over the world. Unloading cargo from the Progress began the day after it docked. Refueling the station began on September 27. The ship was loaded with trash and undocked October 23, and made a normal destructive reentry.[184]

☆ ☆ ☆ ☆ ☆

On October 6, the crew performed routine maintenance on the station's systems. Medical experiments, Earth observation, and astrophysical observations using the Kvant telescopes continued throughout the first half of the month. On October 18, Titov and Manarov began preparations for the second EVA to repair the British TTM X-ray telescope. The crew went over the plans with mission control and checked out their tools and new spacesuits. The previous EVA had failed to repair the telescope when a tool broke while attempting to remove a brass clamp holding the detector unit to the module. Different tools had been developed offering a range of capabilities to break the clamp holding the detector unit and a sample of the clamp and detector was used by the cosmonauts so they could study the problem before starting the EVA. The cosmonauts' new spacesuits had improved features over the older suits developed in 1985. The new suits were totally self-contained and could be operated with the Soviet manned maneuvering unit without communications and reserve power lines used by earlier suits. The suits were made with a hard torso section, and sleeves and legs made of a softer more elastic fabric and were removable to facilitate cleaning and repair. The gloves, critical parts of a spacesuit, were also improved and made more flexible than the earlier versions, which pinched blood flow to the hands. The environmental control system was also improved with stronger ventilation motors and new temperature controls. Cardiograms and respiration data also was measured by the new suit.

On October 20, Polyakov moved into the Soyuz TM-6 while Titov and Manarov depressurized the airlock. This precaution was taken because if there was a problem during the EVA and the crew did not return to close the airlock hatch, Polyakov would be stranded in the wrong end of the station, with the return capsule on the other side of the open airlock. The airlock was opened at 9:59 a.m. and the cosmonauts moved to the end of the Kvant module. They started by removing the insulating blankets they had cut open on the previous EVA. They then successfully removed the brass clamp holding a band around the detector with the new tool delivered by Progress 38. After removing the faulty 50-cm diameter, 40-kg detector unit, they placed the new and improved detector and connected it to the station's power and control lines. They then replaced the insulation covering over the detector. The cosmonauts also had other details to attend to during their rare excursion. Among their tasks was to try to clean dust off of two portholes on the end of the Kvant module using a soft copper brush. The portholes were used to observe approaching spacecraft during docking and undocking operations. They also installed an amateur radio antenna that would be used by the cosmonauts to talk to ham radio operators all over the world as had been done on previous NASA shuttle missions. The experiment was sponsored by a Soviet amateur radio organization. The narrow band FM radio operated at 144.5 MHz and the cosmonauts could speak some English, which improved contacts outside the U.S.S.R. They also installed a foot restraint device on the multiple docking adapter, which would be used during the French EVA in December. The cosmonauts then returned to the airlock and entered the station. On October 25, the crew started observations using the Mariya spectrometer. They also photographed regions of the U.S.S.R., and Polyakov continued medical tests of the cosmonauts' cardiovascular systems.[185,186]

VKK-1 Buran
First Test Flight of Space Shuttle
Launched: November 15, 1988, 6:00 p.m.
Landed: November 15, 1988, 9:44 p.m.
Altitude: 252×256 km @ 51.6°
Crew: none

In March 1988, rumors circulated that a U.S. photo-reconnaissance satellite detected an Energia booster being placed on the launch pad and then it was removed (see box). The Soviets acknowledged that they were testing and retesting systems for the shuttle launch. The booster was being tested with launch pad systems to ensure there would be fewer problems when the shuttle was taken to the pad. In May 1988, the chairman of Glavkosmos stated that the second Energia launch would carry the Soviet shuttle and that it would be the only Energia launch of the year.

K-Type Booster

The type K booster (Energia) is a versatile launch vehicle that can launch either a cargo canister or a space shuttle orbiter into orbit. The Energia was designed by a military design bureau under the direction of Energia designer Boris Gubanov. When using the 8-meter diameter payload canister, the booster can deliver greater weights into orbit, rather than launching the extra weight of a shuttle orbiter that displaces payload weight. The Energia could lift about 100,000 kg to a 180-km circular orbit. When launching a payload on the shuttle orbiter, the payload was estimated at 33,000 kg.

The booster consisted of a core stage that had four LOX-hydrogen high-pressure pre-burning main engines, and four J type first stages as strap-on boosters. The core engines produced from 148,000 kg to 200,000 kg vacuum thrust each, and the strap-on RD-170 engines produced about 740,000 kg to 806,000 kg vacuum each. Total thrust at launch was 3,600,000 kg. The booster was designed to withstand the failure of any three strap-on or core engines and achieve a safe abort of the flight. Each strap-on booster was 41.6 meters tall and about 4 meters wide. The core stage was 61.3 meters long and 8 meters wide. Maximum width of the assembled booster at the base was 16 meters. Dry weight of the entire booster was estimated at 55,000 kg, and 2,400,000 kg fully fueled and loaded.

The Energia/shuttle launch complex at Baykonur consists of three launch pads. Two were modified G-type launch pads originally built in the late 1960s. This can clearly be seen since the service structure for both pads is much taller than necessary to service a shuttle or Energia booster. The two old launch pads are equipped to launch an Energia/shuttle combination. The third launch pad, called the Multipurpose Launcher Testbed, was built specifically to launch Energia boosters without shuttles.

The Energia was first launched on May 15, 1987, at 9:30 p.m. The booster carried only a test article satellite payload with its upper stage booster pods. The strap-on boosters comprising the first stage were jettisoned in pairs, after 2.5 minutes. The core stage fired for 9 minutes total, when it reached an altitude of 80 to 95 km at an orbital inclination of 65°. After separating from the core, the orientation system for the upper stage pods failed to operate correctly. The core stage and the upper stage and payload reentered and burned up over the Pacific Ocean.

In September, Radio Moscow reported that cosmonauts were undergoing shuttle training in simulators, practicing takeoff, maneuvering, and landing, fueling rumors that a manned flight might be attempted soon. Vladimir Dzhanibekov reported that there were six cosmonauts in training for the two positions on the first manned flight of the shuttle, whenever it would occur. In the last week of September, rumors circulated that a U.S. photo-reconnaissance satellite had detected the shuttle being moved to the launch pad (see Figure 6-13).[187–190]

On April 29, 1988, the Soviets announced that their shuttle would be launched shortly on an Energia booster. Pictures released of the Buran orbiter being prepared for flight showed a cylindrical module mounted in the cargo bay, similar in size to the Kvant module. There was no explanation for the purpose of the module, but it probably carried instrumentation to measure the launch and reentry conditions inside the cargo bay that future spacecraft would have to withstand (see Figure 6-14).[191]

The launch was originally scheduled for October 29 at 7:30 a.m. As the countdown proceeded into its final hours, a fault occurred in the ignition system and required the countdown to be delayed for four hours. After recycling the countdown, the count continued to T minus 51 seconds when it was stopped again because the crew access platform did not retract as fast as expected. Even though there was no crew in the orbiter, the crew access and escape arm also provides electrical connections from the ground to the orbiter. Specifically, the orbiter's guidance gyroscopes were updated with accurate ground information. The access platform should have retracted in 3 seconds, but required 38 seconds. The design of the hinge mechanism for the platform was said to be inadequate. Review of problems and corrections would take about two weeks.

On or about November 11, the next launch attempt was set to November 15. The Soviet's announced that live television coverage would be provided of the launch, but as with the previous launch attempt, no live coverage was provided. As launch time approached, launch officials met to consider the worsening weather at Baykonur. The temperature was 4°C and the cloudy weather was predicted to grow worse as a storm moved from the Aral Sea toward Baykonur. The countdown was allowed to continue and workers cleared the pad at 12:00 a.m. November 15, as hydrogen loading of the Energia core stage began. During the final preparations, shuttle cosmonauts flew MiG-25 launch-observation aircraft and a Tu-154 shuttle training aircraft making landings at the shuttle recovery runway to test abort landing conditions.

At 4:49 a.m., the shuttle was switched to an internal launch sequencer, and about 8 seconds before lift-off, the core stage main engines started followed by the 4 strap-on boosters. Lift-off occurred on schedule at 6:00 a.m. After 2.75 minutes, the strap-on boosters were jettisoned in pairs as their propellant was depleted at 60 km altitude. The core stage continued firing, carrying the orbiter

Figure 6-13. The Soviet shuttle used launch facilities originally built for the G-type rocket in the 1960s. The launch tower to the left is new and supports the needs of the shuttle and provides crew access. This photo was taken from on top of the main rotating service tower, which rotates to surround the orbiter on the curved track on the ground. The shuttle is being raised to the vertical position by its erector/transporter for placement on the launch pad. (Source: Sovfoto.)

Space Shuttle

The Soviet shuttle orbiter was 36 meters long, 24 meters wide, with a cargo bay that was 4.7 meters wide and 18.3 meters long. The shuttle orbiter weighed 75,000 kg empty. Payload capability to low Earth orbit was 30,000 kg and return payload was about 20,000 kg, which was roughly the weight of a Salyut, Mir, or Star module. By the time of the first flight of the shuttle in 1988, the Soviets had planned to built at least three or four space-worthy

(continued on next page)

(continued from page 351)

shuttle orbiters. At the time of the first launch, at least five orbiters had been built including two space versions, one boilerplate test article, a structural test article, and one atmospheric, landing test and training orbiter equipped with jet engines. The Soviet shuttle used for atmospheric testing and crew training was equipped with four 9,090 kg thrust Lyulka jet engines. Two engines were installed next to the base of the tail and two on both sides of the boat tail section. The Soviets had intended to install the jets on the first shuttle orbiter, and built the first few test vehicles and models with two jet engines before removing them from the final design.

Soviet space workers have admitted that the orbiter was a copy of the NASA Rockwell orbiter design with some modifications. Differences included the lack of main engines, a single orbital maneuvering system pod, forward landing gear location, and double-layered crew entry hatches. The crew compartment was almost identical to the NASA orbiters. Up to four cosmonauts could be carried in the cockpit and up to six could be carried in the mid deck level. The Soviet orbital maneuvering system propellant tanks held up to 14,000 kg to be used for orbital insertion after separation from the Energia booster, orbital maneuvering, and retrofire. The first orbiter (Buran) was covered by 38,000 insulating tiles similar to NASA orbiter ceramic tiles weighing 9,000 kg. The Soviet shuttle was capable of a 2,000-km cross range maneuver after reentry enabling landing up to 2,000 km away from the orbital flight path.

The final manned version of the shuttle was to be equipped with two modified MiG-25 ejection seats to enable crew ejection. Three crews of test pilots were formed from a group recruited in 1978 to fly the shuttle orbiter. These included Boroday, Stankavichyus, Levchenko, Shchukin, Bachurin, Kononenko, and cosmonaut Volk. Several pilots came and went from the shuttle project from 1978 to 1988. As of 1989, the shuttle pilots included Stankavichyus, Tolboyev, Sultanov, Zabolotskiy, Tresvyatskiy, Sheffer, Volk, and two others.

toward orbit. The core stage shut down 8 minutes after launch and separated from the orbiter at 160 km altitude. The trajectory of the booster and orbiter were both sub-orbital, descending into the atmosphere over the Pacific. The core stage would continue on that path and make a destructive reentry. Two and a half minutes after separation, the orbiter fired its orbital maneuvering engines for 67 seconds to boost the trajectory to about 250 km and avoid falling into the atmosphere. Over the Pacific at 6:47 a.m., the orbiter made another maneuver for 42 seconds circularizing the orbit to 252 × 256 km.

Figure 6-14. The Soviet shuttle is shown here after mating with its Energia booster in the old G-type assembly building built in the late 1960s. Service platforms can be lowered or swung into position to service the booster and shuttle. The booster is resting on its mobile erector/transporter, which is mounted on dual rail lines. (Source: Sovfoto.)

The launch was announced more than an hour later as the orbiter was on its first orbit. The orbiter was in communication with mission control in Kaliningrad during the entire mission using a combination of tracking ships and satellites. The ships *Volkov* and *Belyayev* were stationed in the south Atlantic. The *Marshall Nedelin* was stationed off the coast of Chile and the *Dobrovolski* was stationed to the west of the *Nedelin's* position. The *Marshal Nedelin* was normally used to support military launches (see Figure 6-15). Two Molniya communications satellites were used to relay information, probably from ships to mission control, and a Gorizont and a Luch satellite were also used. During the flight, mission control received television pictures of Earth taken from cameras mounted in the shuttle cockpit.

The orbiter made its first orbit over the Pacific, South America, the South Atlantic, Africa, the U.S.S.R. and back to the Pacific. The second time over the South Pacific, the orbiter turned its tail into the direction of flight and performed retrofire at 8:20 a.m. The orbiter then turned around and coasted toward

Figure 6-15. The *Marshal Nedelin* is usually used to monitor military rocket launches and space missions, but recently has been used to support manned spaceflights, relaying communications and telemetry to mission control. (Source: U.S. Department of Defense.)

reentry. The orbiter touched the fringes of the upper atmosphere at 122 km altitude. For the next 20 minutes, the orbiter was in radio blackout as aerodynamic braking created a plasma shield under the spacecraft. As the orbiter flew through about 40 km altitude, it had completed altering its flight path to the East by about 1,000 km to head toward the Baykonur Cosmodrome. Before the arrival of the shuttle, the Tu-154 trainer was flown on approaches and landings to determine weather conditions at the landing site. The orbiter was also intercepted by two MiG-25 chase planes flown by shuttle test pilots Ural Sultanov and Magomed Tolboyev. They relayed television pictures of the orbiter as it made its final approach to the runway. As the orbiter approached the ground, winds of 64 km/hr (40 mph) blowing 30° to the runway made for a crosswind of 55 km/hr (34 mph), which was well above acceptance values for NASA shuttle landings. The orbiter touched down at 9:25 a.m., traveling about 180 knots (207 mph) with the main landing gear only 1.5 meters from the runway center line. Three braking parachutes were deployed to help slow the orbiter to a stop after traveling about 1,150 meters down the runway.

Inspection revealed that only five of the delicate 38,000 heatshield tiles had fallen off the orbiter during the flight. After the landing, the orbiter was parked outside the orbiter processing facility for initial inspections and propellant removal. The orbiter was scheduled to have some of its major systems disassembled for inspection, but this was still not done by the next June when the orbiter was flown to Paris on the new An-225 carrier aircraft for display at the Paris air show.[192–194]

☆ ☆ ☆ ☆ ☆

In late November, the Mir crew's normal 8.5-hour workday was reduced by one hour. The cosmonauts still had two exercise periods of one hour each to use the ergometer and treadmill, respectively. In addition, conditioning of their cardiovascular systems was begun by periodically using the Chibis low-pressure suit.

Soyuz TM-7

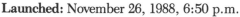

Fifteenth International
Mission—France
Long-Duration Return
and Replacement

Launched: November 26, 1988, 6:50 p.m.
Landed: April 27, 1989, 2:59 a.m.
Crew: Aleksandr Volkov, Sergei Krikalev, and Jean-Loup Chretien
Backup: Aleksandr Victorneko, Aleksandr Serebrov, and Michel Tognini
Altitude: 194 × 235 km @ 51.6°
Call Sign: Donbass

In early 1988, the Soviets hoped that the second large add-on module could be launched to Mir before the French international mission. The module was still undergoing ground testing in December 1988, with the launch scheduled for April 1989. That date would continue to slip into the future and be delayed by other problems.[195] At one time, Chretien said that the space station trainer he was using was already attached to four additional modules of the current space station. The schedule slipped just as it had for the Kvant add-on module. The Soviets also hoped to launch a new module to Mir every five months until the four lateral docking ports were filled.[196] The French/Soviet agreement that led to the mission to fly a Frenchman to Mir included a fee of $30 million. The two countries also agreed that French experiments would be installed on future space station modules and made provisions for a French cosmonaut to visit the station on a month-long mission about every two years.[197]

The launch of the Soyuz TM-7 was delayed itself for five days because of a late decision for French President Mitterrand to view the launch with his 200-member delegation, which had been meeting in Moscow with Soviet leader Gorbachev. The French delegation was also shown the shuttle orbiter Buran and an Energia booster while at the cosmodrome. In a change from previous practice, the Soviets had wanted to sell advertising space on the walls of the space station and cosmonauts' flight suits to Western, especially French, companies for the mission. The French did not agree with the idea although the Soviets did place signs around the Kaliningrad mission control center where they would be seen by television cameras.

Eight minutes after launch, the Soyuz was placed into an initial orbit at 194 × 235 km. The cosmonauts performed the usual checks of the spacecraft and removed their pressure suits. After four orbits the Soyuz boosted its orbit to 256 × 291 km, and on the fifth orbit another boost made the transfer orbit

253 × 305 km. On the next day another maneuver boosted the Soyuz so that it would meet the space station in another day. On November 28, the Soyuz approached the rear docking port and after pausing a few hundred meters from the station the automatic Kurs system soft docked to Mir's aft port at 8:16 p.m. in orbit at 337 × 369 km. Hard docking came 9 minutes later and the crew opened the hatches at about 10:00 p.m. after powering down the Soyuz.

On November 29, the joint crews began working on the French experiment program. French experiments included using the Matra As de Coeur echo-cardiograph, the Viminal and Physalie vision and motor action experiments, Circe radiation detector, and the Ercos experiment testing very large-scale integrated circuits (VLSI chips) exposure to space radiation. On the first day, Chretien started out with the echo-cardiograph, making measurements of his own heart and recording the results on video tape. Medical experiments also included taking urine and blood samples for later analysis on Earth to compare to pre-flight and post-flight samples. The second day of experiments was similar to the first, and Chretien used the cardiograph while wearing the Chibis low-pressure suit to stress his cardiovascular system. He also performed the Physalie experiment measuring the body's motor action. The experiment involved monitoring electrical activity in the body including the limbs, eyes, and heart. Chretien was videotaped from two directions while covered with a mass of wires and electrical boxes and making selected body movements. Chretien commented that some of the experiments required many hours of preparation, and was particularly annoyed at the complexity of the Physalie experiment.

Meanwhile, the other cosmonauts either helped with the French experiments or continued the long-duration crew's work. The Kvant's telescopes continued their observations of various objects, and Titov and Manarov started well planned preparations for their return to Earth at the end of the month. Their exercise intensity would be increased and they began to wear Penguin elastic exercise suits for several hours a day. They also started drinking saltwater solutions to increase their blood volume.[198]

The joint French mission also included an EVA to test the deployment of a space structure with the assistance of cosmonaut Volkov. The EVA was initially planned for December 12, but was moved to December 9 to allow for a second EVA if problems developed during the first EVA. Lighting conditions would have been poor for a second EVA if the original date was not changed.[199] The task of Volkov and Chretien was to deploy a 239-kg experimental carbon fiber structure measuring 3.8 meters long, 3.6 meters wide, and 1 meter thick. Preparations began on December 8 as the cosmonauts prepared the spacesuits and rehearsed their duties.

On December 9 at 12:57 p.m., Chretien and Volkov opened the airlock's hatch and prepared to perform two tasks. The event also marked the first non-American or Soviet to make a spacewalk. Chretien first leaned out the hatch and secured hand holds to the side of the station. He then placed the Enchantillons space exposure experiment on the side of the station. The 15-kg experiment would be retrieved by a later crew and returned to Earth. It would expose composite materials, adhesives, optics, paint, and collect dust for later analysis on Earth.

Chretien and Volkov then exited the airlock and setup equipment outside the station between the docking adapter and the station's work compartment. The cosmonauts first attached a mount to the station's hand rails. The mount was connected to a control panel inside Mir, which would activate the experiment. After securing the 0.6 m × 1 m canister containing the collapsed structure to the mount, Krikalev commanded the experiment to open from inside Mir, but it did not move. Volkov and Chretien shook the unit but it held tightly together. As mission controllers tried to determine the problem, Mir passed out of communications range, leaving the cosmonauts to consider the situation on their own. With the EVA running behind schedule, mission control was about to advise that the experiment be jettisoned. Volkov decided to try kicking the canister, against orders from mission control. After a few good kicks the structure was coaxed to deploy while still out of communications with mission control, and unknown to mission controllers.[200] Video of the event was recorded on Mir and relayed later to mission control. Engineers speculated that the structure was held together by humidity in the air that had frozen the structure together when the airlock was depressurized. Sealing the structure hermetically or keeping the humidity lower would have prevented the problem. The EVA had to be extended by 1.5 hours to last a total of 6 hours because of the deployment problem and problems encountered in attaching the Echantillons materials exposure experiment to the outside of the station. Before the end of the EVA, the structure was jettisoned as planned since it would block one of Mir's side docking ports. By the end of the EVA, the cosmonauts reentered the multiple docking adapter. Chretien had some difficulty closing the hatch, because his face plate was covered with condensed water caused by high humidity in the suit from sweating during the hard work of the EVA. The EVA had lasted 5 hours and 57 minutes. The spacesuits were limited to 6 hours of use at a time, making this the longest Soviet EVA ever, and a record that will not be easily broken for some time to come.[201,202]

As the cosmonauts on Mir prepared to exchange places, ground facilities that had been in use for months also were relieved by fresh units. In the middle of December, the tracking ship *Nevel* left its South American post to be replaced by the *Morzhovets*. The *Borovichi* also left the Caribbean to be replaced by the *Komarov* temporarily, and then permanently by the *Kegostrov* to continue supporting the new Mir crew.

On December 14, Chretien continued experiments including testing deployment of a 28-kg model solar array inside the station. The Physalie and Vinimal experiments were continued and the operation of the Mir Gamma medical installation was checked. The crew also participated in a television conference with children in France. The cosmonauts also used the station's cameras to take pictures of Armenia's recent earthquake damage.

Titov and Manarov had been exercising for about two hours a day, like previous long-duration crews to keep in shape for their return to Earth. The cosmonauts even felt that they could have stayed on the station longer if the mission had called for it. On December 19, the cosmonauts checked out the Soyuz TM-6 and began powering up its systems for the trip home. Landing was scheduled for 9:48 a.m. December 21, with Chretien returning with the long-duration crew, Titov and Manarov. The next day they continued preparing to return, packing the Soyuz with experiment results including film, video tape, and biological specimens. Volkov and Krikalev also checked out the Soyuz TM-7 in preparation for the transfer to the forward docking port after the TM-6 landing.

The next day, Titov, Manarov, and Chretien undocked from Mir in the Soyuz TM-6 at 6:33 a.m. For the first time since the Soyuz 40 flight, the Soyuz did not jettison its orbital module before retrofire. After the Soyuz TM-5 retrofire mishap, the Soviets wanted to keep the orbital module attached in case of a retrofire problem. While no vital equipment was in the orbital module, hygiene facilities and additional water and food were available. Retrofire was scheduled for 9:00 a.m. but was delayed for about three hours when new orientation computer programs, written to avoid any problems with the infrared horizon sensor like those experienced during the TM-5 retrofire, malfunctioned. After consulting with mission control, the crew switched to the Soyuz backup computer and waited for the next available retrofire time. The delay caused the landing point to be moved about 300 km south from the original spot.

Retrofire successfully began at about 12:09 p.m. and 280 seconds later, the orbital module separated. As the spacecraft fell to 140 km altitude at 12:33 p.m., the service module was jettisoned. At 10 km altitude, the parachute was opened and antennas unfurled to enable communications with the recovery forces. The capsule landed 180 km southwest of Dzhezkazgan, in foggy conditions and sub-freezing temperatures at 12:57 p.m. Three hours later, the crew was flown to Dzhezkazgan, and Titov and Manarov walked away from the helicopter with some aid. Two days later the cosmonauts were able to move without help, although they were far from recovered. During their mission, Titov and Manarov conducted about 2,500 experiment sessions using about 150 different procedures and apparatuses.[203,204]

Titov had lost 3 kg of weight and lost 20% muscle volume in the lower legs, and Manarov had gained almost 2 kg. Chretien also lost almost 1 kg during his 25 days in space. These results were not significantly different that previous long-duration flights, and some cosmonauts have returned in worse condition after much shorter flights. The difference is made up by the Soviet advances in understanding how to maintain health by exercise over long periods. The cosmonauts' bodies loss of calcium was thought to be less than in previous flights, and loss of potassium in the blood was 5 to 6%. Five days after landing, Titov and Manarov were walking up to 3 to 4 km a day and swimming 4,500 meters. They then started a training program to increase their back and leg strength. By January 11, the cosmonauts had regained their lost weight and muscle volume. In the middle of the month, Titov and Manarov went to a resort to rest and regain their preflight strength. It was estimated that by March they would be fully recovered.[205]

On December 22 at 9:45 a.m., the new Mir crew of Volkov, Krikalev, and Polyakov undocked the Soyuz TM-7 from the rear docking port and the station was commanded to enable its forward docking port. The station rotated 180° and Volkov redocked to the forward port at 9:59 a.m., clearing the aft port for the next Progress flight.

The new crew's mission was to last five months. They were to stay in space for five months and receive three Progress craft and the "D" or auxiliary Mir expansion module (see Figure 6-16). Until the new module arrived the crew would continue the previous crew's experiments and observations.

Progress 39

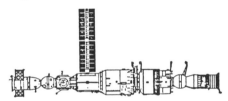

Launched: December 25, 1989,
        7:12 a.m.
Reentry: February 7, 1989
Altitude: 187×237 km @ $51.6°$

Progress 39 entered an initial orbit at 187×237 km after separation from its upper stage. The spacecraft then boosted its orbit to 237×338 km to transfer to Mir's orbit. Two days later, the transfer orbit brought the Progress near the station. Progress 39 then adjusted its orbit to 325×353 km, and approached Mir's aft docking port and docked at 8:55 a.m. on December 27. The Progress delivered mail; food, including fresh vegetables and fruit; propellant; water; replacement parts; and equipment for the new long-duration crew. On January 24, the Progress booster the station's orbit to 340×376 km. Progress 39 undocked from the station on February 7 at 9:46 a.m.

☆ ☆ ☆ ☆ ☆

On January 2, the crew started a new series of Earth observations of Siberia and the eastern Soviet Union. The Bulgarian Rozhen photometer and the Parallax-Zagorka device were used to observe astronomical targets and Earth's atmosphere. The Kvant telescopes continued to be operated to make observations in X-ray wavelengths. Dr. Polyakov continued medical investigations of the new cosmonauts using the French instruments from the last mission. He

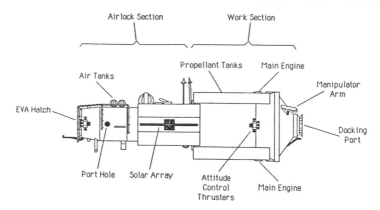

Figure 6-16. The auxiliary "D" module is planned for launch in 1989 and will provide new experiments and a large airlock with a manned maneuvering unit to the Mir station. It is planned that a similar technology "T" module will also dock to Mir at about the same time.

also used a blood analysis device called Retroflon, which was developed by a West German company. There were also reports that on January 17 the cosmonauts were preparing for an EVA. They were probably servicing the spacesuits just used for the French EVA in anticipation for their scheduled EVA's later in the mission. On January 24, the crew began maintenance operations to replace a hydraulic control unit in the station's environmental control system. The work continued over the next few days.

Routine observations using the X-ray telescopes in the Kvant module continued until the end of the month. The cosmonauts also started a new series of experiments using the Yantar device, which deposited thin coatings of silver-palladium and tungsten-aluminum on polymer films. In the beginning of February, the Soviets announced that a planned EVA had been canceled due to problems with preparing the new space station module for launch. On February 7, Progress 39 was filled with the last of the trash and cast off. It reentered the atmosphere later the same day. The aft port was then clear for the Progress 40 flight being prepared for launch at Baykonur.

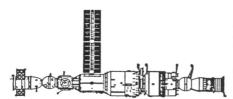

Progress 40

Launched: February 10, 1989,
    11:54 a.m.
Reentry: March 3, 1989
Altitude: 193 × 262 km @ 51.6°

Progress 40 entered an initial orbit at 193 × 262 km after separating from the booster's upper stage. The spacecraft then used a transfer orbit to meet with the Mir space station two days later. On February 12, the spacecraft approached and docked at the aft port at 1:30 p.m. in orbit at 347 × 364 km. The flight carried the usual cargo of food including cucumbers, honey and fresh fruit, water, propellant, experiments, and replacement parts. On February 24 the Progress boosted the station's orbit to 358 × 386 km. The Progress finished refueling by the end of February.

The cargo ship undocked on March 2 at 4:46 a.m. and began an unusual test program. The Progress usually deorbited soon after docking, but Progress 40 carried an experiment that deployed two large multi-segmented metal structures. The structures were folded up against the spacecraft until after undocking when electric heaters mounted on the structures were activated. The heat caused the metal segments of the structure to extend and resume their originally manufactured shape. The Progress continued orbital flight for the next two days, continuing tests of the metal structures before retrofire on March 5 at 4:08 a.m.[206]

☆ ☆ ☆ ☆ ☆

The Mir crew continued with routine work and started new experiments including Diagram, which was a magnetic sensor deployed outside the station through the scientific airlock. The experiment investigated aerodynamic flow around the space station. On March 2, the crew finished the loading of trash and the Progress undocked the next day at 4:46 a.m. to begin its test program before reentry. The cosmonauts video taped Progress 40 as its experimental metal structures unfurled themselves.

After the departure of Progress 40, the crew continued observations with the Kvant's telescopes and use of the Mariye spectrometer. The spectrometer was being used to measure high-energy particles and access any relationship to seismic activity. These observations continued until the middle of March when the next Progress flight was being prepared.

Progress 41

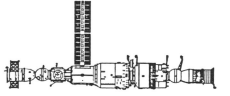

Launched: March 16, 1989,
9:54 p.m.
Reentry: April 25, 1989, 4:02 p.m.
Altitude: 193 × 260 km @ 51.6°

Progress 41 separated from its upper stage and entered an initial orbit of 193 × 260 km. Two days later the Progress docked to the aft docking port of Mir at 11:51 p.m., in orbit at 349 × 363 km. The Progress carried replacement units for the station's power systems, replacement batteries, propellant, food, and water for the crew. On April 10, the Progress boosted the station's orbit to 372 × 400 km. On April 21, the Progress undocked and flew in orbit for five days before reentering the atmosphere. It may have been performing experiments like Progress 40, or it may not have performed a retrofire maneuver for some reason. The spacecraft reentered on April 25 at 4:02 p.m.

☆ ☆ ☆ ☆ ☆

After the docking of Progress 41, the crew began the unloading operations that would last several days. At the same time, routine experiments were continued. The Bulgarian Spektr-256 was used to measure Earth's atmosphere and the French Circe unit was used to measure radiation in the space station. On March 26, the crew relayed their votes in the Soviet Congress of Peoples Deputies elections by radio to mission control. Several cosmonauts were among the candidates for the congress and Viktor Savinykh, Svetlana Savitskaya, and Valeriy Ryumin were elected. A Soviet news report the same day reported that cosmonaut Krikalev had received several notices from the Soviet Army to report for reserve duty, even though he was currently in orbit on Mir! The Soviet press notably referred to the responsible officials as dim-witted bureaucrats.

Normal operations continued through the beginning of April with Earth operations occupying most of the cosmonauts' time. They also replaced several power supply units in Mir as preventive maintenance after three years in orbit. Power supplies were used by various pieces of equipment to convert power from the station's distribution system's alternating current to direct current for electronics in experiments and apparatus. Progress 41 had also delivered new storage batteries for the space station. The station's batteries had been deteriorating as they also had on Salyut 7. The cosmonauts installed a few batteries, but they could not significantly improve the power situation. More batteries would have to be added in the future to keep the normal experiments operating. Also in early April, the crew was treated to large displays of aurora during a period of increased solar activity. The increased radiation from the sun was not a problem for the Mir crew since their orbit was within Earth's magnetic field, which shielded them from most radiation.

On April 10, the Progress boosted the station's orbit to 372×400 km in preparation to leave the station unattended. It had been decided that the replacement crew of Aleksandr Viktorenko and Aleksandr Balandin would not be launched as planned on April 19. They would continue training to receive the large expansion modules that were rescheduled for launch in the last half of 1989.

The Mir crew would return to Earth, leaving the station unmanned for the first time in years for several reasons. The development of the new modules was delayed, just as the Kvant module had been in 1987. The present crew had been trained to operate the new module and its equipment, and were not prepared or trained for a long mission using Mir's current, and in some cases broken, equipment. There was little new research to do on the station until new equipment and experiments were delivered on the new modules. There were also reports that the Mir station had developed some problems with its power system, making it difficult to maintain electrical power continuously to the station's experiments.[207] This made the power-hungry materials processing experiments difficult and probably impractical to operate, and lessened work for the cosmonauts. To bring the station back to full capacity new equipment would have to be taken to Mir or the generating capacity increased by adding the new modules. The problem had to be resolved before the launch of the modules, and probably delayed the modules' launch even further. Vladimir Dzhanibekov, no stranger to power problems since his Soyuz T-13 flight, was put in charge to research the power problems on Mir. The decision to leave Mir unmanned for the first time in two years seemed to be developing as early as February 1989, when there was indecision as to which crew would to be launched in April.

The Soviets expected to leave the station unmanned for at least three months when it was expected that the new modules would be ready for launch. Meanwhile, the cosmonauts had a lot of work to do getting the station in order for unmanned operation. They inventoried all the station's equipment, stowed experiments and instruments, and at the same time prepared for landing using the Chibis suit to stress their cardiovascular systems. The packed trash into Progress 41 for disposal after it undocked on April 21. The crew also had to pack their experiment results from the last four months into the Soyuz TM-7 capsule and power up its systems.

The crew finished their mothballing of the station and closed the hatch to Mir the night of April 26. At 3:28 a.m., April 27, the Soyuz undocked and backed away from the space station. Two orbits later, the Soyuz approached South America and prepared for retrofire. At 6:00 a.m., retrofire began, sending the spacecraft toward Earth. At 6:59 a.m. the capsule landed 140 km northeast of Dzhezkazgan. As the cosmonauts were being lifted out of the capsule, Krikalev hurt his leg but said it was nothing serious. The cosmonauts were flown back to Baykonur as usual, and within two days they were regaining their strength and moving about without trouble.[208]

In the future, crews on the station will have to learn how to operate a space station nearly twice as large as the original Mir and learn how to operate with the new Soviet space shuttle docked to the station. At the same time they will have to host more planned international guests from Austria, Japan, Spain, West Germany, France, and England.

Meanwhile, Soviet scientists and engineers have begun initial concept studies for a Mars transport spacecraft capable of carrying four crewmen. The spacecraft is planned to be propelled by either conventional, nuclear, or electric rockets after being assembled in Earth orbit after a few launches of an Energia booster (see Figure 6-3). The ideas are not new, NASA studies from the late 1960s proposed the same things, but the Soviets now have the capability with the Energia, shuttle, and space station to carry out the plan whenever they wish.[209]

References

1. Kidger, Neville, "Salyut 6 Mission Report: Part 4," *Spaceflight*, Vol. 22, No. 11–12, Nov.–Dec. 1980, p. 343.
2. Foreign Broadcast Information Service, U.S.S.R., Space, JPRS-USP-87-006, Nov. 24, 1987, Joint Publications Research Service, p. 45.
3. "Soviet Scene," *Spaceflight*, Vol. 28, No. 3, March 1986, p. 111.
4. Congressional Research Service, The Library of Congress, *Soviet Space Programs 1976–80, Manned Space Programs and Life Sciences, Part 2*, Washington: Government Printing Office, 1984, p. 467.
5. Foreign Broadcast Information Service, U.S.S.R., Space, JPRS-USP-87-004, July 1987, Joint Publications Research Service, p. 112.
6. Bond, Peter R., "The Soviet Snowstorm," *Spaceflight*, Vol. 31, Feb. 1989, p. 51.
7. Pirard, Theo, "Dunayev: Shuttle Will Fly on Second Energia," *Spaceflight*, Vol. 30, May 1988, p. 186.
8. Foreign Broadcast Information Service, U.S.S.R., Space, JPRS-USP-88-002, April 6, 1988, Joint Publications Research Service, p. 34.
9. Radio Moscow, North American Service, Jan. 24, 1989.
10. Radio Moscow, North American Service, Feb. 2, 1989.
11. Neville, Kidger, "First Launch of Carrier Rocket," *Spaceflight*, Vol. 29, July 1987, p. 5.
12. Foreign Broadcast Information Service, U.S.S.R., Space, JPRS-USP-86-005, September 12, 1986, Joint Publications Research Service, p. 202.
13. Ibid. JPRS-USP-88-002, April 6, 1988, Joint Publications Research Service, p. 8.
14. "Soviets Consider Varied Concepts for 1994 Mars Exploration Flight," *Aviation Week & ST*, July 18, 1988, p. 19.
15. Oberg, James E., *New Race for Space*, Stackpole Books, 1984, pp. 71–74.
16. Foreign Broadcast Information Service, U.S.S.R., Space, JPRS-83430, May 1983, Joint Publications Research Service, p. 46.
17. "Science and Engineering," North American Service, Radio Moscow, June 30, 1988.
18. "Cosmonaut Criticizes Space Effort," *Flight International*, Jan. 2/9, 1988, p. 16.
19. Foreign Broadcast Information Service, U.S.S.R., Space, JPRS-USP-87-006, Nov. 24, 1987, Joint Publications Research Service, p. 45.
20. Turnhill, Reginald (ed.), *Janes Spaceflight Directory 1986*, London: Janes, 1986, p. 225.
21. Foreign Broadcast Information Service, U.S.S.R., Space, JPRS-USP-87-003, April 1987, Joint Publications Research Service, p. 9.

22. Ibid. JPRS-USP-86-004, April 1986, Joint Publications Research Service, p. 22.
23. King-Hele, D. B., Walker, D. M. C., Pilkington, J. A., Winterbottom, A. N., Hiller, H., and Perry, G. E., *The R.A.E. Table of Earth Satellites 1957–1986*, New York: Stockton Press, 1987, p. 864.
24. Foreign Broadcast Information Service, U.S.S.R., Space, JPRS-USP-86-004, April 1986, Joint Publications Research Service, p. 1.
25. "Busy Schedule for Cosmonauts," *Spaceflight*, Vol. 28, July–Aug. 1986, p. 296.
26. Foreign Broadcast Information Service, U.S.S.R., Space, JPRS-USP-86-005, Sept. 1986, Joint Publications Research Service, p. 11.
27. Ibid. JPRS-USP-87-003, April 1987, Joint Publications Research Service, p. 10.
28. Ibid. JPRS-USP-86-005, Sept. 1986, Joint Publications Research Service, pp. 13, 16.
29. "Busy Schedule for Cosmonauts," op. cit. p. 297.
30. Foreign Broadcast Information Service, U.S.S.R., Space, JPRS-USP-86-005, Sept. 1986, Joint Publications Research Service, pp. 19–21, 35.
31. *U.S.S.R. Space Life Sciences Digests*, NASA CR-3922(12), Issue 10, p. 90.
32. Foreign Broadcast Information Service, U.S.S.R., Space, JPRS-USP-86-005, Sept. 1986, Joint Publications Research Service, pp. 24–26.
33. "Busy Schedule for Cosmonauts," op. cit. p. 297.
34. Foreign Broadcast Information Service, U.S.S.R., Space, JPRS-USP-87-003, April 1987, Joint Publications Research Service, p. 9.
35. Ibid. JPRS-USP-86-005, Sept. 1986, Joint Publications Research Service, pp. 27–28, 34.
36. Ibid. JPRS-USP-86-006, Nov. 1986, Joint Publications Research Service, pp. 16–18, 36–37.
37. Kidger, Neville, "New Structure for Mir," *Spaceflight*, Vol. 28, Sept.–Oct. 1986, p. 348.
38. Foreign Broadcast Information Service, U.S.S.R., Space, JPRS-USP-86-005, Sept. 1986, Joint Publications Research Service, p. 38.
39. Ibid. JPRS-USP-86-005, Sept. 1986, Joint Publications Research Service, p. 39.
40. Congressional Research Service, The Library of Congress, *Soviet Space Programs 1981–87, Part 1*, Washington: Government Printing Office, May 1988, p. 467.
41. Foreign Broadcast Information Service, U.S.S.R., Space, JPRS-USP-86-005, Sept. 1986, Joint Publications Research Service, p. 42.
42. Congressional Research Service, The Library of Congress, *Soviet Space Programs 1981–87, Part 1*, Washington: Government Printing Office, May 1988, p. 70.

43. Foreign Broadcast Information Service, U.S.S.R., Space, JPRS-USP-86-005, Sept. 1986, Joint Publications Research Service, pp. 43–46.
44. Ibid. JPRS-USP-86-005, Sept. 1986, Joint Publications Research Service, pp. 52–53.
45. Ibid. JPRS-USP-86-006, Nov. 1986, Joint Publications Research Service, p. 5.
46. Kidger, "New Structure for Mir," op. cit. p. 346.
47. Ibid.
48. Foreign Broadcast Information Service, U.S.S.R., Space, JPRS-USP-86-006, Nov. 1986, Joint Publications Research Service, p. 33.
49. Ibid. JPRS-USP-86-006, Nov. 1986, Joint Publications Research Service, pp. 1–3.
50. Kidger, "New Structure for Mir," op. cit. p. 347.
51. Congressional Research Service, The Library of Congress, *Soviet Space Programs 1981–87, Part 1*, Washington: Government Printing Office, May 1988, pp. 38, 72.
52. Foreign Broadcast Information Service, U.S.S.R., Space, JPRS-USP-86-006, Nov. 1986, Joint Publications Research Service, pp. 6–9.
53. Ibid. pp. 6–7.
54. Ibid. pp. 12–16.
55. Ibid. pp. 18–23.
56. Ibid. JPRS-USP-87-003, April 1987, Joint Publications Research Service, p. 12.
57. Kidger, Neville, "Soviet Scene," *Spaceflight*, Vol. 28, Dec. 1986, p. 425.
58. Foreign Broadcast Information Service, U.S.S.R., Space, JPRS-USP-86-006, Nov. 1986, Joint Publications Research Service, p. 20.
59. Congressional Research Service, The Library of Congress, *Soviet Space Programs 1981–87, Part 1*, Washington: Government Printing Office, May 1988, p. 73.
60. Foreign Broadcast Information Service, U.S.S.R., Space, JPRS-USP-86-006, Nov. 1986, Joint Publications Research Service, pp. 22–23.
61. Kidger, "Soviet Scene," op. cit. p. 425.
62. Foreign Broadcast Information Service, U.S.S.R., Space, JPRS-USP-86-006, Nov. 1986, Joint Publications Research Service, pp. 29, 37.
63. Clark, "Soviet Space Activity, 1985–1986," op. cit. p. 209.
64. Foreign Broadcast Information Service, U.S.S.R., Space, JPRS-USP-86-006, Nov. 1986, Joint Publications Research Service, pp. 38–39.
65. Foreign Broadcast Information Service, FBIS-SOV-89-018 Jan. 30, 1989, pp. 85–86.
66. Foreign Broadcast Information Service, U.S.S.R., Space, JPRS-USP-86-005, Sept. 1986, Joint Publications Research Service, p. 41.

67. Ibid. JPRS-USP-86-006, Nov. 1986, Joint Publications Research Service, pp. 38–39.
68. Ibid. JPRS-USP-87-003, April 1987, Joint Publications Research Service, pp. 27–28.
69. "Mir Mission: Third Solar Array Installed," *Spaceflight*, Vol. 29, Aug. 1987, p. 282.
70. Christy, Robert D., "Satellite Digest—202," *Spaceflight*, Vol. 29, May 1987, p. 186.
71. Foreign Broadcast Information Service, U.S.S.R., Space, JPRS-USP-87-003, April 1987, Joint Publications Research Service, pp. 28–33.
72. Kidger, Neville, "Space Walk Saves Mission," *Spaceflight*, Vol. 29, June 1987, p. 236.
73. "Mir Mission: Third Solar Array Installed," op. cit. p. 284.
74. Foreign Broadcast Information Service, U.S.S.R., Space, JPRS-USP-88-002, April 6, 1988, Joint Publications Research Service, p. 2.
75. Kidger, Neville, "Mir in Action," *Spaceflight*, Vol. 29, No. 4, April 1987, pp. 136–137.
76. Foreign Broadcast Information Service, U.S.S.R., Space, JPRS-USP-87-003, April 1987, Joint Publications Research Service, pp. 23, 26.
77. Ibid. pp. 27.
78. Kidger, "Space Walk Saves Mission," op. cit. p. 236.
79. Kidger, "Mir in Action," op. cit. p. 137.
80. Kidger, "Space Walk Saves Mission," op. cit. p. 236.
81. Foreign Broadcast Information Service, U.S.S.R., Space, JPRS-USP-87-003, April 1987, Joint Publications Research Service, pp. 39, 41.
82. Ibid. p. 41.
83. Ibid. JPRS-USP-87-004, July 1987, Joint Publications Research Service, pp. 1–5.
84. Kidger, "Space Walk Saves Mission," op. cit. p. 237.
85. Foreign Broadcast Information Service, U.S.S.R., Space, JPRS-USP-87-003, April 1987, Joint Publications Research Service, pp. 43–45.
86. Kidger, "Space Walk Saves Mission," op. cit. p. 237.
87. Foreign Broadcast Information Service, U.S.S.R., Space, JPRS-USP-88-002, April 6, 1988, Joint Publications Research Service, p. 40.
88. Ibid. JPRS-USP-87-003, April 1987, Joint Publications Research Service, p. 45.
89. Ibid. JPRS-USP-87-004, July 1987, Joint Publications Research Service, pp. 2–3.
90. Ibid. p. 5.
91. Ibid. p. 7.
92. Ibid. p. 15.
93. "Mir Mission: Third Solar Array Installed," op. cit. p. 282.

94. Ibid. pp. 282–284.
95. Foreign Broadcast Information Service, U.S.S.R., Space, JPRS-USP-87-004, July 1987, Joint Publications Research Service, p. 17.
96. Ibid. p. 18.
97. "Mir Mission: Third Solar Array Installed," op. cit. p. 284.
98. Foreign Broadcast Information Service, U.S.S.R., Space, JPRS-USP-87-004, July 1987, Joint Publications Research Service, p. 20.
99. "Mir Mission: Third Solar Array Installed," op. cit. p. 284.
100. Foreign Broadcast Information Service, U.S.S.R., Space, JPRS-USP-87-004, July 1987, Joint Publications Research Service, p. 22.
101. "Mir Mission: Third Solar Array Installed," op. cit. p. 284.
102. Foreign Broadcast Information Service, U.S.S.R., Space, JPRS-USP-87-004, July 1987, Joint Publications Research Service, pp. 26, 27, 35.
103. "Mir Mission: Third Solar Array Installed," op. cit. p. 284.
104. Kidger, "Space Walk Saves Mission," op. cit. p. 238.
105. Foreign Broadcast Information Service, U.S.S.R., Space, JPRS-USP-87-004, July 1987, Joint Publications Research Service, pp. 28–33.
106. Ibid. pp. 30–34.
107. Ibid. JPRS-USP-87-005, Aug. 1987, Joint Publications Research Service, pp. 5, 24, 26.
108. Ibid. pp. 6–8.
109. Ibid. pp. 8–10.
110. Ibid. p. 11.
111. Kidger, Neville, "Endurance Record Broken," *Spaceflight*, Vol. 29, Nov. 1987, p. 373.
112. Foreign Broadcast Information Service, U.S.S.R., Space, JPRS-USP-87-005, Aug. 1987, Joint Publications Research Service, p. 11.
113. Kidger, "Endurance Record Broken," op. cit. p. 373.
114. Kidger, "Endurance Record Broken," op. cit. p. 374.
115. Foreign Broadcast Information Service, U.S.S.R., Space, JPRS-USP-87-005, Aug. 1987, Joint Publications Research Service, pp. 14–19.
116. Kidger, "Endurance Record Broken," op. cit. p. 374.
117. Foreign Broadcast Information Service, U.S.S.R., Space, JPRS-USP-87-005, Aug. 1987, Joint Publications Research Service, pp. 21–23.
118. Ibid. JPRS-USP-87-006, Nov. 24, 1987, Joint Publications Research Service, p. 10.
119. Kidger, "Endurance Record Broken," op. cit. p. 374.
120. Foreign Broadcast Information Service, U.S.S.R., Space, JPRS-USP-87-001, Feb. 1987, Joint Publications Research Service, p. 2.
121. Ibid. JPRS-USP-87-006, Nov. 24, 1987, Joint Publications Research Service, pp. 1, 6.
122. Kidger, "Endurance Record Broken," op. cit. p. 376.

123. Foreign Broadcast Information Service, U.S.S.R., Space, JPRS-USP-88-002, April 6, 1988, Joint Publications Research Service, p. 4.

124. Ibid. JPRS-USP-87-006, Nov. 24, 1987, Joint Publications Research Service, pp. 7-13.

125. Kidger, "Endurance Record Broken," op. cit. p. 376.

126. Foreign Broadcast Information Service, U.S.S.R., Space, JPRS-USP-87-006, Nov. 24, 1987, Joint Publications Research Service, pp. 7, 12.

127. Ibid. p. 19.

128. Kidger, "Endurance Record Broken," op. cit. p. 376.

129. Foreign Broadcast Information Service, U.S.S.R., Space, JPRS-USP-87-006, Nov. 24, 1987, Joint Publications Research Service, p. 20.

130. Kidger, "Endurance Record Broken," op. cit. p. 377.

131. Foreign Broadcast Information Service, U.S.S.R., Space, JPRS-USP-87-006, Nov. 24, 1987, Joint Publications Research Service, p. 17.

132. Ibid. JPRS-USP-88-002, April 6, 1988, Joint Publications Research Service, p. 2.

133. Kidger, "Endurance Record Broken," op. cit. p. 377.

134. Foreign Broadcast Information Service, U.S.S.R., Space, JPRS-USP-87-006, Nov. 24, 1987, Joint Publications Research Service, p. 21.

135. Kidger, "Endurance Record Broken," op. cit. p. 377.

136. Foreign Broadcast Information Service, U.S.S.R., Space, JPRS-USP-87-006, Nov. 24, 1987, Joint Publications Research Service, p. 22.

137. Ibid. p. 41.

138. Ibid. pp. 22–41.

139. Kidger, "Endurance Record Broken," op. cit. p. 377.

140. Foreign Broadcast Information Service, U.S.S.R., Space, JPRS-USP-87-006, Nov. 24, 1987, Joint Publications Research Service, pp. 32, 40.

141. "Busy Routine for Mir Crew," *Soviet Spaceflight Report*, Kappesser, Peter J. (ed.), No. 5, Sept./Oct. 1987, pp. 1–3.

142. Foreign Broadcast Information Service, U.S.S.R., Space, JPRS-USP-87-006, Nov. 24, 1987, Joint Publications Research Service, p. 36.

143. Ibid. p. 42.

144. Kidger, Neville, "Cosmonauts Observe Supernova," *Spaceflight*, Vol. 30, March 1988, p. 114.

145. Kappesser, op. cit. p. 3.

146. "Record Soviet Manned Space Flight Raises Human Endurance Questions," *Aviation Week & ST*, Jan. 4, 1988, p. 25.

147. Kidger, "Cosmonauts Observe Supernova," op. cit. p. 113.

148. Foreign Broadcast Information Service, U.S.S.R., Space, JPRS-USP-88-002, April 6, 1988, Joint Publications Research Service, p. 8.

149. Christy, Robert D., "Satellite Digest—210," *Spaceflight*, Vol. 30, March 1988, p. 98.

150. *Spaceflight*, Vol. 30, May 1988, p. 192.

151. Foreign Broadcast Information Service, U.S.S.R., Space, JPRS-USP-88-002, April 6, 1988, Joint Publications Research Service, p. 1.

152. Christy, Robert D., "Satellite Digest—211," *Spaceflight*, Vol. 30, April 1988, p. 146.

153. Kidger, "Cosmonauts Observe Supernova," op. cit. p. 113.

154. Foreign Broadcast Information Service, U.S.S.R., Space, JPRS-USP-88-002, April 6, 1988, Joint Publications Research Service, pp. 3, 5.

155. Kidger, "Cosmonauts Observe Supernova," op. cit. p. 117.

156. Foreign Broadcast Information Service, U.S.S.R., Space, JPRS-USP-88-002, April 6, 1988, Joint Publications Research Service, p. 6.

157. "Record Soviet Manned Space Flight Raises Human Endurance Questions," *Aviation Week & ST*, Jan. 4, 1988, p. 25.

158. Foreign Broadcast Information Service, U.S.S.R., Space, JPRS-USP-88-002, April 6, 1988, Joint Publications Research Service, pp. 2–3.

159. Kidger, "Cosmonauts Observe Supernova," op. cit. p. 117.

160. Radio Moscow, North American Service, Jan. 12, 1988.

161. Kidger, Neville, "Bulgarian Set for Mir Visit," *Spaceflight*, Vol. 30, June 1988, p. 229.

162. "Mir Cosmonauts Prepare for First EVA To Install Advanced Array," *Aviation Week & ST*, Feb. 29, 1988, p. 18.

163. Kidger, "Bulgarian Set for Mir Visit," op. cit. p. 228.

164. Radio Moscow, North American Service, March 16, 1988.

165. Radio Moscow, North American Service, March 22, 1988.

166. Kidger, "Bulgarian Set for Mir Visit," op. cit. p. 228.

167. Kidger, "Mir Mission Report," op. cit. p. 395.

168. Christy, Robert D., "Satellite Digest—215," *Spaceflight*, Vol. 30, Oct. 1988, p. 384.

169. Kidger, "Bulgarian Set for Mir Visit," op. cit. p. 228.

170. Foreign Broadcast Information Service, U.S.S.R., Space, JPRS-USP-89-004, Feb. 16, 1989, Joint Publications Research Service, pp. 20–27.

171. Kidger, "Mir Mission Report," op. cit. p. 396.

172. North American Service, Radio Moscow, June 21, 1988.

173. "Soviet Cosmonauts On Mir Fail to Repair Science Instrument," *Aviation Week & ST*, July 19, 1988, p. 27.

174. "Busy Routine for Mir Crew," *Soviet Spaceflight Report*, Kappesser, Peter J. (ed.), No. 5, Sept./Oct. 1987, p. 13.

175. Kidger, "Mir Mission Report," op. cit. p. 396.

176. Ibid. p. 454.

177. North American Service Radio Moscow, Aug. 29, 1988.

178. *Spaceflight*, Vol. 31, Feb. 1989, p. 44.

179. "Soviet Spaceflight Update," *Soviet Spaceflight Report*, Kappesser, Peter J. (ed.), No. 5, Sept./Oct. 1987, p. 13.
180. Lenorovitz, Jeffrey M., "Soviet Long-Duration Cosmonauts Readapt Rapidly to Earth Environment," *Aviation Week & ST*, Jan. 29, 1989, p. 38.
181. "Soyuz TM-5: What Went Wrong," *Flight International*, Oct. 15, 1988, p. 42.
182. "Soviet Spaceflight Update," *Soviet Spaceflight Report*, Kappesser, Peter J. (ed.), No. 5, Sept./Oct. 1987, p. 13.
183. Foreign Broadcast Information Service, U.S.S.R., Space, JPRS-USP-89-004, Feb. 16, 1989, Joint Publications Research Service, pp. 33–35.
184. *Spaceflight*, Vol. 31, Feb. 1989, p. 44.
185. Kidger, "Mir Mission Report," op. cit. pp. 64–65.
186. Foreign Broadcast Information Service, U.S.S.R., Space, JPRS-USP-89-004, Feb. 16, 1989, Joint Publications Research Service, pp. 33–35.
187. Furniss, Tim, "Soviet Shuttle Will Fly on Energia 2," *Flight International*, March 12, 1988, p. 49.
188. "Soviet Spaceflight Update," *Soviet Spaceflight Report*, Kappesser, Peter J. (ed.), No. 5, Sept./Oct. 1987, p. 14.
189. ABC News, May 16, 1988.
190. National Space Society, *Space Hotline*, Sept. 23, 1988.
191. Science and Engineering, Radio Moscow, North American Service, April 29, 1988.
192. Lenorovitz, Jeffrey M., "Soviets Planning Manned Shuttle Mission for 1989," *Aviation Week & ST*, Jan. 16, 1989, p. 34.
193. Kidger, Neville, "Soviet Shuttle," *Spaceflight*, Vol. 31, Jan. 1989, p. 5.
194. Furniss, Tim, "Energia and Buran, The Soviet Space Union," *Flight International*, Feb. 1989, p. 26.
195. Lenorovitz, Jeffrey M., "Long-Term Space Plan Will Lead to Soviet Orbital Infrastructure," *Aviation Week & ST*, Dec. 12, 1988, p. 44.
196. Pirard, Theo, "Dunayev: Shuttle Will Fly on Second Energia," *Spaceflight*, Vol. 30, May 1988, p. 185.
197. Lenorovitz, Jeffrey M., "U.S.S.R. Ground Tests First Building Block Module for Mir Manned Space Station," *Aviation Week & ST*, Dec. 5, 1988, p. 30.
198. Kidger, "Mir Mission Report," op. cit. p. 77.
199. Lenorovitz, Jeffrey M., "International Crew Working on Soviet Mir Space Station," *Aviation Week & ST*, Dec. 12, 1988, p. 31.
200. *Aviation Week & ST*, Dec. 19, 1988, p. 32.
201. Lenorovitz, Jeffrey M., "Soviets Will Tether Maneuvering Unit To Space Station for Initial Missions," *Aviation Week & ST*, Jan. 23, 1989, p. 63.

202. Furniss, Tim, "Soviets Set Space Endurance Records," *Flight International*, Nov. 19, 1988, p. 13.
203. Lenorovitz, "Soviet Long-Duration Cosmonauts Readapt Rapidly to Earth Environment," op. cit. p. 38.
204. Furniss, Tim, "Progress 39 Docks with Mir," *Flight International*, Jan. 14, 1989, p. 13.
205. Foreign Broadcast Information Service, U.S.S.R., Space, JPRS-USP-89-004, Feb. 16, 1989, Joint Publications Research Service, pp. 33–35.
206. Kidger, "Mir Mission Report," op. cit. pp. 192–193.
207. Furniss, Tim, "Mir Repair Mission Scheduled," *Flight International*, May 6, 1989, p. 13.
208. Kidger, "Mir Mission Report," op. cit. pp. 193–194.
209. Lenorovitz, Jeffrey M., "Soviets Plan Medium-Duration Missions on Board Mir This Year," *Aviation Week & ST*, Jan. 9, 1989, p. 23.

Appendix

Specifications of Soviet Space Launch Vehicles, Ballistic Missiles, and Rocket Engines

Soviet Launch Vehicles

Launch sites: B—Baykonur, P—Plesetsk, K—Kapustin Yar
LEO—Low Earth Orbit; GEO—Geostationary Earth Orbit

Type	Launch Sites	Payload	Distance	Year	Comments
A	B	1,500 kg	LEO	1957	SL-1, SL-2, based on SS-6 Sapwood
A-1	B P	5,000 kg	LEO	1959	SL-3, Vostok
		440 kg	moon	1959	Luna
A-2	B P	7,500 kg	LEO	1963	SL-4, Soyuz, Standard, R-7
A-2e	B P	2,100 kg	Molniya	1960	SL-5, SL-6, Molniya communications satellites
		1,800 kg	moon	1962	Luna landers and orbiters
		1,200 kg	Mars	1962	Mars and Zond probes
		1,200 kg	Venus	1962	Venera and Zond probes
B-1	P K	450 kg	LEO	1962	SL-7, based on SS-4, phased out in 1977
C-1	P K B	1,700 kg	LEO	1964	SL-8, Vertical, Kosmos, based on SS-5, ASAT
D-1	B	12,200 kg	LEO	1965	SL-9, Proton 1, 2, 3 test flights
D-1h	B	20,000 kg	LEO	1970	SL-13, Salyut, Star modules
D-1h′	B	21,000 kg	LEO	1986	Uprated D-1h, Mir, Kvant

Engine	Type	Mass	Destination	Year	Description
D-1e	B	1,500 kg	GEO	1967	SL-12, communications satellites
		6,300 kg	moon	1967	Zond 4-8, Luna landers and orbiters
		4,700 kg	Mars	1971	Mars probes
		5,100 kg	Venus	1975	Venera probes
D-1e'	B	2,000 kg	GEO	1986	SL-15, uses Block-D prime 4th stage
F-1m	B P	5,000 kg	LEO	1966	SL-11, based on SS-9, FOBS tests
F-1r	B P	4,500 kg	LEO	1966	RORSAT launcher
F-1s	B P	5,000 kg	LEO	1974	EORSAT launcher
F-2	B P	5,500 kg	LEO	1977	SL-14, Tsyklon (Cyclone), Elint and Meteor
G-1	B	135,000 kg	LEO	---	SL-15, TT-5, N-I, abandoned in 1974
G-1e	B	50,000 kg	moon	---	Abandoned in 1974
J-1	B	15,000 kg	LEO	1982	SL-16, Zenith
K-1	B	100,000 kg	LEO	1987	SL-17, Energia, (Mir 2)
		18,000 kg	GEO	?	
		32,000 kg	moon	?	
		20,000 kg	Mars	?	
		33,000 kg	LEO	1988	Soviet shuttle

Source: Johnson, Nicholas L. Handbook of Soviet Manned Spaceflight, American Astronautical Society: San Diego, 1980, p. 405.

Major Soviet Ballistic Missiles

NATO Name	Deployed	Type	Propellant	Range km	Comments
SS-1 Scunner	1947	MRBM	Alcohol/LOX	350	V-2 derivative sounding rocket, RD-100 engine
SS-1B Scud A	1957	MRBM	UDMH/IRFNA	130	V-2 derivative, mobile
SS-1C Scud B	1957	MRBM	UDMH/IRFNA	280	Modified versions range was 480 km
SS-2 Sibiling	1948	MRBM	Alcohol/LOX		Modified SS-1, used RD-101 engine
SS-3 Shyster	1955	MRBM	Kerosene/LOX	900	Also named V5V, T-1, M-101, Victory, mobile
SS-4 Sandal	1959	IRBM	Kerosene/nitric acid	2,000	Lengthened SS-3, also B-type booster
SS-5 Skean	1961	LRBM	UDMH/nitric acid	4,100	Also C-type booster, used 2 RD-216 engines
SS-6 Sapwood	1957	ICBM	Kerosene/LOX	8,500	Also A-type booster, retired by 1967
SS-7 Saddler	1961	ICBM	Liquid	10,500	Designed for new lighter H-bomb
SS-8 Sasin	1963	ICBM	Liquid	12,000	2-stage, backup design to SS-7
SS-9 Scarp	1965	ICBM	Storable liquid	12,000	2-stage, 5 versions, F-type booster
SS-10 Scrag	1965	ICBM	Kerosene/LOX	12,000	Backup design for SS-9. Not Deployed.
SS-11 Sego	1966	LRBM	Storable liquid	10,500	4 versions, 2-stage. Anti-ship/Biological missions
SS-12 Scaleboard	1968	MRBM	Solid	750	Single-stage, mobile launcher
SS-13 Savage	1968	ICBM	Solid	9,400	3-stage missile
SS-14 Scapegoat	1968	LRBM	Solid	4,000	SS-13 2nd and 3rd stages, mobile
SS-15 Scrooge	1969	LRBM	Solid	5,600	Modified SS-13, replaced SS-14
SS-16 Spinner	1977	ICBM	Solid	9,000	3-stage, not deployed?
SS-17 Spanker	1975	ICBM	Storable liquid	11,000	2-stage, MIRV warheads, replaced SS-11

Designation	Year	Type	Propellant	Range	Notes
SS-18 Satan	1976	ICBM	Storable liquid	16,000	2-stage, MIRV warheads, replaced SS-9
SS-19 Stiletto	1975	ICBM	Storable liquid	0,000	2-stage, replaced SS-11's
SS-20 Saber	1976	LRBM	Solid	5,000	SS-16 2nd & 3rd stages, replaced SS-3, 4, 5 and 15
SS-21 Scarab	1979	SRBM	Solid	120	Replaced Frog type Surface to Surface Missile
SS-22 Scaleboard	1979	MRBM	Solid	900	Replaced SS-12's
SS-23	1986	ICBM	Liquid	11,000	Replaced SS-N-18's
SS-24	1986	ICBM	Solid		Silo or mobile launch
SS-25	1986	ICBM	Solid	9,000	Modified SS-13. Mobile launcher
SS-N-4 Sark	1956	MRBM	Liquid	600	Single-stage, 2-3 carried per submarine
SS-N-5 Serb	1963	MRBM	Liquid		Single-stage, 3 carried per submarine
SS-N-6 Sawfly	1968	LRBM	Liquid	3,000	2-stage, 16 carried per submarine
SS-N-8	1973	ICBM	Liquid	9,100	2-stage, 16-18 carried per submarine
SS-N-17		LRBM	Solid	3,900	Single stage
SS-N-18		ICBM	Liquid	8,000	2-stage, up to 7 MIRV warheads
SS-N-20	1983	ICBM	Solid	8,300	Carried up to 9 MIRV warheads
SS-N-23	1986	ICBM	Liquid	11,000	Replaced SS-N-18's

Sources: Johnson, Nicholas L. Handbook of Soviet Manned Spaceflight, *American Astronautical Society: San Diego, 1980, p. 405.*
Peebles, Curtis. Guardians: Strategic Reconnaissance Satellites, *Novato, CA: Presidio Press, 1987, p. 86.*
"Soviet Rocket Engines—Some New Details," *Spaceflight magazine, Vol. 18, No. 6, June 1976, p. 224.*
Baker, David. The Rocket, *London: New Cavendish Books, 1978, pp. 226–230.*
Additional data courtesy of Nicholas L. Johnson.

Soviet Rocket Engines

Type	Propellant	Development Year	Nozzles	Vacuum Thrust	Vacuum Isp (m/s)	Chamber Pressure	Expansion Ratio
RD-7	Alcohol/LOX	19??–59	1 main 4 verniers	5,617 kg	3,195		
RD-100	Alcohol/LOX	1945–48	1 main	31,358 kg	2,325	16.3 atm	
RD-101	Alcohol/LOX	1947–49	1 main	37,487 + kg			
RD-103	Alcohol/LOX	1952–53	1 main	51,072 kg	2,430	24.5 atm	
RD-107	Kerosene/LOX	1954–57	4 main 2 verniers	102,145 kg	3,077	60 atm	150
RD-108	Kerosene/LOX	1954–57	4 main 4 verniers	96,118 kg	3,087	52.3 atm	150
RD-111	Kerosene/LOX	1959–62	4 main	166,292 kg	3,107	80.5 atm	
RD-119	UDMH/LOX	1958–62	1 main 4 verniers	10,725 kg	3,450	80.9 atm	1,350

RD-170	Kerosene/LOX	19??–87	4 main	806,000 kg	3,293	224 atm	
RD-214	Kerosene/nitric acid	1955–57	4 main	74,656 kg	2,587	44.7 atm	64
RD-216	UDMH/nitric acid	1958–60	2 main	176,506 kg	2,857	75.4 atm	268
RD-219	UDMH/nitric acid	1958–61	2 main	90,194 kg	2,871	75.4 atm	
RD-253	UDMH/N_2O_4	1961–65	1 main	168,845 kg	3,100	150.8 atm	
RD-301	Ammonia/fluorine			10,010 kg	3,928	121 atm	
RD-461	Kerosene/LOX	19??–63	4 main / 4 verniers	30,030 kg	3,234		
RD-?*	Kerosene/LOX	19??–67	1 main	8,382 kg	3,430		
RD-?**	Hydrogen/LOX	19??–87	1 main	200,000 kg	4,165 +		

* Proton escape stage engine
** Energia main engine

Sources: Baker, David. The Rocket, London: New Cavendish Books, 1978, pp. 115–119, 164–165.
Clark, Phillip S. "Soviet Launch Vehicles: An Overview," Journal of the British Interplanetary Society, Vol. 35, No. 2, Feb. 1982, pp. 53–56.
Pavo, H. "Soviet Scene," Spaceflight, Vol. 28, No. 6, June 1986, p. 250.
Kenden, Anthony. "Soviet Rocket Engines—Some New Details," Spaceflight, Vol. 18, No. 6, June 1976, p. 223.
Johnson, Nicholas L. Handbook of Soviet Manned Spaceflight, American Astronautical Society: San Diego, 1980, p. 396.
Additional data courtesy of Nicholas L. Johnson.

Index